Introduction to Urban Water Distribution

Second Edition

Volume 2: Problems & Exercises

IHE Delft Lecture Note Series

ISSN: 1567-7052

Introduction to Urban Water Distribution

Problems & Exercises

Second Edition

Nemanja Trifunović

CRC Press
Taylor & Francis Group
Boca Raton London New York

CRC Press is an imprint of the
Taylor & Francis Group, an **informa** business

United Nations
Educational, Scientific and
Cultural Organization

Institute for
Water Education
in partnership with UNESCO

Second edition published 2020

CRC Press/Balkema is an imprint of the Taylor & Francis Group, an informa business

© 2020 Taylor & Francis Group, London, UK

Typeset by Apex CoVantage, LLC

Library of Congress Cataloging-in-Publication Data
Applied for

Published by: CRC Press/Balkema
              Schipholweg 107C, 2316 XC Leiden, The Netherlands
              e-mail: Pub.NL@taylorandfrancis.com
              www.crcpress.com – www.taylorandfrancis.com

Volume 1

ISBN: 978-0-367-50301-7 (Hbk)
ISBN: 978-0-367-50445-8 (Pbk)
ISBN: 978-1-003-04985-2 (eBook)
DOI: https://doi.org/10.1201/9781003049852

Volume 2

ISBN: 978-0-367-50302-4 (Hbk)
ISBN: 978-0-367-50448-9 (Pbk)
ISBN: 978-1-003-04986-9 (eBook)
DOI: https://doi.org/10.1201/9781003049869

Two-volume set

ISBN: 978-0-367-50295-9 (Hbk)
ISBN: 978-0-367-50443-4 (Pbk)
ISBN: 978-1-003-04941-8 (eBook)

In theory the sky is the limit, but in practice it is time.

Contents

Preface to the second edition

In 2016, precisely ten years after the book 'Introduction to Urban Water Distribution' was first published by Taylor & Francis, I decided to start work on a second edition. I was inspired by numerous positive reactions from my students and peers, a few of whom started using my materials in their educational programmes. However, some of the contents gradually became outdated and an update with newer information became inevitable.

Despite a lot of enthusiasm for this ambition, it took me nearly three years to complete the work. This was because I took the opportunity to strengthen some sections in the book with better elaborated principles, further expand the contents with missing developments in the field, and also add new exercises tested in the classrooms during the last 10–15 years. The result is this material which grew from approximately 500 to 800 pages, with over 700 figures comprising various diagrams, drawings, and computer layouts, including some 200 photos largely from my own collection made during student fieldtrips, project missions, conference exhibitions, or simply by observing interesting water distribution practices while biking on a sunny day around my area. The practical part of the book covered in the appendices has been nearly doubled, by adding two more computer exercises with detailed tutorials, and the selection of 15 solved examination problems and true-false tests, given to the students in Delft in the period 2004 to 2019. Consequently, the electronic materials accompanying this book have also been upgraded with additional spreadsheet and computer modelling applications discussed in the workshop problems and exercises.

Following significant expansion, this book has been published in two volumes: Volume 1 covering the theory (referred to in the table of contents as chapters 1 to 6), and Volume 2 covering the workshop problems and computer modelling exercises (referred to in the table of contents as appendices 1 to 9). Although each of the volumes can be studied separately, in many places in the chapters there are references to the appendices, as well as some clarifications in the appendices that contain references to the chapters/sections. Moreover, all electronic materials have been attached to Volume 2, some of which are also mentioned in Volume 1. Hence, studying both volumes in parallel is the obvious and most effective approach.

This book is used in several water distribution-related specialist modules of the Master of Science programme in Urban Water and Sanitation at IHE Delft Institute for Water Education, in Delft, the Netherlands. In addition, it is the core material in the online version of the short course 'Water Transport and Distribution' and is also used in various training programmes in capacity development projects of IHE Delft. Participants in all these programmes are professionals with various backgrounds and experience, mostly engineers, working in the water supply sector from over forty, predominantly developing, countries from all parts of the world.

This current version is the summary of 25 years of development of IHE educational materials now put at the disposal of both students and teachers in the field of water transport and distribution. On that long road I learned continuously from consultations with experts from the Dutch water sector acting as guest lecturers in our institute and also helping us to organise field visits for our students. A few of the most important water companies in that respect are WATERNET Amsterdam, EVIDES Rotterdam, WMD Assen, and also KWR research institute in Nieuwegein. Furthermore, in my IHE career I have had numerous opportunities to interact with the participants of my training programmes conducted abroad, who have brought to my attention many applications that differ from Dutch and European practice. All these ingredients have helped me tremendously to arrive at the result that will hopefully satisfy the target audience.

Nemanja Trifunović

Introduction

This book was written with the idea of elaborating general principles and practices in water transport and distribution in a straightforward way. Most of its readers are expected to be those who know little or nothing about the subject. However, experts dealing with advanced problems can also use it as a refresher of their knowledge, while the lecturers in this field may wish to use some of the content in their educational programmes.

The general focus in the book is on understanding the hydraulics of distribution networks, which has become increasingly relevant since the large-scale introduction of computers and the exponential growth of computer model applications, also in developing countries. This core is handled in Chapter 3 which discusses the basic hydraulics of pressurised flow, and Chapter 4 which talks about the principles of hydraulic design and computer modelling applied in water transport and distribution. Exercises and tutorials resulting from these chapters are given in appendices 1 to 4.

The main purpose of the exercises is to develop a temporal and spatial perception of the main hydraulic parameters in the system for a given layout and demand scenarios. The workshop problems in Appendix 1 are a collection of calculi tackling various supply schemes and network configurations in a vertical cross-section. Manual calculation is advised here, whilst the spreadsheet lessons illustrated in Appendix 7 can help in checking the results and generating new problems. On the other hand, the tutorials in appendices 2 to 4 discuss, step by step, a computer-aided network design and renovation looking at the network layout in a plan i.e. from a horizontal perspective. Each of these exercises has been formulated with individual data sets that allows attempts with many different source/terrain configurations and demand scenarios. To facilitate the calculation process, the EPANET software of the US Environmental Protection Agency has been used as a network modelling tool. This programme has become popular amongst researchers and practitioners worldwide, owing to its excellent features, simplicity and free distribution via the Internet.

Furthermore, the book contains a relatively detailed discussion on water demand (Chapter 2), which is a fundamental element of any network analysis, and chapters on network construction (Chapter 5) and operation and maintenance (Chapter 6).

Complementary to these contents, more on the maintenance programmes and management issues in water distribution is taught in the Water Governance programmes at IHE Delft. Furthermore, the separate subjects on geographical information systems, water quality and transient flows, all with appropriate lecture notes, make an integral part of the six-week programme on water transport and distribution, which explains the absence of these topics from the scope of this book.

The book comes with a selection of electronic materials containing the spreadsheet hydraulic lessons, a copy of the EPANET software (Version 2.12) and the entire batch of computer model input files and spreadsheet applications mentioned in appendices 1 to 4. Hence, studying with a PC will certainly help to master the contents faster.

The author and IHE Delft are not responsible and assume no liability whatsoever for any results or any use made of the results obtained based on the contents of this book, including the accompanying electronic materials. However, any notification of possible errors or suggestion for improvement will be highly appreciated. Furthermore, any equipment shown in photographs is for illustrative purposes only and is not endorsed or recommended by IHE Delft.

Workshop problems

1.1 Water demand

Problem 1.1.1

Determine the production capacity of a treatment installation for a city with a population of 1,250,000. Assume a specific consumption per capita of 150 l/d, non-domestic water use of 30,000,000 m³/y and 12% of water production as physical loss.

Answer:
Q_{avg} = 111,861 million m³/y or 3.55 m³/s.

Problem 1.1.2

A water supply company delivers an annual quantity of 15,000,000 m³ to a distribution area of 150,000 consumers. At the same time, the collected revenue is 6,000,000 EUR, at an average water tariff of 0.5 EUR/m³. Determine:

a) the delivery on an average consumption day,
b) the percentage of non-revenue water (*NRW*),
c) the specific consumption per capita per day, assuming 60% of the total delivery is for domestic use.

Note:
b) Express the *NRW* as a percentage of the delivered water.

Answers:
a) Q_{avg} = 41,096 m³/day, or 1712 m³/h, or 475.6 l/s
b) *NRW* = 20.0%
c) q = 164 l/c/d

Problem 1.1.3

A family of four pays for annual water consumption of 185 m³. Determine:

a)   the specific consumption per capita per day,
b)   the instantaneous peak factor at a flow of 300 l/h.

Answers:
a) $q = 127$ l/c/d
b) $pf_{ins} = 14$

Problem 1.1.4

An apartment building of 76 occupants pays for an annual water consumption of 4770 m³. Determine:

a)   the specific consumption per capita per day,
b)   the instantaneous peak factor during the maximum consumption flow of 5.5 m³/h.

Answers:
a) $q = 172$ l/c/d
b) $pf_{ins} = 10$

Problem 1.1.5

A residential area of 1200 inhabitants is supplied with an annual water quantity of 63,800 m³, which includes leakage estimated at 10 % of the total supply. During the same period, the maximum flow registered by the district flow meter is 25.4 m³/h. Determine:

a)   the specific consumption per capita per day,
b)   the maximum instantaneous peak factor.

Note:
a)   Specific consumption should not include leakage.
b)   Peak factors include leakage unless the flow is measured at the service connection.

Answers:
a) $q = 131$ l/c/d
b) $pf_{ins} = 3.5$

Problem 1.1.6

A water supply company delivers an annual volume of 13,350,000 m³. The maximum daily demand of 42,420 m³ was observed on 26 July. The minimum, observed on 30 January, was 27,360 m³. The following delivery was registered on 11 March:

Table 1.1 Hourly flows on March 11 – Problem 1.1.6

Hour	1	2	3	4	5	6	7	8	9	10	11	12
m³	433	562	644	835	1450	1644	1856	1922	1936	1887	1721	1712

Hour	13	14	15	16	17	18	19	20	21	22	23	24
m³	1634	1656	1789	1925	2087	2055	1944	1453	1218	813	676	602

Determine:

a) delivery on an average consumption day and the range of seasonal peak factors,
b) the diurnal peak factor diagram,
c) the expected annual range of peak flows supplied to the area.

Answers:
a) Q_{avg} = 36,575 m³/day; $pf_{seasonal}$ = 0.748 – 1.160
b) Q_{avg} = 1435.6 m³/h (on 11 March)

Table 1.2 Diurnal peak factors – Problem 1.1.6

Hour	1	2	3	4	5	6	7	8	9	10	11	12
pf_h	0.302	0.391	0.449	0.582	1.010	1.145	1.293	1.339	1.349	1.314	1.199	1.193

Hour	13	14	15	16	17	18	19	20	21	22	23	24
pf_h	1.138	1.154	1.246	1.341	1.454	1.431	1.354	1.012	0.848	0.566	0.471	0.419

Note that 11 March is not an average consumption day. The average flow derived from the annual quantity is Q_{avg} = 1524 m³/h.

c) Q_{max} = 2563 m³/h; Q_{min} = 343 m³/h

Problem 1.1.7

Estimated leakage in the area from Problem 1.1.6 is 20% of the daily supply, on average. The leakage level is assumed to be constant over 24 hours. Calculate the hourly peak factors for the actual consumption on 11 March.

Note:
Leakage of 20% means a constant flow (loss) of 287.1 m³/h on 11 March.

Answer:
Q_{avg} = 1148.5 m³/h (the consumption)

Table 1.3 Diurnal peak factors – Problem 1.1.7

Hour	1	2	3	4	5	6	7	8	9	10	11	12
pf_h	0.127	0.239	0.311	0.477	1.013	1.181	1.366	1.424	1.436	1.393	1.249	1.241

Hour	13	14	15	16	17	18	19	20	21	22	23	24
pf_h	1.173	1.192	1.308	1.426	1.567	1.539	1.443	1.015	0.811	0.458	0.339	0.274

Problem 1.1.8

The consumption calculated in Problem 1.1.7 consists of three categories: domestic, industrial and commercial. The industrial category contributes to the overall consumption with a constant flow of 300 m³/h, between 8:00 and 20:00 hours. The commercial category requires a flow of 100 m³/h, between 8:00 and 16:00 hours.

a) Determine the hourly peak factors for the domestic consumption category.
b) Assuming the industrial and commercial consumption to be constant through-out the whole year, calculate the average domestic consumption per capita if there are 200,000 people in the area.

Answers:
a) Q_{avg} = 965.2 m³/h (the domestic consumption)

Table 1.4 Diurnal peak factors – Problem 1.1.8

Hour	1	2	3	4	5	6	7	8	9	10	11	12
pf_h	0.151	0.285	0.370	0.568	1.205	1.406	1.626	1.279	1.294	1.243	1.071	1.062

Hour	13	14	15	16	17	18	19	20	21	22	23	24
pf_h	0.981	1.004	1.142	1.386	1.554	1.521	1.406	1.208	0.965	0.545	0.403	0.326

b) q = 124 l/c/d

Problem 1.1.9

The registered annual domestic consumption is presently 38.2 million m³. Determine:

a) the consumption after the first 10 years, assuming an annual population growth of 3.8 %,
b) the consumption after the following 10 years (11 to 20) assuming an annual population growth of 2.2 %.

Compare the results of the Linear and Exponential models discussed in Section 2.5.

Answers:
a) In 10 years from now: Q_{lin} = 52.7 million m³; Q_{exp} = 55.5 million m³
b) In 20 years from now: Q_{lin} = 64.3 million m³; Q_{exp} = 69.0 million m³

Problem 1.1.10

The following annual consumptions were registered in the period 2010–2015 (in million m^3):

Table 1.5 Consumption 2010 to 2015 – Problem 1.1.10

Year	2010	2011	2012	2013	2014	2015
Q (10^6 m^3)	125.4	131.8	138.2	145.4	152.6	159.9

Make a forecast for the year 2025.

Answer:
Q_{2025} = 260.7 million m^3 (exponential growth of 5 %)

1.2 Single pipe calculation

Problem 1.2.1

A pipe of length L = 500 m, diameter D = 300 mm and absolute roughness k = 0.02 mm transports a flow Q = 456 m^3/h. Determine the hydraulic gradient by using the Darcy-Weisbach formula. The water temperature can be assumed to be 10° C. Check the result by using the hydraulic tables in Appendix 6.

Answer:
By using the Darcy-Weisbach formula, S = 0.0079.
From the tables for k = 0.01 mm, S = 0.007 if Q = 434.1 m^3/h. If S = 0.010, Q = 526.9 m^3/h. By linear interpolation: S = 0.0077, which is close to the calculated result.

Problem 1.2.2

A pipe of length L = 275 m, diameter D = 150 mm and absolute roughness k = 0.1 mm transports a flow Q = 80 m^3/h. Determine the hydraulic gradient by using the Darcy-Weisbach formula. The water temperature can be assumed to be 15° C. Check the result by using the hydraulic tables in Appendix 6.

Answer:
S = 0.0108;
From the tables for k = 0.1 mm, S = 0.010 if Q = 76.7 m^3/h.

Problem 1.2.3

A pipe of length L = 1000 m and diameter D = 800 mm transports a flow Q = 1.2 m^3/s. Determine the hydraulic gradient:

a) by using the Darcy-Weisbach formula for k = 0.2 mm,
b) the Hazen-Williams formula for C_{hw} = 130,
c) the Manning formula for N = 0.010 m$^{-1/3}$s.

The water temperature can be assumed to be 10° C.

Answers:
a) $S = 0.0055$
b) $S = 0.0054$
c) $S = 0.0049$

Problem 1.2.4

Determine the maximum capacity of a pipe where $D = 400$ mm and $k = 0.5$ mm at the maximum-allowed hydraulic gradient $S_{max} = 0.0025$. The water temperature equals 10° C. Check the result by using the hydraulic tables in Appendix 6.

Answer:
$Q_{max} = 429.8$ m³/h
From the tables for $k = 0.5$ mm, $Q = 384.9$ m³/h if $S = 0.002$ and 473.2 m³/h for $S = 0.003$. By linear interpolation: $Q_{max} = 429.1$ m³/h.

Problem 1.2.5

Determine the maximum capacity of a pipe where $D = 200$ mm at the maximum-allowed hydraulic gradient $S_{max} = 0.005$:

a)   if $k = 0.01$ mm,
b)   if $k = 1$ mm.

The water temperature equals 10° C.

Answers:
a) $Q_{max} = 123.1$ m³/h
b) $Q_{max} = 89.8$ m³/h

Problem 1.2.6

Determine the maximum capacity of a pipe where $D = 1200$ mm and $k = 0.05$ mm at the maximum-allowed hydraulic gradient:

a) $S_{max} = 0.001$,
b) $S_{max} = 0.005$.

The water temperature equals 10° C.

Answers:
a) $Q_{max} = 5669$ m³/h
b) $Q_{max} = 13,178$ m³/h

Problem 1.2.7

Determine the maximum capacity of a pipe where $D = 100$ mm and $k = 0.4$ mm at the maximum-allowed hydraulic gradient $S_{max} = 0.01$. Use the Moody diagram shown in Figure 3.9. The water temperature equals $10°$ C.

Answer:
$Q_{max} = 22.6$ m³/h.

Problem 1.2.8

Determine the pipe diameter that can transport flow $Q = 720$ m³/h at the maximum-allowed hydraulic gradient $S_{max} = 0.002$. The pipe roughness $k = 0.05$ mm. Assume the water temperature to be $12°$ C. Check the result by using the hydraulic tables in Appendix 6.

Answer:
$D = 477$ mm; the first higher manufactured diameter $D = 500$ mm delivers 820.0 m³/h.
From the tables for $k = 0.05$ mm and $S = 0.002$, $Q = 818.2$ m³/h for $D = 500$ mm.

Problem 1.2.9

A pipe, $L = 450$ m, $D = 300$ mm and $k = 0.3$ mm, conveys flow $Q = 100$ l/s. An increase in flow to 300 l/s is planned. Determine:

a) the diameter of the pipe laid in parallel to the existing pipe,
b) the pipe diameter if, instead of laying a second pipe, the existing pipe is replaced by a larger one,
c) the pipe diameter if the existing pipe is replaced by two equal pipes.

For all new pipes, $k = 0.01$ mm. Assume the water temperature to be $10°$ C.

Note:
The present hydraulic gradient has to be maintained in all three options.

Answers:
For $S = 0.007$

a) $Q_2 = 200$ l/s; $D_2 = 363$ mm (adopted $D = 400$ mm)
b) $Q = 300$ l/s; $D = 423$ mm (adopted $D = 500$ mm)
c) $Q_1 = Q_2 = 150$ l/s; $D_1 = D_2 = 326$ mm (adopted $D = 350$ mm)

Problem 1.2.10

Find the equivalent diameters of two pipes connected in parallel, where $L = 850$ m and $k = 0.05$ mm, in the following cases:

a) $D_1 = D_2 = 200$ mm; $Q_1 = Q_2 = 20$ l/s,
b) $D_1 = D_2 = 400$ mm; $Q_1 = Q_2 = 100$ l/s,
c) $D_1 = D_2 = 800$ mm; $Q_1 = Q_2 = 800$ l/s.

The water temperature equals $10°$ C.

Answer:
For $Q = Q_1 + Q_2$

a) $S = 0.0020$; $D = 259$ mm (adopted $D = 300$ mm)
b) $S = 0.0013$; $D = 520$ mm (adopted $D = 600$ mm)
c) $S = 0.0021$; $D = 1042$ mm (adopted $D = 1100$ mm)

Problem 1.2.11

Find the equivalent diameters of two pipes connected in series, where $L_1 = 460$ m, $L_2 = 240$ m, in the following cases:

a) $D_1 = 400$ mm, $D_2 = 200$ mm; $Q = 80$ l/s,
b) $D_1 = 200$ mm, $D_2 = 400$ mm; $Q = 80$ l/s,
c) $D_1 = 600$ mm, $D_2 = 300$ mm; $Q = 400$ l/s.

Assume for all pipes that $k = 0.01$ mm and the water temperature is $10°$ C.

Answer:
For $L = 700$ m

a) $S = 0.0087$; $D = 246$ mm (adopted $D = 250$ mm)
b) $S = 0.0159$; $D = 217$ mm (adopted $D = 250$ mm)
c) $S = 0.0239$; $D = 368$ mm (adopted $D = 400$ mm)

1.3   Branched systems

Problem 1.3.1

For the branched system shown in Figure 1.1, calculate the pipe flows and nodal pressures for a surface level (msl) in the reservoir that can maintain a minimum network pressure of 20 mwc. Assume for all pipes that $k = 1$ mm and the water temperature is $10°$ C.

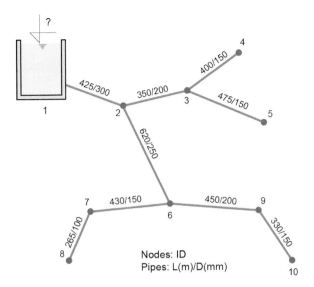

Figure 1.1 Network layout – Problem 1.3.1

Table 1.6 Nodal elevations and demands – Problem 1.3.1

Node	1	2	3	4	5	6	7	8	9	10
Z (msl)	-	18.2	26.5	16.2	13.6	16.3	14.8	13.1	11.3	12.8
Q (l/s)	-79.0	4.5	12.4	11.4	9.9	5.2	11.1	3.3	10.4	10.8

Answer:
The surface elevation of 52.2 msl at node 1 results in the pressures shown in Figure 1.2. The minimum pressure appears to be in node 3 (20.0 mwc).

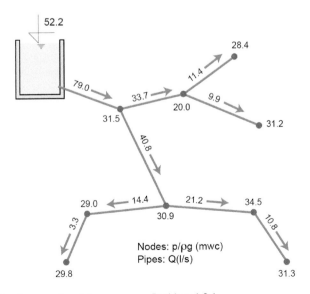

Figure 1.2 Pipe flows and nodal pressures – Problem 1.3.1

Problem 1.3.2

The minimum pressure criterion for the branched system shown in Figure 1.3 is 25 mwc. Determine the surface level of the reservoir in node 1 that can supply a flow of 50 l/s. What will be the water level in the second tank in this scenario? Calculate the pressures and flows in the system. Assume for all pipes that $k = 0.5$ mm and the water temperature is 10° C.

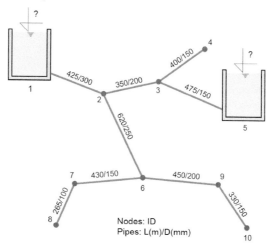

Figure 1.3 Network layout – Problem 1.3.2

Table 1.7 Nodal elevations and demands – Problem 1.3.2

Node	1	2	3	4	5	6	7	8	9	10
Z (msl)	-	18.2	26.5	16.2	-	16.3	14.8	13.1	11.3	12.8
Q (l/s)	-50.0	7.6	16.4	9.2	-34.9	15.2	11.1	9.3	8.3	7.8

Answer:
See Figure 1.4.

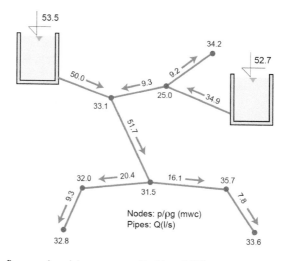

Figure 1.4 Pipe flows and nodal pressures – Problem 1.3.2

Problem 1.3.3

For the same system as in Problem 1.3.2 and the same surface levels in the reservoirs as shown in Figure 1.4, determine the pressures and flows if the demand in node 8 has increased for 10 l/s and in node 10 for 20 l/s.

Note:
Flows in the pipes on the route 1-2-3-5 depend on the water surface elevation difference between the reservoirs, which reflects the total head loss along this route. A trial-and-error process can be applied for the exact flow distribution. This is done until the correct position of the hydraulic grade line connecting the reservoirs has been obtained from the friction loss calculation for each of the three pipes. The two branches emerging from nodes 2 and 3 have fixed flow distribution based on the downstream nodal demands, and their friction losses can be calculated afterwards. The nodal pressures in these branches will be influenced by the reservoir elevations through the heads in nodes 2 and 3.

Answer:
See Figure 1.5.

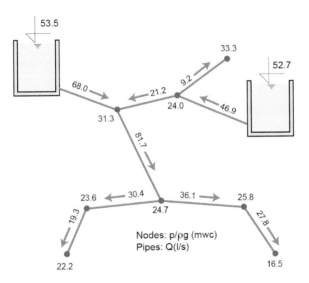

Figure 1.5 Pipe flows and nodal pressures – Problem 1.3.3

Due to the increase in demand, the minimum pressure point has moved from node 3 to node 10.

Problem 1.3.4

Determine the pipe diameters for the layout shown in Figure 1.6, if the maximum-allowed hydraulic gradient $S_{max} = 0.005$. Determine the surface level of the reservoir at the supply point, which can maintain a minimum pressure of 20 mwc. Assume for all pipes that $k = 0.05$ mm and the water temperature is $10°$ C.

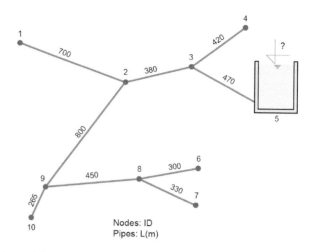

Figure 1.6 Network layout – Problem 1.3.4

Table 1.8 Nodal elevations and demands – Problem 1.3.4

Node	1	2	3	4	5	6	7	8	9	10
Z (msl)	17.6	18.2	16.0	21.4	-	18.0	16.5	19.0	20.4	22.7
Q (l/s)	5.0	7.6	9.9	3.5	-58.1	3.8	4.4	10.5	9.2	4.2

Answer:
Calculation of the pipe diameters based on $S = 0.005$ in each pipe gives the results as shown in Figure 1.7. All nodal pressures in the branch emerging from node 2 are below 20 mwc, thus insufficient.

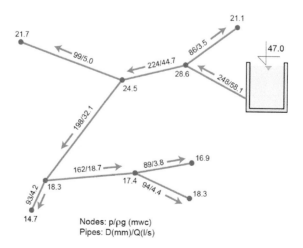

Figure 1.7 Pipe diameters/flows and nodal pressures for $S = 0.005$ in each pipe – Problem 1.3.4

The effect of rounding the pipe diameters to the first higher value can be seen in Figure 1.8. The improvement of the nodal pressures is visible but is still not sufficient.

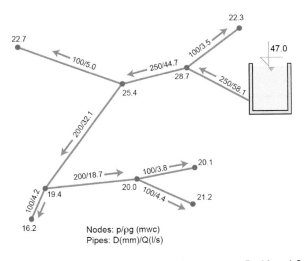

Figure 1.8 Rounded pipe diameters/flows and nodal pressures – Problem 1.3.4

Consequently, a few diameters need to be further enlarged to bring all the pressures above 20 mwc, which is shown in the final solution in Figure 1.9.

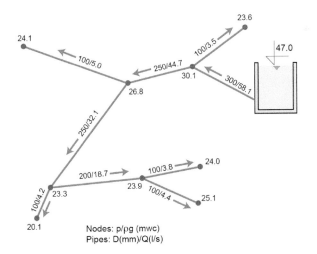

Figure 1.9 Final pipe diameters/flows and nodal pressures – Problem 1.3.4

Problem 1.3.5

For the same system as in Problem 1.3.4 and the same surface level in the reservoir as shown in Figure 1.7, determine the pressures and flows if the demand in nodes 6 and 7 has increased for 10 l/s. Change the pipe diameters where necessary in order to meet the design criteria (S_{max} and $p_{min}/\rho g$).

Answer:

By increasing the demand in nodes 6 and 7 to 13.8 and 14.4 l/s respectively, the pressures in the network will be as shown in Figure 1.10. Nodes 6 to 10 have pressure below 20 mwc. To satisfy the design pressure and hydraulic gradient, pipes 3–2, 2–9, 9–10, 9–8, 8–6 and 8–7 have to be enlarged (see Figure 1.11).

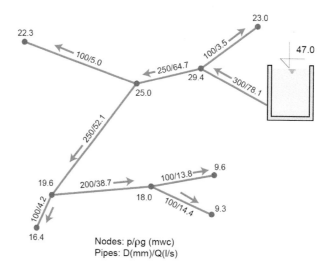

Figure 1.10 Pipe diameters/flows and nodal pressures – Problem 1.3.5

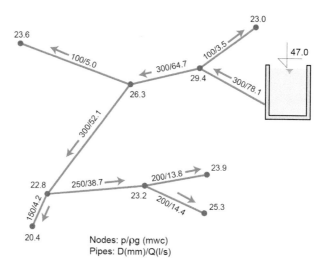

Figure 1.11 Pipe diameters/flows and nodal pressures for increased *D* – Problem 1.3.5

1.4   Looped systems

Problem 1.4.1

For the same system as in Problem 1.3.3 and the same surface levels in the reservoirs as shown in Figure 1.5, determine the pressures and flows if nodes 3 and 9 are connected with a pipe, where $L = 780$ m, $D = 200$ mm and $k = 0.05$ mm.

Note:
Remove all the branches and add their demands to the nodes of the loop 2-3-9-6. A 'dummy' loop, 1-2-3-5, should be formed to determine the flows from the tanks. $\Delta H_{1-5} = 53.5–52.7 = 0.8$ mwc is kept fixed while balancing the heads throughout the iterative calculation.

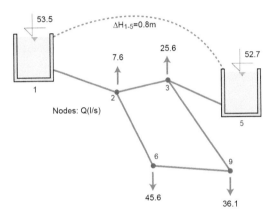

Figure 1.12 Network layout and nodal demands – Problem 1.4.1

Answer:
See Figure 1.13. Adding an extra pipe to create the loop significantly improves the pressures as can be seen in comparison with Figure 1.5.

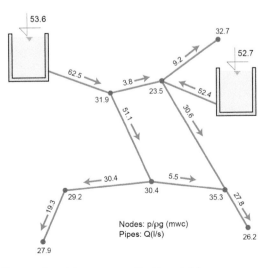

Figure 1.13 Pipe flows and nodal pressures – Problem 1.4.1

Problem 1.4.2

For the same system as in Problem 1.3.5 and keeping the layout as shown in Figure 1.11, determine the pressures and flows if nodes 2 and 8, and 3 and 6 are connected with pipes with respective lengths of 680 m and 470 m. For both pipes $D = 150$ mm and $k = 0.05$ mm.

Answer:
See Figure 1.14. As the figure shows, the pressures in the network will improve by creating loops, compared to those shown in Figure 1.11.

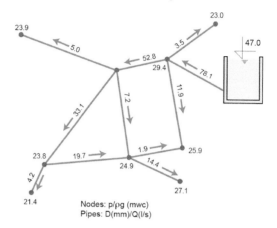

Figure 1.14 Pipe flows and nodal pressures – Problem 1.4.2

Problem 1.4.3

For the layout shown in Figure 1.14, analyse the pressure in the system:

a) after the failure of pipe 9–8, and
b) after the failure of pipe 2–3.

What is the deficit of pressure to be provided at the supply point, in both cases?

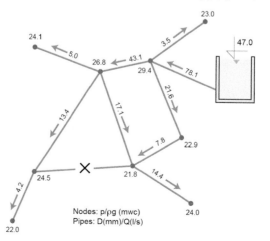

Figure 1.15 Pipe flows and nodal pressures – Problem 1.4.3a

Answers:

a) There is no pressure deficit in the system caused by the failure of pipe 9–8 (Figure 1.15).

b) In this case, Figure 1.16 shows a severe drop of pressure in the system. The observed maximum deficit is 37.3 mwc in node 8 (for $p_{min}/\rho g = 20$ mwc).

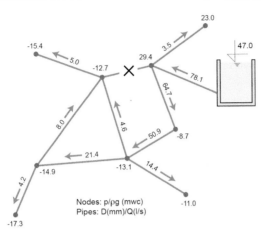

Figure 1.16 Pipe flows and nodal pressures – Problem 1.4.3b

1.5   Hydraulics of storage and pumps

Problem 1.5.1

For the gravity system shown in Figure 1.17, find the maximum capacity of the transport pipe, when $L = 3000$ m, $D = 800$ mm and $k = 0.5$ mm, which can be delivered with a pressure of 35 mwc at the entrance of the city. Assume the water temperature to be 10° C.

Answer:
$Q_{max} = 3782$ m³/h

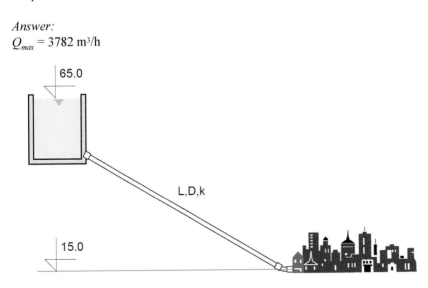

Figure 1.17 Distribution scheme – Problem 1.5.1

Problem 5.2

For the same system as in Problem 1.5.1, a pumping station is built next to the reservoir, as shown in Figure 1.18. The pump characteristics valid during the operation of all the pumps are shown in Figure 1.19.

Determine:
a) the maximum flow of the transport system that can be delivered to the city with the same pressure as in Problem 1.5.1, and
b) the pressure at the entrance of the city if the pumping station delivers the same flow as in Problem 1.5.1.

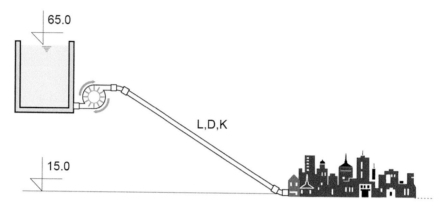

Figure 1.18 Distribution scheme – Problem 1.5.2

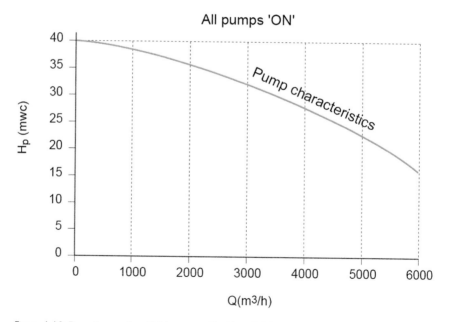

Figure 1.19 Pumping station Q/H curve – Problem 1.5.2

Answers:

a) From the graph in Figure 1.20, the pump delivers a maximum capacity of ±5630 m³/h. The pumping head of ±18 mwc is used in this case to cover the friction loss increase.

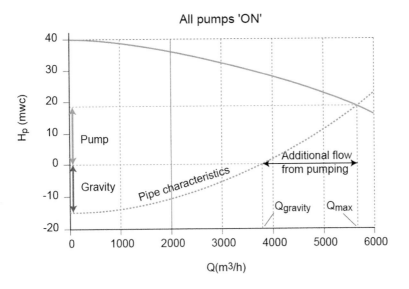

Figure 1.20 Pumping station operation – Problem 1.5.2a

b) From the graph in Figure 1.21, the pump delivers a head of ±29 mwc. As the entire friction loss is covered by gravity, the pumping head will be utilized to deliver the pressure at the entrance of the city. Thus, $p_{entr}/\rho g = 35+29 = 64$ mwc.

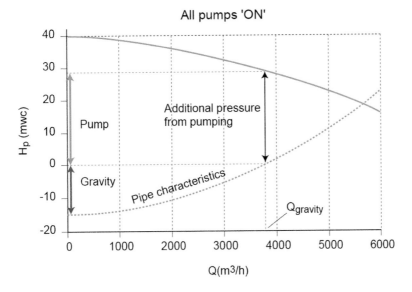

Figure 1.21 Pumping station operation – Problem 1.5.2b

The hydraulic grade lines for both modes of operation are shown in Figure 1.22.

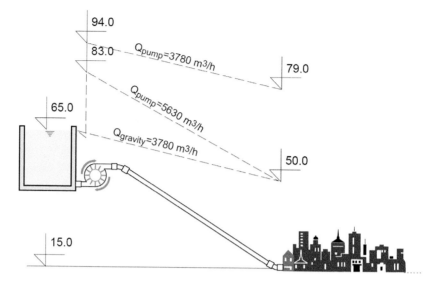

Figure 1.22 Hydraulic grade lines – Problem 1.5.2

Problem 1.5.3

For the combined system shown in Figure 1.23, find the maximum capacity and corresponding pressure at the entrance of the city. Avoid negative pressures along the route. The pipes are:

A-B: $L = 2000$ m, $D = 600$ mm, $k = 1.0$ mm,
B-C: $L = 1200$ m, $D = 700$ mm, $k = 0.1$ mm.

The pumping station operates according to the curve in Figure 1.19. The water temperature can be assumed to be 10° C.

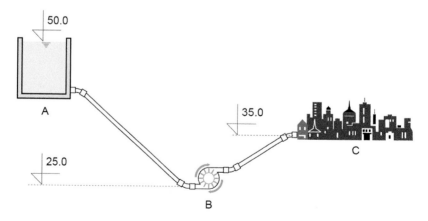

Figure 1.23 Distribution scheme – Problem 1.5.3

Note:
The theoretical maximum flow, without negative pressures, is reached for ΔH_{A-B} = 50–25 = 25 mwc.

Answer:
Q_{max} = 2596 m³/h. The pumping head for this flow is ±34 mwc. Consequently, the calculated ΔH_{B-C} = 4.3 mwc leads to a $p_C/\rho g$ of 19.7 mwc.

Problem 1.5.4

For the system shown in Figure 1.24, determine the pressure at the entrance of the city for a flow of 800 m³/h. The pipes are as follows:

A-B: L = 1500 m, D = 500 mm, k = 0.5 mm,
B-C: L = 1200 m, D = 400 mm, k = 0.1 mm.

Both pumping stations in A and B operate according to the curve shown in Figure 1.25. The water temperature can be assumed to be 10° C.

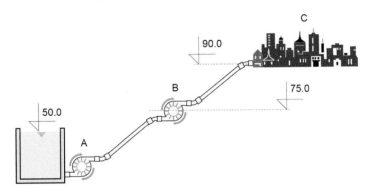

Figure 1.24 Distribution scheme – Problem 1.5.4

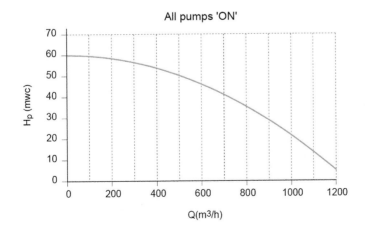

Figure 1.25 Pumping station Q/H curve – Problem 1.5.4

Answer:

For $Q = 800$ m³/h, the total pumping head $H_p = 70.4$ mwc. $\Delta H_{A-B} + \Delta H_{B-C} = 4.0 + 7.6 = 11.6$ mwc. Thus, $p_C/\rho g = 18.8$ mwc.

Problem 1.5.5

For the system shown in Figure 1.26, determine the maximum flow that can be pumped from reservoir A to reservoir B. If the same capacity has to be transported by gravity, find the pressure at the entrance of the city. The pipes are as follows:

    A-B: $L = 1350$ m, $D = 450$ mm, $k = 0.1$ mm,
    B-C: $L = 1800$ m, $D = 500$ mm, $k = 0.1$ mm.

The pumping station operates according to the curve shown in Figure 1.25. The water temperature may be assumed to be 10° C.

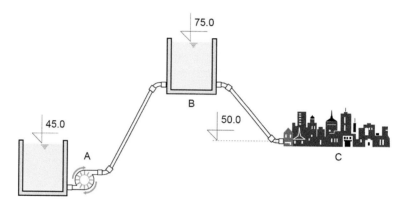

Figure 1.26 Distribution scheme – Problem 1.5.5

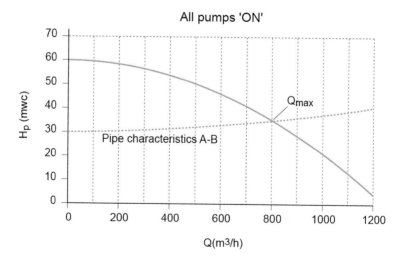

Figure 1.27 Pump operation at section A-B – Problem 1.5.5

Answer:

The maximum pumping capacity Q = 805 m³/h (see Figure 1.27). This flow is delivered by gravity when pressure $p_c/\rho g$ = 21.2 mwc.

Problem 1.5.6

Pumping station B in Figure 1.28 supplies distribution area C from reservoir A through a pipe, where L =1000 m, D = 600 mm, k = 1 mm. The pump characteristics of one pump unit are shown in Figure 1.29. The water temperature can be assumed to be 10° C.

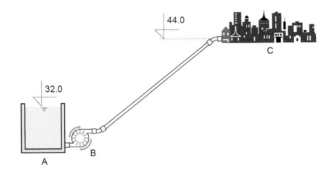

Figure 1.28 Distribution scheme – Problem 1.5.6

The demand of the distribution area registered on the maximum consumption day was 28,008 m³. The demand variation pattern during 24 hours is given in the following table:

Table 1.9 Diurnal peak factors – Problem 1.5.6

Hour	1	2	3	4	5	6	7	8	9	10	11	12
pf_h	0.28	0.30	0.33	0.51	1.07	1.32	1.31	1.38	1.40	1.39	1.36	1.22

Hour	13	14	15	16	17	18	19	20	21	22	23	24
pf_h	1.07	1.04	1.12	1.18	1.29	1.37	1.30	1.33	1.06	0.60	0.44	0.33

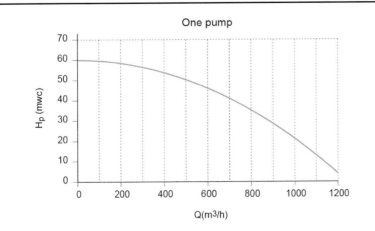

Figure 1.29 Pump characteristics – Problem 1.5.6

Determine:

a) the balancing volume of the reservoir assuming a constant (average) inflow over 24 hours,

b) the required number of pumps arranged in parallel, which can provide the minimum-required pressure of $p_{min}/\rho g = 30$ mwc at the pipe end, during the maximum consumption hour,

c) the same as in b) but for the minimum consumption hour instead,

d) the excessive pumping energy during the maximum and minimum consumption hours; the overall efficiency of the pumping station $\eta_{pst} = 0.65$.

Answers:

a) $Q_{avg} = 1167$ m³/h, $V_{bal} = 4.21Q_{avg} = 4913$ m³.

b) The maximum demand occurs at 9:00 hours, when $Q_9 = 1.40 \times 1167 = 1634$ m³/h. For this flow, $\Delta H_9 = 4.97 \approx 5$ mwc. The head required by one pump, $H_{p,9} = 12 + 5 + 30 = 47$ mwc, is reached for a flow of ± 580 m³/h. Thus, three pumps are necessary.

c) The minimum consumption occurs at 1:00 hours, when $Q_1 = 0.28 \times 1167 = 327$ m³/h. $\Delta H_1 = 0.2$ mwc, $H_{p,1} = 12 + 0.2 + 30 = 42.2$ mwc, for $Q_{p,1} = \pm 680$ m³/h. Hence, one pump is sufficient.

d) The actual pumping head during the maximum consumption hour is ± 48.5 mwc (for the flow 1634 / 3 = 544.7 m³/h). The excessive head is 1.5 mwc and the wasted energy $E_w = 3.43$ kWh per single unit. Hence, for three units $E_w = 10.28$ kWh. During the minimum supply conditions, a flow of 327 m³/h will be pumped against a head of ± 56 mwc. Thus, the excessive head is 13.8 mwc and the wasted energy $E_w = 18.92$ kWh when there is one unit in operation.

Problem 1.5.7

Pumping station B in Figure 1.30 supplies the distribution area C from reservoir A through a pipe $L = 1100$ m, $D = 250$ mm, $k = 0.05$ mm. The pump characteristics of one pump unit are shown in Figure 1.31. The minimum-required pressure at the entrance of the city is 25 mwc. The water temperature can be assumed to be $10°$ C.

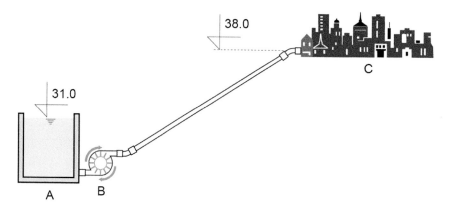

Figure 1.30 Distribution scheme – Problem 1.5.7

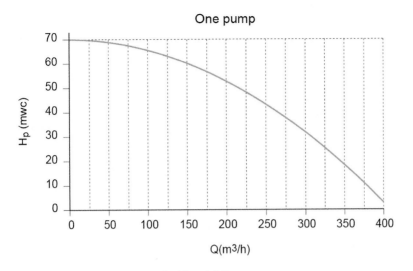

Figure 1.31 Pump characteristics – Problem 1.5.7

Determine:
a)   the maximum flow that can be supplied when one pump is in operation,
b)   the maximum flow that can be supplied if two parallel pumps are in operation,
c)   the maximum flow that can be supplied in cases a) and b), if another pipe with
     $D = 250$ mm is laid in parallel.

Answers:
a) $Q_{max} = 260$ m³/h (see Figure 1.32)

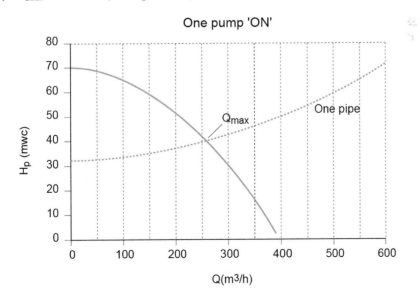

Figure 1.32 Pump operation, one pump – Problem 1.5.7a

b) $Q_{max} = 410$ m³/h (see Figure 1.33).

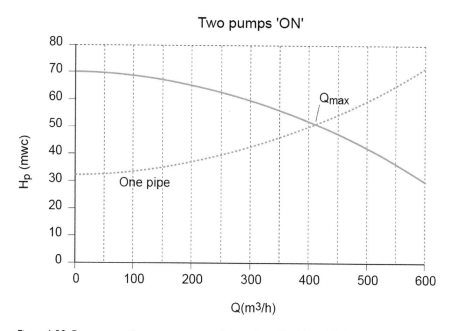

Figure 1.33 Pump operation, two pumps and one pipe – Problem 1.5.7b

c)   By laying the second pipe where $D = 250$ mm, each pipe will transport half of the initial flow, which reduces the friction losses. The composite system characteristics is shown in Figure 1.34. From the graph: $Q_{max,1} = 280$ m³/h for one pump in operation, and $Q_{max,2} = 520$ m³/h, for two pumps.

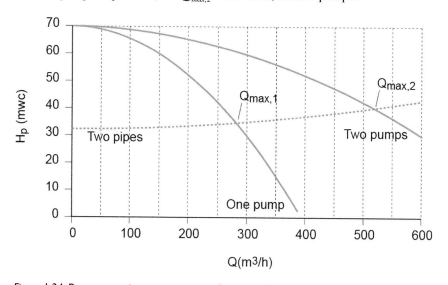

Figure 1.34 Pump operation, two pumps and two parallel pipes – Problem 1.5.7c

Problem 1.5.8

The distribution area C in Figure 1.35 is supplied by gravity through a pipe where $L = 750$ m, $D = 500$ mm, $k = 0.5$ mm. The volume of reservoir B is recovered by pumping A from a well field. The water temperature $T = 10°$ C.

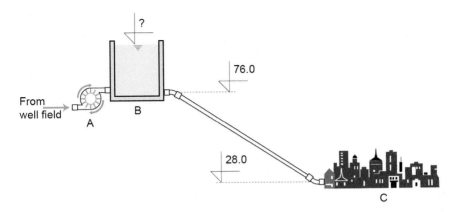

Figure 1.35 Distribution scheme – Problem 1.5.8

The demand of the distribution area that was registered on the maximum consumption day was 30,480 m³. The demand variation pattern during 24 hours is given in the following table:

Table 1.10 Diurnal peak factors – Problem 1.5.8

Hour	1	2	3	4	5	6	7	8	9	10	11	12
pf_h	0.71	0.75	0.77	0.79	0.96	1.14	1.15	1.18	1.20	1.19	1.17	1.07

Hour	13	14	15	16	17	18	19	20	21	22	23	24
pf_h	0.96	0.94	1.00	1.04	1.12	1.17	1.16	1.14	0.96	0.87	0.79	0.76

Determine:
a) the balancing volume of the reservoir, assuming constant (average) pumping over 24 hours,
b) the 24-hour water level variation in the tank, assuming the tank has a cross-section area of 1000 m² and provision for all other purposes of 60 % of the total volume,
c) the range of pressures that appear over 24 hours at the pipe end.

Answers:
a) $Q_{avg} = 1270$ m³/h, $V_{bal} = 1.64 Q_{avg} = 2085$ m³
b) The total volume $V_{tot} = 5215$ m³. The available tank depth is 5.22 m.

Table 1.11 Water depth variation in the tank – Problem 1.5.8

Hour	pf_{out}	pf_{in}	pf_{in}-pf_{out}	Sum	Depth (m)
1	0.71	1	0.29	0.29	4.29
2	0.75	1	0.25	0.54	4.61
3	0.77	1	0.23	0.77	4.90
4	0.79	1	0.21	0.99	5.17
5	0.96	1	0.04	**1.02**	**5.22**
6	1.14	1	-0.14	0.88	5.04
7	1.15	1	-0.15	0.73	4.85
8	1.18	1	-0.18	0.55	4.62
9	1.20	1	-0.20	0.36	4.37
10	1.19	1	-0.19	0.17	4.13
11	1.17	1	-0.17	0.00	3.92
12	1.07	1	-0.07	-0.07	3.83
13	0.96	1	0.04	-0.04	3.87
14	0.94	1	0.06	0.02	3.94
15	1.00	1	0.00	0.02	3.94
16	1.04	1	-0.04	-0.02	3.89
17	1.12	1	-0.12	-0.14	3.74
18	1.17	1	-0.17	-0.32	3.52
19	1.16	1	-0.16	-0.48	3.31
20	1.14	1	-0.14	**-0.62**	**3.13**
21	0.96	1	0.04	-0.58	3.18
22	0.87	1	0.13	-0.46	3.34
23	0.79	1	0.21	-0.24	3.61
24	0.76	1	0.24	0.00	3.92

At midnight, $V_0 = 0.6V_{tot} + 0.62Q_{avg} = 3917$ m³ which corresponds to the depth of 3.92 m (see Figure 1.36).

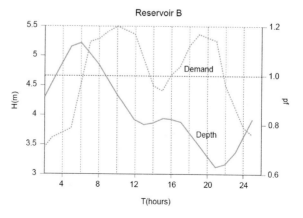

Figure 1.36 Water depth variation vs. demand pattern – Problem 1.5.8

c) The maximum demand occurs at 9:00 hours, where Q_9 = 1524 m³/h. For this flow, ΔH_9 = 7.14 mwc and the remaining pressure in C, $p_{C,9}/\rho g$ = 76 + 4.37 − 7.14−28 ≈ 45 mwc. For the minimum at 1:00 hours, Q_1 = 902 m³/h. ΔH_1 = 2.54 mwc and $p_{C,1}/\rho g$ ≈ 50 mwc.

Problem 1.5.9

For the same problem as in 1.5.8 determine the balancing volume of the reservoir assuming constant pumping of twice the average flow during 12 hours:

a) night time: between 20:00 and 8:00 hours,
b) daytime: from 9:00 to 15:00 hours, and night time: from 23:00 until 5:00 hours.

Answers:
a) V_{bal} = 13.21Q_{avg} = 16,780 m³
b) V_{bal} = 8.47Q_{avg} =10,760 m³

Problem 1.5.10

A distribution area is supplied, as shown in Figure 1.37. The following is a typical demand variation:

Table 1.12 Diurnal peak factors – Problem 1.5.10

Hour	1	2	3	4	5	6	7	8	9	10	11	12
pf$_h$	0.79	0.82	0.83	0.83	0.94	1.06	1.09	1.08	1.11	1.12	1.11	1.05

Hour	13	14	15	16	17	18	19	20	21	22	23	24
pf$_h$	1.02	1.00	1.04	1.07	1.08	1.08	1.09	1.09	1.01	0.96	0.89	0.83

On the maximum consumption day, the pumping station was working at constant (average) capacity of 1800 m³/h against a head of 34 mwc.

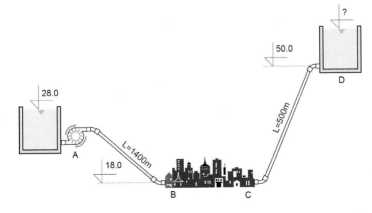

Figure 1.37 Distribution scheme – Problem 1.5.10

Determine:

a) the balancing volume of the tank located in point D,

b) the 24-hour water level variation in the tank, assuming the tank has a cross-section area of 1650 m² and there is provision for all other purposes of 70 % of the total volume,

c) pipe diameters A-B and C-D, providing the minimum pressures at B and C of 35 mwc, during the maximum consumption hour,

d) pressures during the minimum consumption hour, required to refill the tank.

For both pipes $k = 0.5$ mm. Assume the water temperature to be 10° C.

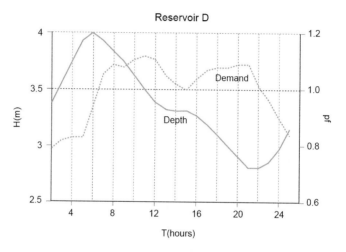

Figure 1.38 Water depth variation in tank D vs. demand pattern – Problem 1.5.10

Answers:

a) $V_{bal} = 1.1Q_{avg} = 1980$ m³

b) The total volume $V_{tot} = 6600$ m³. The available tank depth is 4.0 m. At midnight, $V_0 = 0.7V_{tot} + 0.32Q_{avg} = 5196$ m³ when the depth is 3.15 m (see Figure 1.38).

c) The maximum demand occurs at 10:00 hours, $Q_{10} = 2016$ m³/h, while the minimum appears at 1:00 hours, $Q_1 = 1422$ m³/h. In both cases the pumping station supplies 1800 m³/h and the difference comes from, or goes into, the tank. Hence, $Q_{A-B} = 1800$ m³/h, $H_{A-B} = (28+34) – (18+35) = 9$ mwc. The optimum diameter for these conditions is $D_{A-B} = 574$ mm (manufactured diameter = 600 mm). At 10:00 hours, the tank supplies the system with 216 m³/h from the elevation 50 + 3.50 = 53.50 msl. $\Delta H_{D-C} = 53.5 – (18+35) = 0.5$ mwc. The optimum diameter for such conditions is $D_{D-C} = 369$ mm (manufactured at 400 mm).

d) During the minimum demand hour, the tank receives 378 m³/h to the elevation of 50 + 3.38 = 53.38 msl. $\Delta H_{C-D} \approx 1.0$ mwc for $D = 400$ mm. Thus, the required pressure in C, $p_C/\rho g = 53.4 + 1–18 = 36.4$ mwc. For $D_{A-B} = 600$ mm, $\Delta H_{A-B} = 7.2$ mwc and $p_B/\rho g = 36.8$ mwc > $p_C/\rho g$. Hence, neglecting the resistance in the distribution area itself would allow water to reach the tank. However, as this is not reasonable to assume, either a higher pumping head in A or an additional booster station is needed.

1.6   Examination problems

The following selection of 15 problems was given as a part of the examinations set at IHE Delft in the period 2004–2019. It is an open book exam and students are encouraged to use the spreadsheet applications elaborated in Appendix 7 for the calculations. Each problem is worth 50 out of the total of 100 points, and the available time to solve it is planned at approximately 90 minutes (the total exam duration is three hours).

For simplification purposes, in all the problems (unless specified differently):

1. The water temperature is assumed to be $T = 10°$ C.
2. The specified water surface elevations in the reservoirs/sources are fixed.
3. All minor losses can be neglected.
4. All the pump curves are drawn based on the indicated duty head/flow (H_d/Q_d). The equation used is $H = c - aQ^2$, where $c = 4H_d/3$ and $a = H_d/(3Q_d^2)$.
5. Minor inaccuracies in plotting the pump curve diagrams to reflect the above equation are acceptable if the reading of the heads and flows is done directly from the graph. The values presented in the solutions are calculated by the formula.
6. The warning #VALUE! indicated in some spreadsheet results refers to the pump power which is not considered in the problems.

Problem 1.6.1

Area C shown in Figure 1.39 is to receive water supplied by the pumping station from tank A. The pumping station consists of two equal pump units connected in parallel, each of them operating according to the curve shown in Figure 1.41 ($H_d = 40$ mwc, $Q_d = 100$ l/s). The information about the connecting route is as follows:

Section	L (m)	k (mm)
A – B	1400	0.1
B – C	1200	0.1

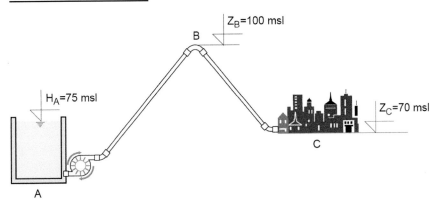

Figure 1.39 Problem 1.6.1 – questions a) and b)

The diurnal pattern of area C on the maximum consumption day is given in Table 1.13. The minimum pressure to be maintained is $p_c/\rho g = 20$ mwc.

Table 1.13 Design diurnal demand of area C – Problem 1.6.1

Hour	1	2	3	4	5	6	7	8	9	10	11	12
Q (m³/h)	250	260	325	445	540	680	720	670	580	550	430	390

Hour	13	14	15	16	17	18	19	20	21	22	23	24
Q (m³/h)	350	370	330	360	440	600	660	550	405	330	295	270

Questions:

a) Calculate the pipe diameter D_{BC} that can maintain the minimum pressure in area C while the minimum pressure in cross section B is $p_B/\rho g = 5$ mwc.

b) What will the pipe diameter D_{AB} be that can supply the most critical flow while maintaining the minimum pressure in area B, as above?

c) In the alternative layout, shown in Figure 1.40, a reservoir is put in position B at the water level of 105 msl in order to simulate the same hydraulic conditions. What is the maximum reduction of the diameter D_{AB} that can be achieved with this action if the pumping station operates with two units?

d) Can the reservoir in B be supplied with only one pump and if so, what would be the diameter D_{AB}?

e) Calculate the balancing volume of the reservoir in B for optimal supply conditions. How is this volume affected by the operation of one or two pumps?

NOTE: All pipe diameters should be calculated with exact values (without rounding!)

Marking matrix:

Question	a	b	c	d	e
Max. points	10	10	10	10	10

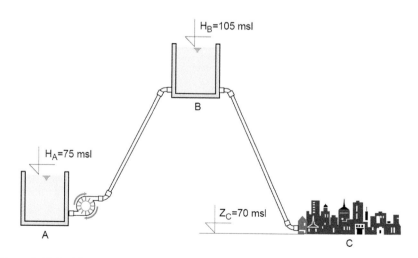

Figure 1.40 Problem 1.6.1 – questions c) to e)

Pump curve

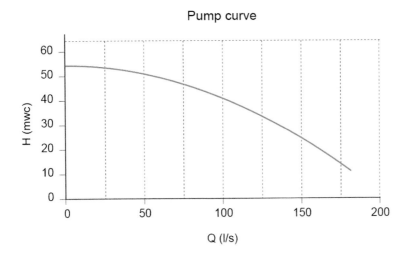

Figure 1.41 Problem 1.6.1 – pump curve for single unit (H_d = 40 mwc, Q_d = 100 l/s)

Answers:

a)   At the pressure $p_C/\rho g$ = 20 mwc, the head H_C = 90 msl. At the pressure $p_B/\rho g$
= 5 mwc, the head H_B = 105 msl. Hence, the available friction loss along B-C
= 15 mwc and the hydraulic gradient S_{BC} = 15/1200 = 0.0125. The maximum
flow in the system appears at 7:00 hours, which is Q_7 = 720 m³/h = 200 l/s
Consequently, D_{BC} = 336 mm (from Spreadsheet Lesson 1–5).

INPUT		OUTPUT	
L (m)	1200	h_f (mwc)	15.00
k (mm)	0.1	u (m²/s)	1.31E-06
Q (l/s)	200	D (mm)	336
S (-)	0.0125	Re (-)	579354
T (°C)	10	λ (-)	0.0162
H₂ (msl)	-	v (m/s)	2.26

b)   With two pumps in parallel, each pump supplies the flow of $Q_{p,7}$ = 100 l/s
at the head of $H_{p,7}$ = 40 mwc. The available friction loss along A-B = 75 +
40–105 = 10 mwc and the gradient S_{AB} = 10/1400 = 0.007. Consequently, D_{AB}
= 375 mm (from Spreadsheet Lesson 1–5).

INPUT		OUTPUT	
L (m)	1400	h_f (mwc)	10.00
k (mm)	0.1	u (m²/s)	1.31E-06
Q (l/s)	200	D (mm)	375
S (-)	0.007143	Re (-)	519627
T (°C)	10	λ (-)	0.0161
H₂ (msl)	-	v (m/s)	1.81

c) The maximum reduction of the diameter can be achieved if $Q_{AB} = 125$ l/s being the average flow for the day (450 m³/h). With two pumps, each pump delivers 62.5 l/s at the head of 48.13 mwc. The available friction loss along A-B is then $= 75 + 48.13 – 105 = 18.13$ mwc and the hydraulic gradient $S_{AB} = 18.13/1400 = 0.013$. Consequently, $D_{AB} = 279$ mm (from Spreadsheet Lesson 1–5).

INPUT		OUTPUT	
L (m)	1400	h_f (mwc)	18.13
k (mm)	0.1	u (m²/s)	1.31E-06
Q (l/s)	125	D (mm)	279
S (-)	0.01295	Re (-)	436122
T (°C)	10	λ (-)	0.0170
H₂ (msl)	-	v (m/s)	2.04

d) If one pump delivers the entire average flow of 125 l/s, the pumping head is then 32.50 mwc. The available friction loss along A-B is consequently $= 75 + 32.50 – 105 = 2.5$ mwc and the hydraulic gradient $S_{AB} = 2.5/1400 = 0.0018$. Consequently, $D_{AB} = 414$ mm (from Spreadsheet Lesson 1–5).

INPUT		OUTPUT	
L (m)	1400	h_f (mwc)	2.50
k (mm)	0.1	u (m²/s)	1.31E-06
Q (l/s)	125	D (mm)	414
S (-)	0.001786	Re (-)	294465
T (°C)	10	λ (-)	0.0167
H₂ (msl)	-	v (m/s)	0.93

e) The balancing volume of the reservoir is 1040 m³ (from Spreadsheet Lesson 8–10), which is not affected by the operation of one or two pumps. In both cases, the average flow is pumped into the reservoir.

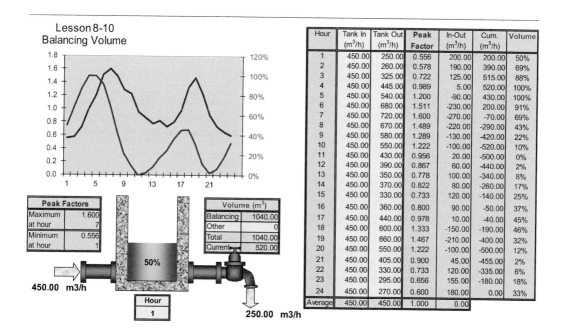

Hour	Tank In (m³/h)	Tank Out (m³/h)	Peak Factor	In-Out (m³/h)	Cum. (m³/h)	Volume
1	450.00	250.00	0.556	200.00	200.00	50%
2	450.00	260.00	0.578	190.00	390.00	69%
3	450.00	325.00	0.722	125.00	515.00	88%
4	450.00	445.00	0.989	5.00	520.00	100%
5	450.00	540.00	1.200	-90.00	430.00	100%
6	450.00	680.00	1.511	-230.00	200.00	91%
7	450.00	720.00	1.600	-270.00	-70.00	69%
8	450.00	670.00	1.489	-220.00	-290.00	43%
9	450.00	580.00	1.289	-130.00	-420.00	22%
10	450.00	550.00	1.222	-100.00	-520.00	10%
11	450.00	430.00	0.956	20.00	-500.00	0%
12	450.00	390.00	0.867	60.00	-440.00	2%
13	450.00	350.00	0.778	100.00	-340.00	8%
14	450.00	370.00	0.822	80.00	-260.00	17%
15	450.00	330.00	0.733	120.00	-140.00	25%
16	450.00	360.00	0.800	90.00	-50.00	37%
17	450.00	440.00	0.978	10.00	-40.00	45%
18	450.00	600.00	1.333	-150.00	-190.00	46%
19	450.00	660.00	1.467	-210.00	-400.00	32%
20	450.00	550.00	1.222	-100.00	-500.00	12%
21	450.00	405.00	0.900	45.00	-455.00	2%
22	450.00	330.00	0.733	120.00	-335.00	6%
23	450.00	295.00	0.656	155.00	-180.00	18%
24	450.00	270.00	0.600	180.00	0.00	33%
Average	450.00	450.00	1.000	0.00		

Lesson 8-10
Balancing Volume

Peak Factors	
Maximum	1.600
at hour	7
Minimum	0.556
at hour	1

Volume (m³)	
Balancing	1040.00
Other	0
Total	1040.00
Current	520.00

450.00 m3/h

50%

Hour
1

250.00 m3/h

Problem 1.6.2

Area C shown in Figure 1.42 receives water by gravity from the reservoir in B, which is supplied by the pumping station at reservoir A. The information about the connecting route is as follows:

Section	L (m)	k (mm)
A – B	800	0.1
B – C	750	0.1

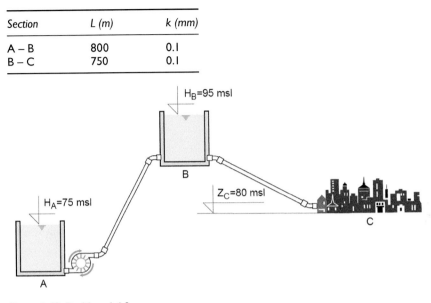

Figure 1.42 Problem 1.6.2

The diurnal pattern of area C on the maximum consumption day is given in Table 1.14.

Table 1.14 Design diurnal demand of area C – Problem 1.6.2

Hour	1	2	3	4	5	6	7	8	9	10	11	12
Q (m³/h)	260	250	320	450	540	680	750	640	580	550	430	370

Hour	13	14	15	16	17	18	19	20	21	22	23	24
Q (m³/h)	350	370	330	360	440	600	660	500	400	330	290	270

Questions:
a) For the diurnal demand in Table 1.14, calculate the balancing volume of the tank in B.
b) A pump unit with the system curve as shown in Figure 1.43 (H_d = 40 mwc, Q_d = 40 l/s) is used to pump the average flow from reservoir A to reservoir B. Calculate the diameter of pipe A-B if three pumps are operated in parallel.
c) Calculate the diameter of pipe B-C that can provide the minimum pressure in area C of $p_c/\rho g$ = 10 mwc. For the selected diameter, calculate the maximum pressure in area C.

Marking matrix:

Question	a	b	c
Max. points	10	20	20

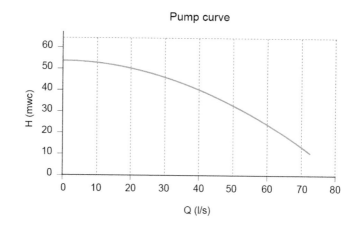

Figure 1.43 Problem 1.6.2 – pump curve for single unit (H_d = 40 mwc, Q_d = 40 l/s)

Answers:

a)   The balancing volume of the reservoir is 1063.31 m³ (from Spreadsheet Lesson 8–10).

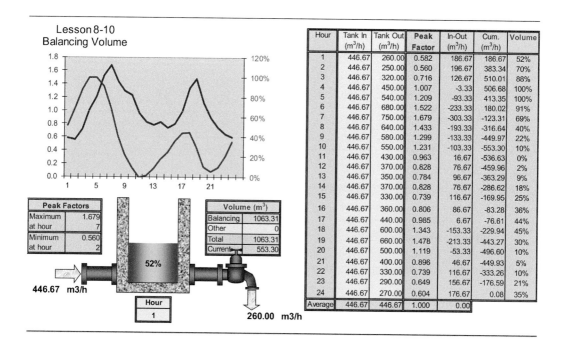

Lesson 8-10
Balancing Volume

Hour	Tank In (m³/h)	Tank Out (m³/h)	Peak Factor	In-Out (m³/h)	Cum. (m³/h)	Volume
1	446.67	260.00	0.582	186.67	186.67	52%
2	446.67	250.00	0.560	196.67	383.34	70%
3	446.67	320.00	0.716	126.67	510.01	88%
4	446.67	450.00	1.007	-3.33	506.68	100%
5	446.67	540.00	1.209	-93.33	413.35	100%
6	446.67	680.00	1.522	-233.33	180.02	91%
7	446.67	750.00	1.679	-303.33	-123.31	69%
8	446.67	640.00	1.433	-193.33	-316.64	40%
9	446.67	580.00	1.299	-133.33	-449.97	22%
10	446.67	550.00	1.231	-103.33	-553.30	10%
11	446.67	430.00	0.963	16.67	-536.63	0%
12	446.67	370.00	0.828	76.67	-459.96	2%
13	446.67	350.00	0.784	96.67	-363.29	9%
14	446.67	370.00	0.828	76.67	-286.62	18%
15	446.67	330.00	0.739	116.67	-169.95	25%
16	446.67	360.00	0.806	86.67	-83.28	36%
17	446.67	440.00	0.985	6.67	-76.61	44%
18	446.67	600.00	1.343	-153.33	-229.94	45%
19	446.67	660.00	1.478	-213.33	-443.27	30%
20	446.67	500.00	1.119	-53.33	-496.60	10%
21	446.67	400.00	0.896	46.67	-449.93	5%
22	446.67	330.00	0.739	116.67	-333.26	10%
23	446.67	290.00	0.649	156.67	-176.59	21%
24	446.67	270.00	0.604	176.67	0.08	35%
Average	446.67	446.67	1.000	0.00		

Peak Factors	
Maximum	1.679
at hour	7
Minimum	0.560
at hour	2

Volume (m³)	
Balancing	1063.31
Other	0
Total	1063.31
Current	553.30

52%

446.67 m3/h

Hour
1

260.00 m3/h

b) The pumping station feeds the reservoir in B with the average flow of Q_{AB} = 446.67 m³/h = 124.08 l/s. Three equal pump units in parallel arrangement then each deliver the flow of $Q_{p,avg}$ = 41.36 l/s at the head of $H_{p,avg}$ = 39.08 mwc. The static head between the two reservoirs is $H_B - H_A = 95-75$ = 20 mwc. Hence, the available friction loss along A-B is 39.08–20 = 19.08 mwc, leading to the hydraulic gradient S_{AB} = 19.08/800 = 0.02385. Consequently, D_{AB} = 247 mm (from Spreadsheet Lesson 1–5), rounded off to 250 mm.

INPUT		OUTPUT	
L (m)	800	h$_f$ (mwc)	19.08
k (mm)	0.1	u (m²/s)	1.31E-06
Q (l/s)	124.08	D (mm)	247
S (-)	0.02385	Re (-)	489597
T (°C)	10	λ (-)	0.0172
H₂ (msl)	-	v (m/s)	2.59

c) At the pressure $p_c/\rho g$ = 10 mwc, the head H_C = 90 msl. Hence, the available friction loss along B-C = 95–90 = 5 mwc and the hydraulic gradient S_{BC} = 5/750 = 0.006667. The maximum flow in the system appears at 7:00 hours, which is Q_7 = 750 m³/h = 208.33 l/s. Consequently, D_{BC} = 386 mm (from Spreadsheet Lesson 1–5), rounded off to 400 mm.

INPUT		OUTPUT	
L (m)	750	h$_f$ (mwc)	5.00
k (mm)	0.1	u (m²/s)	1.31E-06
Q (l/s)	208.33	D (mm)	386
S (-)	0.006667	Re (-)	525925
T (°C)	10	λ (-)	0.0160
H₂ (msl)	-	v (m/s)	1.78

With D_{BC} = 400 mm, the maximum pressure in area C will appear during the peak minimum hour, at 2:00 hours, when the demand Q_2 = 250 m³/h = 69.44 l/s. Consequently, the friction loss h_{BC} = 0.52 mwc (from Spreadsheet Lesson 1–2), and the maximum pressure $p_c/\rho g$ = 95–0.52–80 = 14.48 mwc.

INPUT		OUTPUT	
L (m)	750	v (m/s)	0.55
D (mm)	400	u (m²/s)	1.31E-06
k (mm)	0.1	Re (-)	169177
Q (l/s)	69.44	λ (-)	0.0179
T (°C)	10	h$_f$ (mwc)	0.52
H₂ (msl)	-	S (-)	0.0007

Problem 1.6.3

Area C shown in Figure 1.44 receives water by gravity from the reservoir in B, which is supplied by the pumping station at reservoir A. The pumping station comprises two equal pump units connected in parallel, each of them operating according to the curve shown in Figure 1.46 (H_d = 20 mwc, Q_d = 100 l/s). The information about the connecting route is as follows:

Section	L (m)	D (mm)	k (mm)
A – B	500	300	0.6
B – C	400	250	1.0

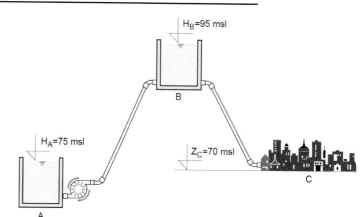

Figure 1.44 Problem 1.6.3 – questions a) and b)

Questions:
a) Calculate the pumping flow for the water levels in the tanks A and B, and, assuming that the equal flows, $Q_{AB} = Q_{BC}$, determine the pressure in area C if one pump unit is in operation.
b) Calculate the pumping flow for the water levels in the tanks A and B, and assuming that the equal flows, $Q_{AB} = Q_{BC}$, determine the pressure in area C if both pump units are in operation.
c) In the alternative layout, shown in Figure 1.45, area C is directly connected to the pumping station in A, bypassing the hill with the reservoir. The pipe section between A and C has the following characteristics: L_{AC} = 1600 m and k_{AC} = 0.2 mm. What will the required diameter of pipe section AC be if the same flow and the pressure at C, as calculated in question b), are to be maintained; the pipe route ABC is closed in this case and the reservoir at B will be disconnected for maintenance purposes.

Marking matrix:

Question	a	b	c
Max. points	15	15	20

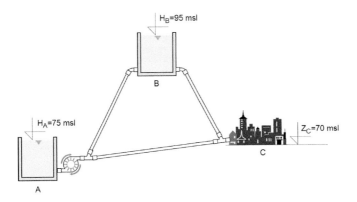

Figure 1.45 Problem 1.6.3 – question c)

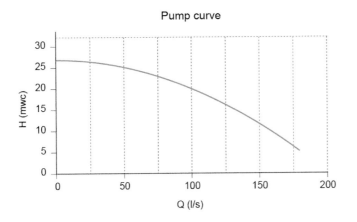

Figure 1.46 Problem 1.6.3 – pump curve for single unit (H_d = 20 mwc, Q_d = 100 l/s)

Answers:

a) The static head between the two reservoirs is $H_B - H_A = 95 - 75 = 20$ mwc. In addition, the pump needs to overcome the friction loss, which is dependent on the flow rate. The solution can be obtained by trying with a few values of (pump) flows to calculate the pipe friction loss that needs to match the pumping head corresponding to the pumping flow. Because of the static head value, the pumping flow for the given pump curve will be less than 100 l/s, as a starting point. The following table can be produced by using Spreadsheet Lesson 1–2 for the friction loss calculation.

$Q_{p,1}$ (l/s)	$H_{p,1}$ (mwc)	$20 + h_f$ (mwc)	ΔH (mwc)
95	20.65	23.71	−3.06
90	21.27	23.33	− 2.06
85	21.85	22.98	− 1.13
80	22.40	22.64	− 0.24
75	22.92	22.33	0.59

The difference between the pumping head and the sum of the static head and the friction loss needs to be zero. The table shows that this is going to happen for the pumping flow between 75 and 80 l/s. A more accurate value can be obtained further refining the flows by repeating the same process.

$Q_{p,1}$ (l/s)	$H_{p,1}$ (mwc)	20 + h_f (mwc)	ΔH (mwc)
79	22.51	22.58	-0.07
78	22.61	22.51	0.10
78.8	22.53	22.56	-0.03
78.6	22.55	22.55	**0**
78.4	22.57	22.54	0.03

Alternatively, the system characteristics can be drawn for pipe A-B, which will show the correct pump flow in the intersection with the pump curve; the spreadsheet lessons 6–2 or 6–3 can be used to facilitate this calculation, with minor adaptations. Afterwards, the pressure in area C can be calculated as explained in the problem 1.6.2-c. Finally, and most easily, using Spreadsheet Lesson 7–2 offers straightforward results, which are $Q_{p,1} = Q_{AB} = 78.6$ l/s and $p_C/\rho g = 18.96$ mwc, for $Q_{AB} = Q_{BC}$. Parameter H_w in the spreadsheet corresponds to ΔH in the above tables.

b) With two pumps in parallel, the combined operation of both pumps can be described by the pump curve with the same duty head and doubled duty flow ($H_d = 20$ mwc, $Q_d = 200$ l/s). The results in this case will be $Q_{p,2} = Q_{AB} = 107.6$ l/s and $p_C/\rho g = 13.72$ mwc, for $Q_{AB} = Q_{BC}$ (from Spreadsheet Lesson 7–2).

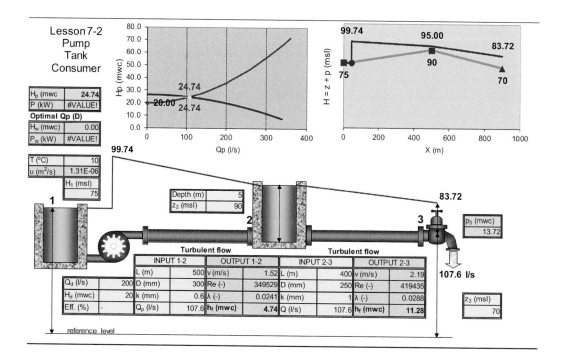

c) Two equal pumps in parallel should supply the flow $Q_{AC} = 107.6$ l/s at the pressure $p_C/\rho g = 13.72$ mwc. Hence, each pump delivers 53.8 l/s at the head of 24.74 mwc. The available friction loss along A-C is then $= 75 + 24.74 - 13.72 - 70 = 16.02$ mwc and the hydraulic gradient $S_{AC} = 16.02/1600 = 0.010013$. Consequently, $D_{AC} = 284$ mm (from Spreadsheet Lesson 1–5), rounded off to 300 mm.

INPUT		OUTPUT	
L (m)	1600	h_f (mwc)	16.02
k (mm)	0.2	u (m²/s)	1.31E-06
Q (l/s)	107.6	D (mm)	284
S (-)	0.010013	Re (-)	369376
T (°C)	10	λ (-)	0.0192
H₂ (msl)	-	v (m/s)	1.70

Problem 1.6.4

Area C and the reservoir in B shown in Figure 1.47 receive water supplied by the pumping station from reservoir A. The water is also supplied by gravity from B to C, at the same time. The pumping station consists of two equal pump units connected in parallel, each of them operating according to the curve shown in Figure 1.48 ($H_d = 40$ mwc, $Qd = 100$ l/s). The information about the connecting routes is as follows:

Section	L (m)	D (mm)	k (mm)
A – B	2000	300	0.5
B – C	2200	250	1.5
A – C	2800	300	2.0

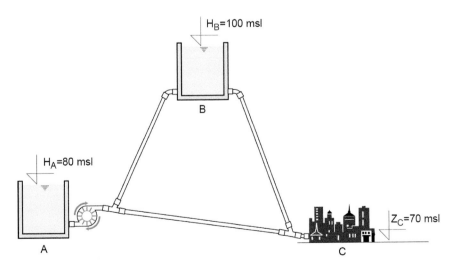

Figure 1.47 Problem 1.6.4

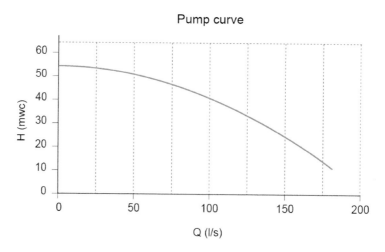

Figure 1.48 Problem 1.6.4 – pump curve for single unit (H_d = 40 mwc, Q_d = 100 l/s)

Questions:
a) The flow registered in the pumping station at a particular moment in time is Q_p = 200 l/s (both pump units are in operation). Calculate the demand and the pressure in area C at that moment. How much of the flow is supplied to the reservoir in B?
b) In the event of maintenance of reservoir B (the route A-B-C is closed), can the entire demand of area C calculated in question a) be supplied through the route A-C and by one pump only? What will the pressure in area C be in such a situation?
c) If the backflow is to be hydraulically prevented from reservoir B to reservoir A, what must be the maximum pumping flow if the pumping station operates with one pump unit? Calculate the diameter of new pipe A-C (at k = 0.1 mm) that can supply this flow with the pressure in area C of $p_c/\rho g$ = 10 mwc. What is the demand of area C at that point in time?

Marking matrix:

Question	a	b	c
Max. points	20	10	20

Answers:
a) At the pumping flow Q_p = 200 l/s, each of the two pumps delivers 100 l/s at the head of 40 mwc. The available friction loss along A-B is then = 80 + 40–100 = 20 mwc and the hydraulic gradient S_{AB} = 20/2000 = 0.01. Consequently, Q_{AB} = 113.03 l/s (from Spreadsheet Lesson 1–4).

INPUT		OUTPUT	
L (m)	2000	h_f (mwc)	20.00
D (mm)	300	u (m²/s)	1.31E-06
k (mm)	0.5	Re (-)	367387
S (-)	0.01	λ (-)	0.0230
T (°C)	10	v (m/s)	1.60
H_2 (msl)	-	Q (l/s)	113.03

$Q_{AC} = Q_p - Q_{AB} = 200-113.03 = 86.97$ l/s. Based on this flow, the friction loss along A-C is 24.19 mwc (from Spreadsheet Lesson 1–2).

INPUT		OUTPUT	
L (m)	2800	v (m/s)	1.23
D (mm)	300	u (m²/s)	1.31E-06
k (mm)	2	Re (-)	282515
Q (l/s)	86.97	λ (-)	0.0336
T (°C)	10	h_f (mwc)	**24.19**
H₂ (msl)	-	S (-)	0.0086

The pressure in C is then $p_C/\rho g = 80 + 40-24.19-70 = 25.81$ mwc. For this pressure, the available friction loss for supply from reservoir B is $100-70-25.81 = 4.19$ mwc. Consequently, $S_{BC} = 4.19/2200 = 0.001905$ and $Q_{BC} = 26.05$ l/s (from Spreadsheet Lesson 1–4).

INPUT		OUTPUT	
L (m)	2200	h_f (mwc)	4.19
D (mm)	250	u (m²/s)	1.31E-06
k (mm)	1.5	Re (-)	101414
S (-)	0.001905	λ (-)	0.0332
T (°C)	10	v (m/s)	0.53
H₂ (msl)	-	Q (l/s)	**26.05**

Therefore, the total supply of area B is $Q_{AC} + Q_{BC} = 86.97 + 26.05 = 113.02$ l/s (incidentally, very similar to Q_{AB}).

b)   For $Q_{AC} = 113.02$ l/s, the friction loss along A-C is 40.74 mwc (from Spreadsheet Lesson 1–2).

INPUT		OUTPUT	
L (m)	2800	v (m/s)	1.60
D (mm)	300	u (m²/s)	1.31E-06
k (mm)	2	Re (-)	367136
Q (l/s)	113.02	λ (-)	0.0335
T (°C)	10	h_f (mwc)	**40.74**
H₂ (msl)	-	S (-)	0.0145

For the same pumping flow delivered by one pump only, the pumping head is $H_p = 36.30$ mwc, and the pressure in area C will be $p_C/\rho g = 80 + 36.30 - 40.74 - 70 = 5.56$ mwc. Hence, the supply of the entire demand by one pump is possible but at rather low pressure.

c) The pumping station supplies the reservoir in B as long the pumping head is bigger than the surface level difference of A and B (the static head), which is $H_B - H_A = 100-80 = 20$ mwc. The flow direction along A-B will reverse at the moment the pumping head drops below this value. Hence, for the critical pumping head $H_p = 20$ mwc, one pump in operation delivers $Q_p = Q_{AC} = 158.11$ l/s ($Q_{AB} = 0$). For the pressure in C of $p_C/\rho g = 10$ mwc, the available friction loss alongside pipe A-C is $H_A - H_C = 80 + 20-10-70 = 20$ mwc. Consequently, $S_{BC} = 20/2800 = 0.007143$ and $D_{AC} = 343$ mm (from Spreadsheet Lesson 1–5).

INPUT		OUTPUT	
L (m)	2800	h_f (mwc)	20.00
k (mm)	0.1	u (m²/s)	1.31E-06
Q (l/s)	158.11	D (mm)	343
S (-)	0.007143	Re (-)	449071
T (°C)	10	λ (-)	0.0165
H₂ (msl)	-	v (m/s)	1.71

At the same time, $H_B - H_C = 100-10-70 = 20$ mwc. Consequently, $S_{BC} = 20/2200 = 0.009091$ and $Q_{BC} = 57.38$ l/s (from Spreadsheet Lesson 1–4).

INPUT		OUTPUT	
L (m)	2200	h_f (mwc)	20.00
D (mm)	250	u (m²/s)	1.31E-06
k (mm)	1.5	Re (-)	223876
S (-)	0.009091	λ (-)	0.0326
T (°C)	10	v (m/s)	1.17
H₂ (msl)	-	Q (l/s)	57.38

Finally, the total demand in area C is $158.11 + 57.38 = 215.49$ l/s.

Problem 1.6.5

Area C shown in Figure 1.49 receives water by gravity from the reservoir in B, which is supplied by the pumping from reservoir A. The pump unit operates according to the curve shown in Figure 1.50 ($H_d = 30$ mwc, $Q_d = 15$ l/s). The information about the connecting route is as follows:

Section	L (m)	D (mm)	k (mm)
A – B	800	?	1.5
B – C	750	200	0.2

Questions:
a) On the day of maximum consumption, the registered demand in C was 675 m³; the maximum hourly peak demand factor is 2.0. What was the pressure in area C during the maximum consumption hour on the maximum consumption day?

b) At the same time, the pump unit is supplying the average flow on the maximum consumption day from reservoir A to reservoir B. Determine the pipe diameter D_{AB} needed to satisfy these conditions.

c) A new residential area in D is to be connected to the system. The planned average specific demand of 3000 inhabitants is 100 l/c/d (including leakage) and the peak seasonal variation factor is 1.5; the maximum hourly peak demand factor is the same as in the case of area C. What is going to be the pressure in area D if connected with area C by the pipe of $L_{CD} = 500$ m, $D_{CD} = 150$ mm and $k_{CD} = 0.1$ mm?

d) Alternatively, what is going to be the pressure in area D if connected directly to the reservoir in B by the pipe of $L_{BD} = 600$ m, $D_{BD} = 100$ mm and $k_{BD} = 0.1$ mm?

e) Can the pump cope with the increase of the demand after the residential area in D has been connected? Explain your answer.

Marking matrix:

Question	a	b	c	d	e
Max. points	10	10	10	10	10

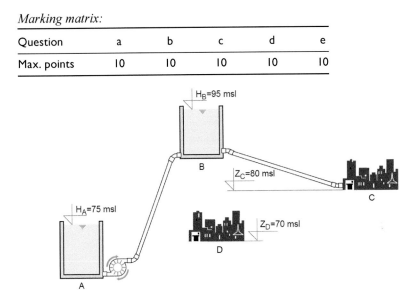

Figure 1.49 Problem 1.6.5

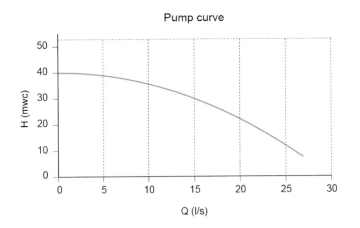

Figure 1.50 Problem 1.6.5 – pump curve for single unit (H_d = 30 mwc, Q_d = 15 l/s)

Answers:

a)   The daily demand of 675 m³ converts into an average flow of 675/24/3.6 = 7.8125 l/s. With the peak demand factor of 2.0, the demand in area C during the maximum consumption hour on the maximum consumption day is 15.63 l/s. For this demand, the friction loss along B-C is 1.09 mwc (from Spreadsheet Lesson 1–2), leading to the pressure $p_C/\rho g = 95 - 1.09 - 80 = 13.91$ mwc.

INPUT		OUTPUT	
L (m)	750	v (m/s)	0.50
D (mm)	200	u (m²/s)	1.31E-06
k (mm)	0.2	Re (-)	76159
Q (l/s)	15.63	λ (-)	0.0230
T (°C)	10	h_f (mwc)	1.09
H₂ (msl)	-	S (-)	0.0014

b)   The average pumping flow $Q_p = 7.8125$ l/s is delivered at the pumping head $H_p = 37.29$ mwc. The available friction loss along A-B is then = 75 + 37.29 - 95 = 17.29 mwc and the hydraulic gradient $S_{AB} = 17.29/800 = 0.021613$. Consequently, $D_{AB} = 101$ mm (from Spreadsheet Lesson 1–5).

INPUT		OUTPUT	
L (m)	800	h_f (mwc)	17.29
k (mm)	1.5	u (m²/s)	1.31E-06
Q (l/s)	7.8125	D (mm)	101
S (-)	0.021613	Re (-)	75569
T (°C)	10	λ (-)	0.0444
H₂ (msl)	-	v (m/s)	0.98

c)   3000 new inhabitants using on average 100 l/c/d convert into an average flow of 3000×100/24/3600 = 3.472 l/s. Applying both the peak seasonal factor and the hourly peak demand factor, the demand in area D during the maximum consumption hour on the maximum consumption day equals 3.472×1.5×2.0 = 10.42 l/s. Connecting this area with the specified pipe to the area in C results in the friction loss along C-D of 1.31 mwc (from Spreadsheet Lesson 1–2), leading to the pressure $p_D/\rho g = 80 + 13.91 - 1.31 - 70 = 22.60$ mwc.

INPUT		OUTPUT	
L (m)	500	v (m/s)	0.59
D (mm)	150	u (m²/s)	1.31E-06
k (mm)	0.1	Re (-)	67697
Q (l/s)	10.42	λ (-)	0.0222
T (°C)	10	h_f (mwc)	1.31
H₂ (msl)	-	S (-)	0.0026

d) Connecting area D with the specified pipe directly to the reservoir in B results in the friction loss along B-D of 12.01 mwc (from Spreadsheet Lesson 1–2), leading to the pressure $p_D/\rho g = 95-12.01-70 = 12.99$ mwc. This is the worse alternative purely because of the reduced pipe diameter generating much higher friction loss than the one calculated in question c).

INPUT		OUTPUT	
L (m)	600	v (m/s)	1.33
D (mm)	100	u (m²/s)	1.31E-06
k (mm)	0.1	Re (-)	101545
Q (l/s)	10.42	λ (-)	0.0223
T (°C)	10	h$_f$ (mwc)	**12.01**
H$_2$ (msl)	-	S (-)	0.0200

e) To maintain appropriate demand balancing by the reservoir in B, after connecting area D, the average pumping from A to B on the maximum consumption day needs to increase. The increased flow $Q_p = 7.81 + 10.42/2.0 = 13.02$ l/s is delivered at the pumping head $H_p = 32.47$ mwc, which is 4.82 mwc below the original pumping head of 37.29 mwc. There are two options to mitigate this problem: (1) to add an additional pump unit in parallel; with the same size, two pumps will deliver the head of $H_p = 38.12 > 37.29$ mwc for the flow $Q_p = 13.02$ l/s; (2) to keep the single pump but further increase the pipe diameter D_{AB}; in this scenario, the available friction loss along A-B is $= 75 + 32.47-95 = 12.47$ mwc and the hydraulic gradient $S_{AB} = 12.47/800 = 0.015588$. Consequently, $D_{AB} = 129$ mm (from Spreadsheet Lesson 1–5). The first option will result in lower investment, but higher operational costs.

INPUT		OUTPUT	
L (m)	800	h$_f$ (mwc)	12.47
k (mm)	1.5	u (m²/s)	1.31E-06
Q (l/s)	13.02	D (mm)	**129**
S (-)	0.015588	Re (-)	98053
T (°C)	10	λ (-)	0.0406
H$_2$ (msl)	-	v (m/s)	0.99

Problem 1.6.6

Area B shown in Figure 1.51 receives water by gravity from the sources in A and C. The information about the connecting route is as follows:

Section	L (m)	D (mm)	k (mm)
A – B	1000	250	0.5
B – C	500	?	0.045

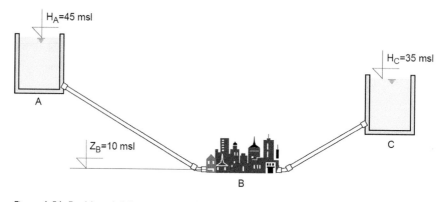

Figure 1.51 Problem 1.6.6 – questions a) to c)

Questions:
a) At a particular point in time during the day, the pressure in area B is measured at $p_B/\rho g = 20$ mwc. Determine the supply from source A for this situation.
b) For the same scenario as in question a), determine the pipe diameter D_{BC} needed to provide the equal supply flow from source C.
c) At a certain moment, during the night time, the pressure in area B is measured at $p_B/\rho g = 26$ mwc. Determine the demand in area B for this situation. The pipe diameter D_{BC} is the same as calculated in question b).
d) Options are considered to improve the supply by installing a pump in source A as shown in Figure 1.52. This pump operates according to the curve shown in Figure 1.53 ($H_d = 40$ mwc, $Q_d = 150$ l/s). What is the demand increase in area B resulting from this intervention if the same pressure of $p_B/\rho g = 20$ mwc is maintained during the daytime?
e) What is the demand increase in area B resulting from the pump installation if the same pressure of $p_B/\rho g = 26$ mwc is maintained during the night time?

Marking matrix:

Question	a	b	c	d	e
Max. points	5	5	10	15	15

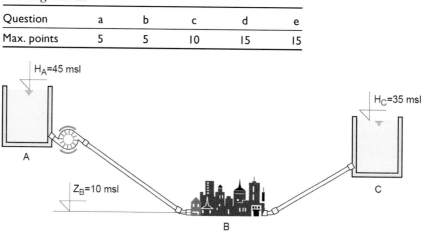

Figure 1.52 Problem 1.6.6 – questions d) and e)

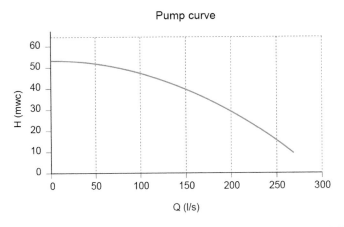

Figure 1.53 Problem 1.6.6 – pump curve for single unit (H_d = 40 mwc, Q_d = 150 l/s)

Answers:

a) For the pressure $p_B/\rho g$ = 20 mwc, the head H_B = 30 msl. The available friction loss for the supply from reservoir A is then 45–30 = 15 mwc. Consequently, S_{AB} = 15/1000 = 0.015 and Q_{AB} = 85.77 l/s (from Spreadsheet Lesson 1–4).

INPUT		OUTPUT	
L (m)	1000	h_f (mwc)	15.00
D (mm)	250	u (m²/s)	1.31E-06
k (mm)	0.5	Re (-)	334858
S (-)	0.015	λ (-)	0.0241
T (°C)	10	v (m/s)	1.75
H₂ (msl)	-	Q (l/s)	85.77

b) The flow Q_{BC} = Q_{AB} = 85.77 l/s. The available friction loss along B-C is then = 35–30 = 5 mwc and the hydraulic gradient S_{BC} = 5/500 = 0.01. Consequently, D_{BC} = 250 mm (from Spreadsheet Lesson 1–5).

INPUT		OUTPUT	
L (m)	500	h_f (mwc)	5.00
k (mm)	0.045	u (m²/s)	1.31E-06
Q (l/s)	85.77	D (mm)	250
S (-)	0.01	Re (-)	334599
T (°C)	10	λ (-)	0.0160
H₂ (msl)	-	v (m/s)	1.75

c)   For the pressure $p_B/\rho g = 26$ mwc, the head $H_B = 36$ msl. The available friction loss for the supply from reservoir A is then $45-36 = 9$ mwc. Consequently, $S_{AB} = 9/1000 = 0.009$ and $Q_{AB} = 66.20$ l/s (from Spreadsheet Lesson 1–4).

INPUT		OUTPUT	
L (m)	1000	h$_f$ (mwc)	9.00
D (mm)	250	u (m^2/s)	1.31E-06
k (mm)	0.5	Re (-)	258319
S (-)	0.009	λ (-)	0.0243
T (°C)	10	v (m/s)	1.35
H$_2$ (msl)	-	Q (l/s)	66.20

At the same time, the available friction loss for the supply from reservoir C is then $35-36 = -1$ mwc, meaning that reservoir A also supplies reservoir C during the night time ($H_B > H_C$). Consequently, $S_{BC} = 1/500 = 0.002$ and $Q_{BC} = 36.37$ l/s (from Spreadsheet Lesson 1–4).

INPUT		OUTPUT	
L (m)	500	h$_f$ (mwc)	1.00
D (mm)	250	u (m^2/s)	1.31E-06
k (mm)	0.045	Re (-)	141597
S (-)	0.002	λ (-)	0.0179
T (°C)	10	v (m/s)	0.74
H$_2$ (msl)	-	Q (l/s)	36.37

Finally, the demand in area B is $Q_B = 66.20-36.37 = 29.83$ l/s.

d)   The pump should deliver the water from the head $H_A = 45$ msl to the head $H_B = 30$ msl. Thus, the static head is negative ($\Delta H = -15$ mwc) because the supply is still partly facilitated by the gravity (a smaller pump needed for that purpose). The further considerations are similar to those in problem 1.6.3-a. The trial-and-error process will include the flows higher than the one delivered by gravity, as calculated in question a). The following table can be produced by using Spreadsheet Lesson 1–2 for the friction loss calculation.

Q_p (l/s)	H_p (mwc)	$-15 + h_f$ (mwc)	ΔH (mwc)
90	48.53	1.50	47.03
110	46.16	9.53	36.63
130	43.32	19.16	24.16
150	40.00	30.37	9.63
170	36.21	43.16	− 6.95

The difference between the pumping head and the sum of the static head and the friction loss along A-B needs to be zero. The table shows that this is going to happen for the pumping flow between 150 and 170 l/s. A more accurate value can be obtained by further refining the flows by repeating the same process. Alternatively, Spreadsheet Lesson 6–2 gives the final result, of $Q_{AB} = 161.91$ l/s. The supply from the reservoir in C is unchanged and the total demand in area B will be $Q_B = 161.91 + 85.77 = 247.68$ l/s, which is an increase of nearly 45 %.

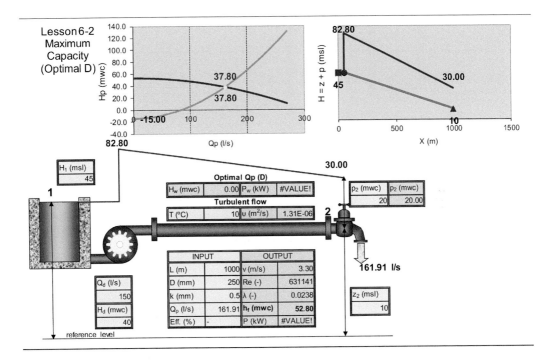

e) For the night time conditions and the pressure $p_B/\rho g = 26$ mwc, the static head will be $\Delta H = -9$ mwc and the flow delivered by the pump is consequently going to be $Q_{AB} = 154.60$ l/s (from Spreadsheet Lesson 6–2). The supply to the reservoir in C remains as calculated in question c), and finally the demand in area B is $Q_B = 154.60 - 36.37 = 118.23$ l/s, which is a significant increase compared to the supply by gravity alone (nearly four times higher).

Problem 1.6.7

A water supply system is to be designed for the supply of area C shown in Figure 1.54. The diurnal demand pattern is expected to be as given in Table 1.15

Table 1.15 Design diurnal demand of area C – Problem 1.6.7

Hour	1	2	3	4	5	6	7	8	9	10	11	12
Q (m³/h)	260	250	330	450	540	680	780	640	580	550	430	370

Hour	13	14	15	16	17	18	19	20	21	22	23	24
Q (m³/h)	350	370	330	360	440	600	660	540	400	330	290	270

The information about the connecting route is as follows:

Section	L (m)	k (mm)
A – B	800	0.5
B – C	550	0.5

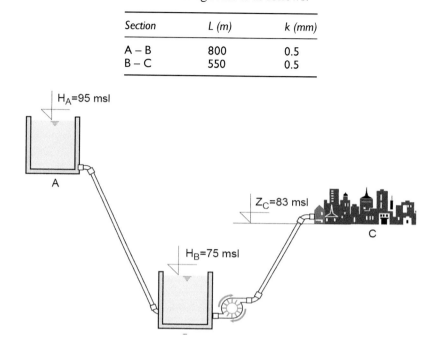

Figure 1.54 Problem 1.6.7

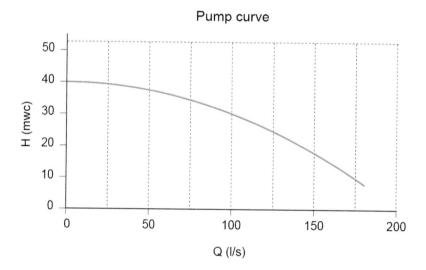

Figure 1.55 Problem 1.6.7 – pump curve for single unit (H_d = 30 mwc, Q_d = 100 l/s)

Questions:

a)   The reservoir in A receives water from the source and supplies the reservoir in B by gravity for 18 hours per day, between the hours 4 and 21. During the night time, between hours 22 and 3, the volume of reservoir A has to be recharged to have enough water for the next supply cycle of B. For such a regime and the demand in Table 1.15, calculate the balancing volumes of the tanks in A and B.

b) Calculate the diameter of pipe A-B that can provide the demand balancing from the reservoir in B. Assume the first higher manufactured value from the following list (in mm): 100, 150, 200, 250, 300, 350, 400, 500, 600, 800 and 1000. Can the duration of supply from the reservoir A be shorter than 18 hours based on the selection of the manufactured diameter? Adapt the balancing volume of the reservoir in B calculated in question a), if necessary.

c) A pump unit with the curve as shown in Figure 1.55 is used to pump the demand in Table 1.15 from the reservoir in B. Calculate the diameter of pipe B-C that can provide the minimum pressure in area C of $p_C/\rho g = 10$ mwc, if three pumps are operated in parallel. For the diameter chosen from the same list as in question b), calculate the maximum pressure in area C using the minimum required number of pumps.

Marking matrix:

Question	a	b	c
Max. points	15	20	15

Answers:
a) The balancing volume of the reservoir in B is 1820 m³ (from Spreadsheet Lesson 8–10).

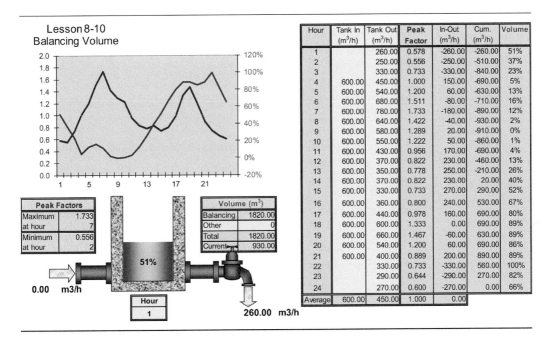

This volume is larger when supply of the reservoir is shorter than 24 hours. The reservoir in A is recharged only when the feeding of B is stopped, therefore the balancing volume needs to be sufficient to receive the entire daily demand, which is $600 \times 18 = 10,800$ m³.

b) The flow $Q_{AB} = 600$ m³/h = 166.67 l/s. The available friction loss along A-B is $H_A - H_B = 95-75 = 20$ mwc and the hydraulic gradient $S_{AB} = 20/800 = 0.025$. Consequently, $D_{AB} = 291$ mm (from Spreadsheet Lesson 1–5).

INPUT		OUTPUT	
L (m)	800	h$_f$ (mwc)	20.00
k (mm)	0.5	u (m²/s)	1.31E-06
Q (l/s)	166.67	D (mm)	291
S (-)	0.025	Re (-)	557489
T (°C)	10	λ (-)	0.0230
H₂ (msl)	-	v (m/s)	2.49

When this diameter is rounded off to the higher value of 300 mm, the pipe can also supply more flow. $Q_{AB} = 179.63$ l/s (from Spreadsheet Lesson 1–4).

INPUT		OUTPUT	
L (m)	800	h$_f$ (mwc)	20.00
D (mm)	300	u (m²/s)	1.31E-06
k (mm)	0.5	Re (-)	583227
S (-)	0.025	λ (-)	0.0228
T (°C)	10	v (m/s)	2.54
H₂ (msl)	25	Q (l/s)	179.63

The total daily demand can therefore be supplied from A to B in a shorter period of time, which is 10,800/(179.63×3.6) = 16.7 hours. Applying a 17-hour supply with an average flow of 10,800/17 = 635.29 m³/h results in the balancing volume of 2278.77 m³ if the supply from A to B starts one hour later in the morning i.e. at 5:00 instead of 4:00 hours, or 2193.93 m³ if the supply from A to B stops one hour earlier in the evening i.e. at 20:00 instead of 21:00 hours (from Spreadsheet Lesson 5–8).

c) The minimum pressure $p_c/\rho g = 10$ mwc is to be maintained at the maximum demand, which occurs at 7:00 hours ($Q_7 = 780$ m³/h = 216.67 l/s). Three equal pumps of the size shown in Figure 1.55 will pump one third of this demand ($Q_p = 72.22$ l/s) each at the head of $H_p = 34.78$ mwc. The available friction loss for the supply from reservoir B is then $75 + 34.78-10-83 = 16.78$ mwc. Consequently, $S_{BC} = 16.78/550 = 0.030509$ and $D_{BC} = 310$ mm (from Spreadsheet Lesson 1–5).

INPUT		OUTPUT	
L (m)	550	h$_f$ (mwc)	16.78
k (mm)	0.5	u (m²/s)	1.31E-06
Q (l/s)	216.67	D (mm)	310
S (-)	0.030509	Re (-)	681048
T (°C)	10	λ (-)	0.0225
H₂ (msl)	-	v (m/s)	2.87

When this diameter is rounded off to 350 mm, the minimum demand, which occurs at 2:00 hours (Q_2 = 250 m³/h = 69.44 l/s) can be supplied with one pump unit delivering the pressure to area C of $p_c/\rho g$ = 26.23 mwc. The straightforward answer can be obtained from Spreadsheet Lesson 6–1.

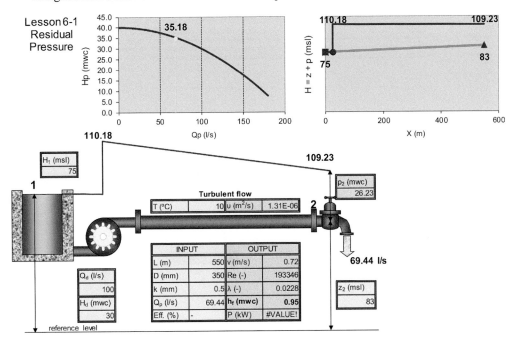

Problem 1.6.8

Area C shown in Figure 1.56 receives water combined from the reservoir in A (by pumping) and the one in C (by gravity), while area B is supplied by direct pumping. The pump unit operates according to the curve shown in Figure 1.57 (H_d = 20 mwc, Q_d = 100 l/s). The information about the connecting route is as follows:

Section	L (m)	D (mm)	k (mm)
A – B	400	250	0.25
A – C	300	200	0.4
C – D	625	150	0.9

Questions:

a) At a particular moment in time, the pressure gauges in the pumping station registered the pumping head H_p = 18 mwc, whilst the average pressure measured in area C was $p_c/\rho g$ = 20 mwc. What is the demand of both areas and the pressure in area B, at the same moment in time?

b) In the alternative layout shown in Figure 1.58, section A-B is disconnected and the entire supply of area B is provided from reservoir D. The pipe section B-D has the following characteristics: L_{BD} = 550 m, D_{BD} = 200 mm and k_{BD} = 0.2 mm. Calculate the demand of area B if the same pressure as determined in question a) needs to be maintained.

c)   What will be the flow that can be delivered from reservoir A to reservoir D if the minimum night flow of $Q_C = 108$ m³/h is to be maintained in area C? What is the pressure in area C at that moment?

Marking matrix:

Question	a	b	c
Max. points	25	10	15

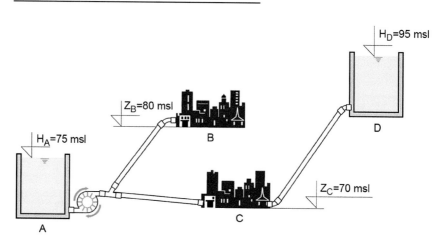

Figure 1.56 Problem 1.6.8 – question a)

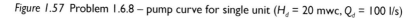

Figure 1.57 Problem 1.6.8 – pump curve for single unit ($H_d = 20$ mwc, $Q_d = 100$ l/s)

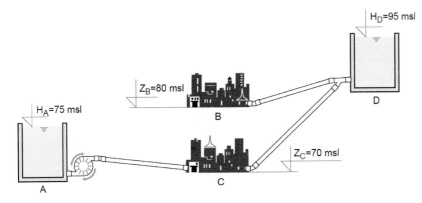

Figure 1.58 Problem 1.6.8 – questions b) and c)

Answers:

a) For a given duty head and flow, the pump delivers Q_p = 114 l/s at H_p = 18
    mwc. For the pressure $p_c/\rho g$ = 20 mwc, the available friction loss along A-C
    is 75 + 18–20–70 = 3 mwc and the hydraulic gradient S_{AC} = 3/300 = 0.01.
    Consequently, Q_{AC} = 39.75 l/s (from Spreadsheet Lesson 1–4).

INPUT		OUTPUT	
L (m)	300	h_f (mwc)	3.00
D (mm)	200	u (m²/s)	1.31E-06
k (mm)	0.4	Re (-)	194409
S (-)	0.01	λ (-)	0.0245
T (°C)	10	v (m/s)	1.27
H₂ (msl)	-	Q (l/s)	39.75

Consequently, Q_{AB} = Q_p – Q_{AC} = 114–39.75 = 74.25 l/s. For this flow, the
friction loss along A-B is $h_{f,AB}$ = 3.88 mwc (from Spreadsheet Lesson 1–2),
leading to the pressure in area B of $p_B/\rho g$ = 75 + 18–3.88–80 = 9.12 mwc.

INPUT		OUTPUT	
L (m)	400	v (m/s)	1.51
D (mm)	250	u (m²/s)	1.31E-06
k (mm)	0.25	Re (-)	289434
Q (l/s)	74.25	λ (-)	0.0208
T (°C)	10	h_f (mwc)	3.88
H₂ (msl)	-	S (-)	0.0097

Finally, the area in C also receives water from the reservoir in D. The available friction loss along D-C is 95–20–70 = 5 mwc and the hydraulic gradient $S_{DC} = 5/625 = 0.008$. Consequently, $Q_{DC} = 14.88$ l/s (from Spreadsheet Lesson 1–4), leading to the total demand of area C of $Q_C = 39.75 + 14.88 = 54.63$ l/s.

INPUT		OUTPUT	
L (m)	625	h$_f$ (mwc)	5.00
D (mm)	150	u (m^2/s)	1.31E-06
k (mm)	0.9	Re (-)	96439
S (-)	0.008	λ (-)	0.0332
T (°C)	10	v (m/s)	0.84
H$_2$ (msl)	-	Q (l/s)	14.88

b) In the alternative layout, the available friction loss along D-B is $H_D - H_B = 95–9.12–80 = 5.88$ mwc and the hydraulic gradient $S_{BD} = 5.88/550 = 0.010691$. Consequently, the supply from the reservoir in D is $Q_{DB} = 44.30$ l/s (from Spreadsheet Lesson 1–4).

INPUT		OUTPUT	
L (m)	550	h$_f$ (mwc)	5.88
D (mm)	200	u (m^2/s)	1.31E-06
k (mm)	0.2	Re (-)	215840
S (-)	0.010691	λ (-)	0.0211
T (°C)	10	v (m/s)	1.41
H$_2$ (msl)	-	Q (l/s)	44.30

c) During the night time, $Q_p = Q_{AC} = Q_C + Q_{CD}$. Thus, the pump needs to pump the flow higher than 108 m³/h = 30 l/s, meaning that the pumping head will be lower than 26.07 mwc. In this situation, the continuity of the hydraulic grade line along the route A-C-D needs to be maintained i.e. $H_A + H_p - h_{f,AC} = H_D + h_{f,CD} = H_C$. The following table can be produced by using Spreadsheet Lesson 1–2 for the friction loss calculation.

Q_{AC} (l/s)	H_p (mwc)	$h_{f,AC}$ (mwc)	$H_{C,left}$ (msl)	Q_{CD} (l/s)	$h_{f,CD}$ (mwc)	$H_{C,right}$ (msl)	ΔH_C (msl)
35	25.85	2.34	98.51	5	0.59	95.59	2.92
40	25.60	3.04	97.56	10	2.29	97.29	0.27
45	25.32	3.83	96.49	15	5.08	100.08	- 3.59

The difference ΔH_C between $H_{C,left} = H_A + H_p - h_{f,AC}$ and $H_{C,right} = H_D + h_{f,CD}$ needs to be zero. The table shows that this is going to happen for the pumping flow between 40 and 45 l/s; actually, very close to 40 l/s. A more accurate value can be obtained by further refining the flows by repeating the same process. Finally, $Q_{CD} = 40.42$ l/s and $H_C = 97.48$ msl leading to the pressure $p_C/\rho g = 27.48$ mwc.

Q_{AC} (l/s)	H_p (mwc)	$h_{f,AC}$ (mwc)	$H_{C,left}$ (msl)	Q_{CD} (l/s)	$h_{f,CD}$ (mwc)	$H_{C,right}$ (msl)	ΔH_C (msl)
40.2	25.59	3.07	97.52	10.2	2.38	97.38	0.14
40.4	25.58	3.10	97.48	10.4	2.47	97.47	0.01
40.42	25.58	3.10	97.48	10.42	2.48	97.48	**0**

Problem 1.6.9

Area B shown in Figure 1.59 receives water by pumping from the sources in A and C. Both pumps operate according to the curve shown in Figure 1.60 ($H_d = 60$ mwc, $Q_d = 120$ l/s).The information about the connecting route is as follows:

Section	L (m)	D (mm)	k (mm)
A – B	500	300	0.3
B – C	400	200	0.4

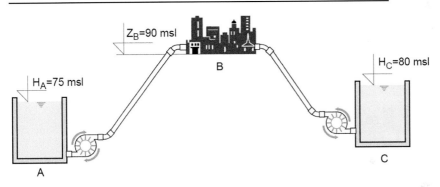

Figure 1.59 Problem 1.6.9 – questions a) and b)

Questions:
a) Calculate the demand of area B at the pressure $p_B/\rho g = 30$ mwc, if both pumps A and C are switched on.
b) Can the same demand, as calculated in question a), be satisfied if the section B-C is temporarily closed due to the maintenance of the pump in C? What will be the pressure at B at that moment if an additional pump of the same size is connected in parallel to the one in A?
c) Area D is planned to be connected to the system and will be supplied by a booster station installed at B, as shown in Figure 1.61 (the same pump curve can be used as for the pumps in A and C). The pipe section B-D has the following characteristics: $L_{BD} = 300$ m, $D_{BD} = 100$ mm and $k_{BD} = 0.1$ mm. To supply area D, the same demand as for area B, as calculated in question a), is to be maintained but at the lower minimum pressure $p_B/\rho g = 20$ mwc, with both pumps A (one unit!) and C in operation. What will be the demand and pressure at D for such a scenario?

Marking matrix:

Question	a	b	c
Max. points	15	15	20

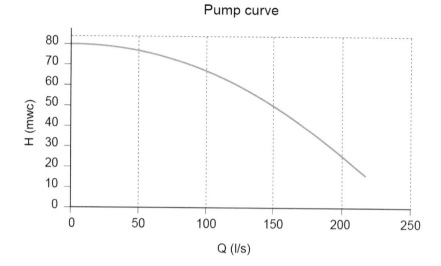

Figure 1.60 Problem 1.6.9 – pump curve for single unit (H_d = 60 mwc, Q_d = 120 l/s)

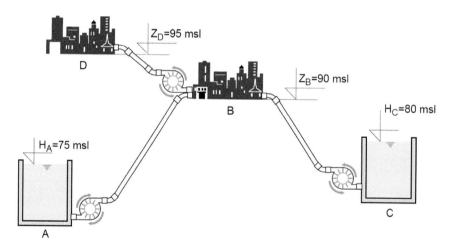

Figure 1.61 Problem 1.6.9 – question c)

Answers:

a)   For the given pressure $p_B/\rho g$ = 30 mwc, H_B = 120 msl. The available friction loss along A-B is 120–75 = 45 mwc, which is the static head to be met by pump A. The available friction loss along C-B is 120–80 = 40 mwc, which is the static head to be met by pump C. Consequently, the flows Q_{AB} and Q_{CB} are equal to 141.99 l/s and 101.84 l/s, respectively (from Spreadsheet Lesson 6–2), leading to the total demand of area B of Q_B = 243.83 l/s.

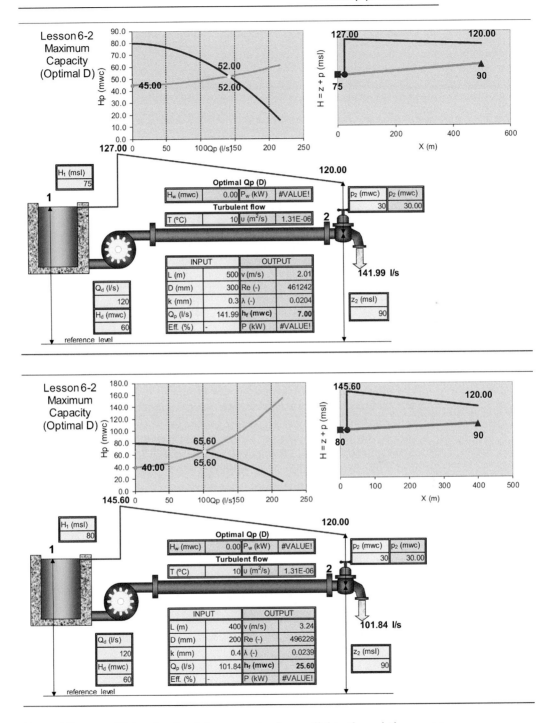

b) Adding an additional unit of the same size in parallel to the existing pump will result in the composite pump curve of $H_d = 60$ mwc, and $Q_d = 240$ l/s. The residual pressure for the demand $Q_B = 243.83$ l/s, will be $p_B/\rho g = 24.02$ mwc (from Spreadsheet Lesson 6–1).

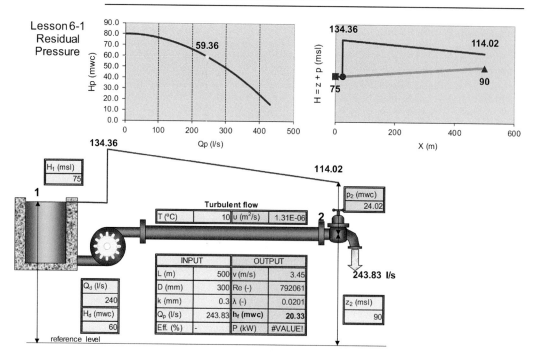

c) For the reduced pressure $p_B/\rho g = 20$ mwc, $H_B = 110$ msl. The available friction loss along A-B is 35 mwc, while the one along C-B is 30 mwc. Consequently, both pumps A and C can deliver more water. Applying the same calculations as in question a), the flows Q_{AB} and Q_{CB} are equal to 161.07 l/s and 113.93 l/s, respectively (from Spreadsheet Lesson 6–2). The demand pumped to the area in D is then $Q_{BD} = 161.07 + 113.93 - 243.83 = 31.17$ l/s. The booster pump in B will deliver this flow at the head $H_p = 78.65$ mwc, and the friction loss will be $h_{f,BD} = 49.94$ (from Spreadsheet Lesson 1–2), resulting in the residual pressure $p_D/\rho g = 110 + 78.65 - 49.94 - 95 = 43.71$ mwc.

INPUT		OUTPUT	
L (m)	300	v (m/s)	3.97
D (mm)	100	u (m²/s)	1.31E-06
k (mm)	0.1	Re (-)	303759
Q (l/s)	31.17	λ (-)	0.0207
T (°C)	10	h_f (mwc)	49.94
H₂ (msl)	-	S (-)	0.1665

Problem 1.6.10

Areas D and E shown in Figure 1.62 receive water combined by pumping from the reservoir in A, and by gravity from the reservoir in C. The pump operates according to the curve shown in Figure 1.63 ($H_d = 30$ mwc, $Q_d = 150$ l/s). The information about the connecting routes is as follows:

Section	L (m)	D (mm)	k (mm)
A – B	400	300	0.5
B – D	600	250	1.0
C – B	500	200	0.5
B – E	500	300	0.5

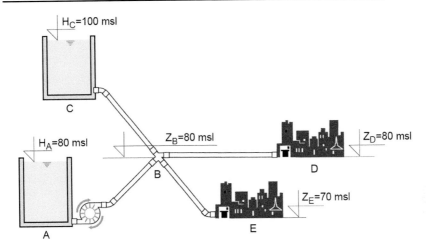

Figure 1.62 Problem 1.6.10

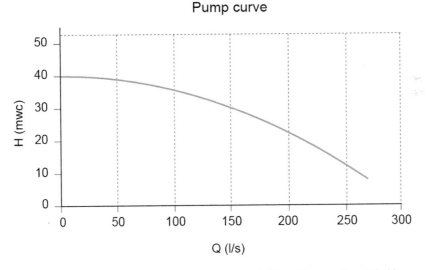

Figure 1.63 Problem 1.6.10 – pump curve for single unit (H_d = 30 mwc, Q_d = 150 l/s)

Questions:
a) During the night time, the minimum observed flow of the pump was Q_p = 150 l/s while the observed pressure in area D was $p_D/\rho g$ = 17 mwc. What was the demand of areas D and E, and the pressure in area E at that moment in time?
b) During the daytime, the maximum observed flow of the pump was Q_A = 190 l/s while the observed pressure in area D was $p_D/\rho g$ = 7 mwc. What was the demand of areas D and E, and the pressure in area E at that moment in time?

c)   A renovation of the section C-B is planned. What should be the enlarged diameter D_{CB} that can enable the reservoir in C to supply, with the pump out of operation, the entire demands and pressures calculated for the supply in question b)?

Marking matrix:

Question	a	b	c
Max. points	20	15	15

Answers:

a)   For the given duty head and flow, the pump delivers Q_p = 150 l/s at H_p = 30 mwc. Furthermore, the friction loss along A-B is $h_{f,AB}$ = 7 mwc (from Spreadsheet Lesson 1–2), leading to the head in the cross section in B of H_B = 80 + 30–7 = 103 msl.

INPUT		OUTPUT	
L (m)	400	v (m/s)	2.12
D (mm)	300	u (m²/s)	1.31E-06
k (mm)	0.5	Re (-)	487262
Q (l/s)	150	λ (-)	0.0229
T (°C)	10	hf (mwc)	7.00
H₂ (msl)	-	S (-)	0.0175

Hence, the available friction loss along B-C is 103–100 = 3 mwc and the hydraulic gradient S_{BC} = 3/500 = 0.006. Consequently, Q_{BC} = 29.83 l/s (Spreadsheet Lesson 1–4). Hence, the reservoir in C receives water.

INPUT		OUTPUT	
L (m)	500	hf (mwc)	3.00
D (mm)	200	u (m²/s)	1.31E-06
k (mm)	0.5	Re (-)	145424
S (-)	0.006	λ (-)	0.0261
T (°C)	10	v (m/s)	0.95
H₂ (msl)	-	Q (l/s)	29.83

For $p_D/\rho g$ = 17 mwc, the available friction loss along B-D is 103–17–80 = 6 mwc and the hydraulic gradient S_{BD} = 6/600 = 0.01. Consequently, Q_{BD} = 63.84 l/s (from Spreadsheet Lesson 1–4).

INPUT		OUTPUT	
L (m)	600	hf (mwc)	6.00
D (mm)	250	u (m²/s)	1.31E-06
k (mm)	1	Re (-)	248752
S (-)	0.01	λ (-)	0.0290
T (°C)	10	v (m/s)	1.30
H₂ (msl)	-	Q (l/s)	63.84

Finally, the demand of area E = 150 l/s – 29.83 l/s (in C) – 63.84 l/s (in D) = 56.33 l/s. For this flow, the friction loss along B-E is $h_{f,BE}$ = 1.27 mwc (from Spreadsheet Lesson 1–2), leading to the pressure in area E of $p_E/\rho g$ = 103–1.27–70 = 31.73 msl.

INPUT		OUTPUT	
L (m)	500	v (m/s)	0.80
D (mm)	300	u (m²/s)	1.31E-06
k (mm)	0.5	Re (-)	182983
Q (l/s)	56.33	λ (-)	0.0236
T (°C)	10	hf (mwc)	**1.27**
H₂ (msl)	-	S (-)	0.0025

b) During the daytime, the pump delivers Q_p = 190 l/s at H_p = 23.96 mwc. The friction loss along A-B is $h_{f,AB}$ = 11.18 mwc (from Spreadsheet Lesson 1–2), leading to the head H_B = 80 + 23.96–11.18 = 92.78 msl.

INPUT		OUTPUT	
L (m)	400	v (m/s)	2.69
D (mm)	300	u (m²/s)	1.31E-06
k (mm)	0.5	Re (-)	617199
Q (l/s)	190	λ (-)	0.0228
T (°C)	10	hf (mwc)	**11.18**
H₂ (msl)	-	S (-)	0.0279

Hence, the available friction loss along C-B is 100–92.78 = 7.22 mwc and the hydraulic gradient S_{CB} = 7.22/500 = 0.01444. Consequently, Q_{CB} = 46.64 l/s (from Spreadsheet Lesson 1–4). Hence, the reservoir in C supplies water.

INPUT		OUTPUT	
L (m)	500	hf (mwc)	7.22
D (mm)	200	u (m²/s)	1.31E-06
k (mm)	0.5	Re (-)	226555
S (-)	0.01444	λ (-)	0.0257
T (°C)	10	v (m/s)	1.48
H₂ (msl)	-	Q (l/s)	**46.64**

For $p_D/\rho g$ = 7 mwc, the available friction loss $h_{f,BD}$ = 92.78–7–80 = 5.78 mwc and the hydraulic gradient S_{BD} = 5.78/600 = 0.009633. Based on this, Q_{BD} = 62.65 l/s (from Spreadsheet Lesson 1–4).

INPUT		OUTPUT	
L (m)	600	hf (mwc)	5.78
D (mm)	250	u (m²/s)	1.31E-06
k (mm)	1	Re (-)	244925
S (-)	0.009633	λ (-)	0.0290
T (°C)	10	v (m/s)	1.28
H₂ (msl)	-	Q (l/s)	**62.65**

Finally, the demand of area E = 190 l/s + 46.64 l/s (in C) – 62.65 l/s (in D) = 173.99 l/s. For this flow, the friction loss along B-E is $h_{f,BE}$ = 11.73 mwc (from Spreadsheet Lesson 1–2), leading to the pressure in area E of $p_E/\rho g$ = 92.78–11.73–70 = 11.05 mwc.

INPUT		OUTPUT	
L (m)	500	v (m/s)	2.46
D (mm)	300	u (m²/s)	1.31E-06
k (mm)	0.5	Re (-)	565192
Q (l/s)	173.99	λ (-)	0.0228
T (°C)	10	h$_f$ (mwc)	11.73
H₂ (msl)	-	S (-)	0.0235

c) During the renovation period, the reservoir in C should deliver $Q_C = Q_D + Q_E$ = 62.65 + 173.99 = 236.64 l/s. To meet the same pressures in D and E as in question b), the head H_B = 92.78 msl needs to be maintained and the available friction loss along C-B is then still $h_{f,CB}$ = 7.22 mwc as well as the hydraulic gradient S_{CB} = 0.01444, leading to the diameter D_{CB} = 370 mm (from Spreadsheet Lesson 1–5), rounded off to 400 mm.

INPUT		OUTPUT	
L (m)	500	h$_f$ (mwc)	7.22
k (mm)	0.5	u (m²/s)	1.31E-06
Q (l/s)	236.64	D (mm)	370
S (-)	0.01444	Re (-)	623151
T (°C)	10	λ (-)	0.0217
H₂ (msl)	-	v (m/s)	2.20

Problem 1.6.11

Area E shown in Figure 1.64 receives water combined by pumping from the reservoir in A, and by gravity from the reservoirs in C and D. The pump operates according to the curve shown in Figure 1.65 (H_d = 40 mwc, Q_d = 150 l/s). The information about the connecting routes is as follows:

Section	L (m)	D (mm)	k (mm)
A – B	400	300	0.5
B – D	1000	250	1.0
C – B	500	150	0.5
B – E	500	350	0.5

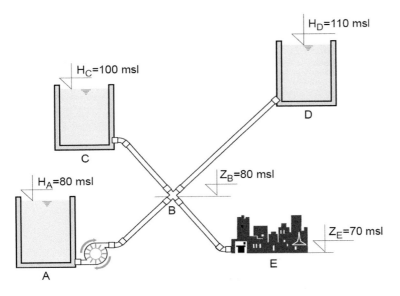

Figure 1.64 Problem 1.6.11

Questions:
a) During the night time, the observed pressure in the intersection at B was $p_B/\rho g = 40$ mwc. What was the demand and pressure in area E at that moment in time?
b) During the daytime, the observed pressure in the intersection at B dropped to $p_B/\rho g = 15$ mwc. What was the demand and pressure in area E at that moment in time?
c) At which flow will the pump supply the reservoir in C if $Q_{BD} = 0$ but pipe B-D is not closed? What will be the demand and pressure in area E at that moment in time?

Marking matrix:

Question	a	b	c
Max. points	20	15	15

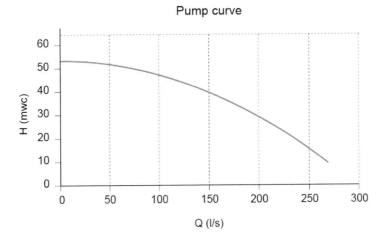

Figure 1.65 Problem 1.6.11 – pump curve for single unit (H_d = 40 mwc, Q_d = 150 l/s)

Answers:

a)   Due to $H_A = Z_B = 80$ msl, the pressure $p_B/\rho g = 40$ mwc is at the same time the static head along A-B. Consequently, the flow $Q_{AB} = 121.38$ l/s (from Spreadsheet Lesson 6–2).

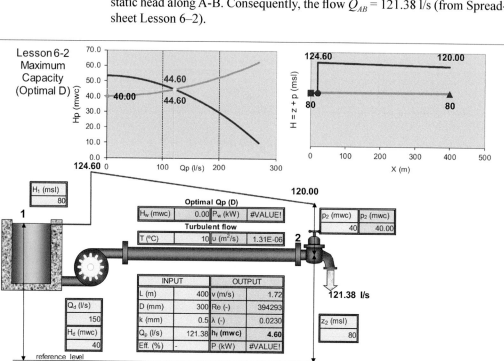

For $H_B = 120$ msl, the available friction loss along B-C is $120{-}100 = 20$ mwc, and along B-D is $120{-}110 = 10$ mwc. The hydraulic gradients are then $S_{BC} = 20/500 = 0.04$, and $S_{BD} = 10/1000 = 0.01$, leading to $Q_{BC} = 36.47$ l/s and $Q_{BD} = 63.84$ l/s (both from Spreadsheet Lesson 1–4). Hence, these reservoirs receive water.

INPUT		OUTPUT	
L (m)	500	h$_f$ (mwc)	20.00
D (mm)	150	u (m^2/s)	1.31E-06
k (mm)	0.5	Re (-)	236505
S (-)	0.04	λ (-)	0.0276
T (°C)	10	v (m/s)	2.06
H$_2$ (msl)	-	Q (l/s)	36.47

INPUT		OUTPUT	
L (m)	1000	h$_f$ (mwc)	10.00
D (mm)	250	u (m^2/s)	1.31E-06
k (mm)	1	Re (-)	248752
S (-)	0.01	λ (-)	0.0290
T (°C)	10	v (m/s)	1.30
H$_2$ (msl)	-	Q (l/s)	63.84

Finally, the demand of area E = 121.38 l/s – 36.47 l/s (in C) – 63.84 l/s (in D) = 21.07 l/s. For this flow, the friction loss along B-E is $h_{f,BE} = 0.09$ mwc (from

Spreadsheet Lesson 1–2), leading to the pressure in area E of $p_E/\rho g = 120–0.09–70 = 49.91$ mwc.

INPUT		OUTPUT	
L (m)	500	v (m/s)	0.22
D (mm)	350	u (m²/s)	1.31E-06
k (mm)	0.5	Re (-)	58666
Q (l/s)	21.07	λ (-)	0.0249
T (°C)	10	h_f (mwc)	0.09
H₂ (msl)	-	S (-)	0.0002

b) During the daytime, the static head along A-B equals the pressure $p_B/\rho g = 15$ mwc. In this scenario, $Q_{AB} = 206.19$ l/s (from Spreadsheet Lesson 6–2).

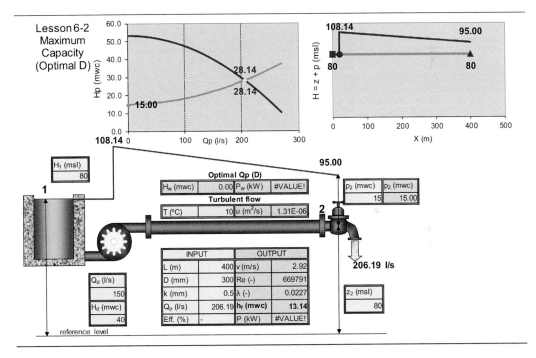

For $H_B = 95$ msl, the available friction loss along C-B is $100–95 = 5$ mwc, and along D-B is $110–95 = 15$ mwc. The hydraulic gradients are then $S_{CB} = 5/500 = 0.01$, and $S_{DB} = 15/1000 = 0.015$, leading to $Q_{CB} = 18.04$ l/s and $Q_{DB} = 78.32$ l/s (both from Spreadsheet Lesson 1–4). Hence, the reservoirs supply water.

INPUT		OUTPUT	
L (m)	500	h_f (mwc)	5.00
D (mm)	150	u (m²/s)	1.31E-06
k (mm)	0.5	Re (-)	117105
S (-)	0.01	λ (-)	0.0282
T (°C)	10	v (m/s)	1.02
H₂ (msl)	-	Q (l/s)	18.04

INPUT		OUTPUT	
L (m)	1000	h_f (mwc)	15.00
D (mm)	250	u (m²/s)	1.31E-06
k (mm)	1	Re (-)	306156
S (-)	0.015	λ (-)	0.0289
T (°C)	10	v (m/s)	1.60
H₂ (msl)	-	Q (l/s)	78.32

Finally, the demand of area E = 206.19 l/s + 18.04 l/s (in C) + 78.32 l/s (in D) = 302.55 l/s. For this flow, the friction loss along B-E is $h_{f,BE}$ = 15.71 mwc (from Spreadsheet Lesson 1–2), leading to the pressure in area E of $p_E/\rho g$ = 95–15.71–70 = 9.29 mwc.

INPUT		OUTPUT	
L (m)	500	v (m/s)	3.14
D (mm)	350	u (m²/s)	1.31E-06
k (mm)	0.5	Re (-)	842407
Q (l/s)	302.55	λ (-)	0.0218
T (°C)	10	**h_f (mwc)**	**15.71**
H₂ (msl)	-	S (-)	0.0314

c) For the condition Q_{BD} = 0, H_B = H_D = 110 mwc, and therefore $p_B/\rho g$ = 30 mwc. As a consequence, Q_{AB} = 160.73 l/s (from Spreadsheet Lesson 6–2).

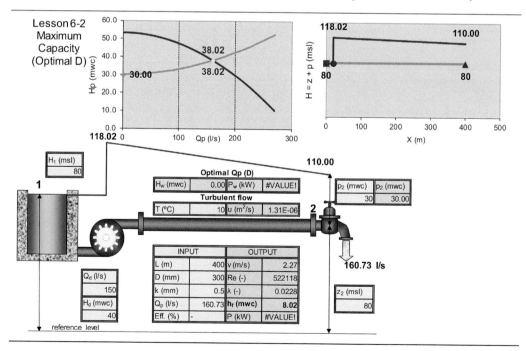

For H_B = 110 msl, the available friction loss along B-C is 110–100 = 10 mwc. The hydraulic gradient is then S_{BC} = 10/500 = 0.02, leading to Q_{BC} = 25.67 l/s (from Spreadsheet Lesson 1–4). Hence, the reservoir in C receives water.

INPUT		OUTPUT	
L (m)	500	h_f (mwc)	10.00
D (mm)	150	u (m²/s)	1.31E-06
k (mm)	0.5	Re (-)	166472
S (-)	0.02	λ (-)	0.0279
T (°C)	10	v (m/s)	1.45
H₂ (msl)	-	Q (l/s)	**25.67**

Finally, the demand of area E = 160.73–25.67 = 135.06 l/s. For this flow, the friction loss along B-E is $h_{f,BE}$ = 3.19 mwc (from Spreadsheet Lesson 1–2), leading to the pressure in area E of $p_E/\rho g$ = 110–3.19–70 = 36.81 mwc.

INPUT		OUTPUT	
L (m)	500	v (m/s)	1.40
D (mm)	350	u (m²/s)	1.31E-06
k (mm)	0.5	Re (-)	376055
Q (l/s)	135.06	λ (-)	0.0222
T (°C)	10	h_f (mwc)	**3.19**
H₂ (msl)	-	S (-)	0.0064

Problem 1.6.12

Areas B and C shown in Figure 1.66 receive water: B by gravity from reservoir A, and C combined by pumping from the same reservoir and by gravity from the reservoir in D. The equal pump units in the pumping station operate according to the curve shown in Figure 1.67 (H_d = 30 mwc, Q_d = 100 l/s). The information about the connecting routes is as follows:

Section	L (m)	D (mm)	k (mm)
A – B	500	400	0.5
B – C	350	250	1.5
C – D	800	350	0.5

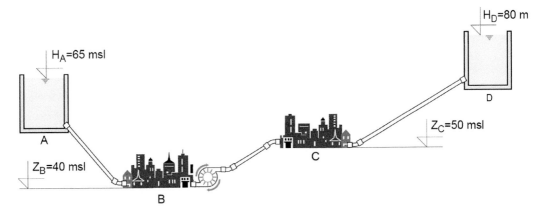

Figure 1.66 Problem 1.6.12

Questions:

a) At a certain moment in time, the pressures in area B, at the suction side of the pumping station, as well as in area C, are measured both at $p_B/\rho g = p_C/\rho g = 20$ mwc. Determine the demands in areas A and B if one pump is in operation.

b) During the night time, the reservoir in D will need to be replenished with the minimum flow of 160 l/s pumped from the reservoir in A, while the required demand in each of the two areas, B and C, should be at least 50 l/s. For this purpose the diameter of section B-C will be increased to 350 mm at a new absolute roughness of 0.5 mm. Find out the least number of pumps in parallel arrangement that are needed to provide this operation.

c) The reservoir in D is temporarily disconnected for maintenance purposes. Can the number of pumps that is determined in question b) supply the entire demand of areas B and C calculated in question a), if the diameter of pipe D_{AB} has also been increased to maintain the same pressure of 20 mwc in area B (as in question a)? Find out the diameter D_{AB} at the same absolute roughness of 0.5 mm and the pressure in area C in this situation.

Marking matrix:

Question	a	b	c
Max. points	20	15	15

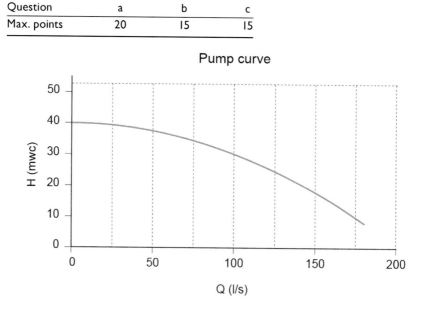

Pump curve

Figure 1.67 Problem 1.6.12 – pump curve for single unit (H_d = 30 mwc, Q_d = 100 l/s)

Answers:

a) For $p_B/\rho g = p_C/\rho g = 20$ mwc, the heads $H_B = 60$ msl and $H_C = 70$ msl creating the static head along B-C of 10 mwc. At the same time, the available friction loss along A-B is $h_{f,AB} = 65-60 = 5$ mwc, while the one along D-C is $h_{f,DC} = 80-70 = 10$ mwc. Consequently, the corresponding hydraulic gradients are then $S_{AB} = 5/500 = 0.01$, and $S_{DC} = 10/800 = 0.0125$, leading to $Q_{AB} = 241.25$ l/s and $Q_{DC} = 189.99$ l/s (both from Spreadsheet Lesson 1–4). Hence, the reservoirs supply water.

INPUT		OUTPUT	
L (m)	500	h_f (mwc)	5.00
D (mm)	400	υ (m²/s)	1.31E-06
k (mm)	0.5	Re (-)	587819
S (-)	0.01	λ (-)	0.0213
T (°C)	10	v (m/s)	1.92
H₂ (msl)	-	**Q (l/s)**	**241.25**

INPUT		OUTPUT	
L (m)	800	h_f (mwc)	10.00
D (mm)	350	υ (m²/s)	1.31E-06
k (mm)	0.5	Re (-)	527736
S (-)	0.0125	λ (-)	0.0220
T (°C)	10	v (m/s)	1.97
H₂ (msl)	-	**Q (l/s)**	**189.99**

Furthermore, with one pump in operation $Q_{BC} = 123.76$ l/s (from Spreadsheet Lesson 6–2).

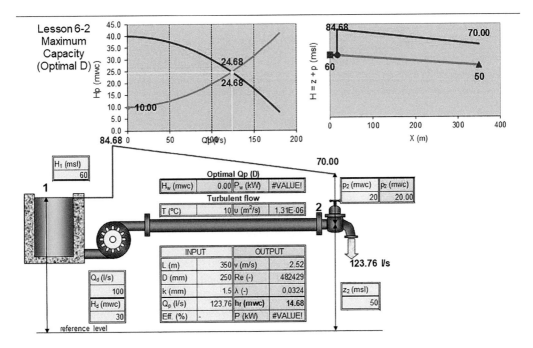

Finally, the demands in the supplied areas are $Q_B = Q_{AB} - Q_{BC} = 241.25 - 123.76 = 117.49$ l/s, and $Q_C = Q_{BC} + Q_{DC} = 123.76 + 189.99 = 313.75$ l/s.

b) During the night time, the minimum demands are $Q_B = Q_C = 50$ l/s while the minimum recharge of the reservoir in D is $Q_{CD} = 160$ l/s. Thus, the pumping station has to provide the minimum flow of $50 + 160 = 210$ l/s, while $Q_{AB} = 260$ l/s. Consequently, the friction losses will be $h_{f,AB} = 5.80$ mwc and $h_{f,CD} = 7.12$ mwc (both from Spreadsheet Lesson 1–2), leading to the heads in area B of $H_B = 65-5.80 = 59.20$ msl, and in area C of $H_C = 80 + 7.12 = 87.12$ msl.

INPUT		OUTPUT	
L (m)	500	v (m/s)	2.07
D (mm)	400	u (m²/s)	1.31E-06
k (mm)	0.5	Re (-)	633441
Q (l/s)	260	λ (-)	0.0213
T (°C)	10	h$_f$ (mwc)	**5.80**
H₂ (msl)	-	S (-)	0.0116

INPUT		OUTPUT	
L (m)	800	v (m/s)	1.66
D (mm)	350	u (m²/s)	1.31E-06
k (mm)	0.5	Re (-)	445497
Q (l/s)	160	λ (-)	0.0221
T (°C)	10	h$_f$ (mwc)	**7.12**
H₂ (msl)	-	S (-)	0.0089

Hence, the static head along B-C is $87.12-59.20 = 27.92$ mwc. At the same time the flow $Q_{BC} = 210$ l/s with the enlarged pipe generates the friction loss of $h_{f,BC} = 5.33$ mwc (from Spreadsheet Lesson 1–2).

INPUT		OUTPUT	
L (m)	350	v (m/s)	2.18
D (mm)	350	u (m²/s)	1.31E-06
k (mm)	0.5	Re (-)	584715
Q (l/s)	210	λ (-)	0.0220
T (°C)	10	h$_f$ (mwc)	**5.33**
H₂ (msl)	-	S (-)	0.0152

This leads to the conclusion that the pumping station needs to pump the minimum flow of $Q_p = 210$ l/s at the minimum pumping head of $H_p = 27.92 + 5.33 = 33.25$ mwc. For the given pump curve, two pumps in parallel deliver the required flow at the head of $H_p = 28.98$ mwc, while three pumps in parallel will deliver the head $H_p = 35.10$ mwc > 33.25 mwc. Hence, three units are necessary.

c) From question a): $Q_B = 117.49$ l/s, and $Q_C = 313.75$ l/s. Thus, $Q_{AB} = 431.24$ l/s and $Q_{BC} = Q_p = 313.75$ l/s. For $p_B/\rho g = 20$ mwc, the available friction loss along A-B is still $h_{f,AB} = 5$ mwc leading to the $S_{AB} = 0.01$ and the diameter $D_{AB} = 498$ mm (from Spreadsheet Lesson 1–5), rounded off to 500 mm.

INPUT		OUTPUT	
L (m)	500	h_f (mwc)	5.00
k (mm)	0.5	u (m²/s)	1.31E-06
Q (l/s)	431.24	D (mm)	498
S (-)	0.01	Re (-)	843128
T (°C)	10	λ (-)	0.0201
H₂ (msl)	-	v (m/s)	2.21

Three pumps will pump $Q_p = 313.75$ l/s at the head $H_p = 29.06$ mwc. At the same time, the friction loss $h_{f,BC} = 11.82$ mwc (from Spreadsheet Lesson 1–2), leading to the pressure in C of $p_C/\rho g = 60 + 29.06 - 11.82 - 50 = 27.24$ mwc.

INPUT		OUTPUT	
L (m)	350	v (m/s)	3.26
D (mm)	350	u (m²/s)	1.31E-06
k (mm)	0.5	Re (-)	873591
Q (l/s)	313.75	λ (-)	0.0218
T (°C)	10	h_f (mwc)	11.82
H₂ (msl)	-	S (-)	0.0338

Problem 1.6.13

Areas D and E shown in Figure 1.68 receive water: D by gravity from reservoir C, and E by serial pumping from the reservoir in A. The equal pump units in A and B operate according to the curve shown in Figure 1.69 ($H_d = 50$ mwc, $Q_d = 200$ l/s). The information about the connecting routes is as follows:

Section	L (m)	D (mm)	k (mm)
A – B	400	300	0.5
B – C	450	250	1.0
C – D	600	150	0.5
B – E	350	200	1.5

Questions:

a) For the supply of area E, $Q_E = 50$ l/s, calculate the flow and head of pump A, and supply of reservoir C from the pump in B. What is the pressure at the suction side of pump B, and what will be the pressure in area E at that moment in time? Furthermore, if the observed pressure in area D is $p_D/\rho g = 14$ mwc, what is the demand of this area?

b) How much water would area E receive, at the pressure $p_E/\rho g = 30$ mwc, if it is supplied by gravity from the reservoir in C (the pumps at A and B are switched off)? Calculate the equivalent diameter for the section C-B-E in this case, for the total length equal to $L_{CBE} = L_{BC} + L_{BE} = 800$ m. The roughness factor for the new pipe can be assumed at $k_{CBE} = 0.1$ mm.

Marking matrix:

Question	a	b
Max. points	30	20

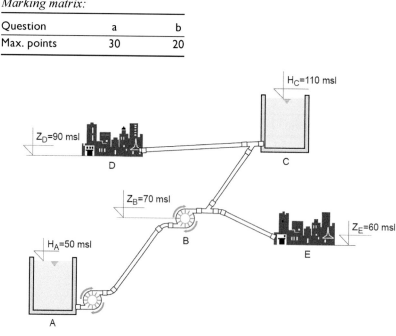

Figure 1.68 Problem 1.6.13

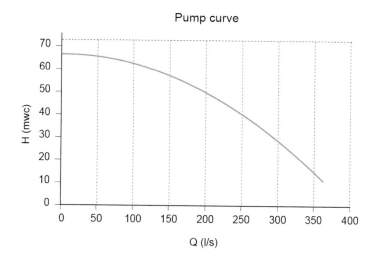

Figure 1.69 Problem 1.6.13 – pump curve for single unit (H_d = 50 mwc, Q_d = 200 l/s)

Answers:

a) For the demand in area E of $Q_E = 50$ l/s, $Q_{AB} = Q_{BC} + 50$ l/s. Both pumps are of the same size and connected in series; hence, they pump the same flow $Q_p = Q_{AB}$ with the same pumping head H_p. Based on this, the integrity of the hydraulic grade line along A-B-C is maintained by $H_A = 50 + H_p - h_{f,AB} + H_p - h_{f,BC} = 110 = H_C$. Thus, $H_p = 30 + (h_{f,AB} + h_{f,BC})/2$. A selected range of pumping flows, leading to the pumping heads above 30 mwc, is used to calculate the friction losses along A-B-C (from Spreadsheet Lesson 1–2), until the equation has been satisfied. The results are shown in the following table.

Q_p (l/s)	$H_{p,curve}$ (mwc)	Q_{AB} (l/s)	$h_{f,AB}$ (mwc)	Q_{BC} (l/s)	$h_{f,BC}$ (mwc)	$H_{p,calc}$ (msl)	ΔH_p (msl)
260	38.50	260	20.83	210	48.03	64.43	25.93
240	42.67	240	17.77	190	39.35	58.56	15.89
220	46.50	220	14.95	170	31.53	53.24	6.74
200	50.00	200	12.37	150	24.57	48.47	−1.53

The difference ΔH_p between the calculated value and the value taken from the pumping curve needs to be zero. The table shows that this is going to happen for the pumping flow between 200 and 220 l/s; actually, close to 200 l/s. A more accurate value can be obtained by further refining the flows by repeating the same process. Finally, with minor interpolation $Q_p = 203.85$ l/s and $H_p = 49.35$ msl leading to the pressure at the suction side of pump B of $p_B/\rho g = 50 + 49.35 - 12.85 - 70 = 16.50$ mwc.

Q_p (l/s)	$H_{p,curve}$ (mwc)	Q_{AB} (l/s)	$h_{f,AB}$ (mwc)	Q_{BC} (l/s)	$h_{f,BC}$ (mwc)	$H_{p,calc}$ (msl)	ΔHp (msl)
202	49.67	202	12.62	152	25.23	48.93	−0.74
204	49.33	204	12.87	154	25.89	49.38	0.05
203.9	49.34	203.9	12.86	153.9	25.86	49.36	0.02
203.8	49.36	203.8	12.84	153.8	25.83	49.34	−0.02

At the same time, the friction loss along B-E will be $h_{f,BE} = 7.88$ mwc (from Spreadsheet Lesson 1–2), leading to the pressure in area E of $p_E/\rho g = 70 + 16.50 + 49.35 - 7.88 - 60 = 67.97$ mwc.

INPUT		OUTPUT	
L (m)	350	v (m/s)	1.59
D (mm)	200	u (m²/s)	1.31E-06
k (mm)	1.5	Re (-)	243631
Q (l/s)	50	λ (-)	0.0349
T (°C)	10	h_f (mwc)	7.88
H₂ (msl)	-	S (-)	0.0225

Finally, for $p_D/\rho g = 14$ mwc, the head $H_D = 104$ msl, creating the available friction loss along C-D of 6 mwc. The corresponding hydraulic gradient is then $S_{DC} = 6/600 = 0.01$, to $Q_{DC} = 18.04$ l/s (from Spreadsheet Lesson 1–4).

INPUT		OUTPUT	
L (m)	600	h_f (mwc)	6.00
D (mm)	150	u (m²/s)	1.31E-06
k (mm)	0.5	Re (-)	117105
S (-)	0.01	λ (-)	0.0282
T (°C)	10	v (m/s)	1.02
H₂ (msl)	-	Q (l/s)	18.04

b) For $p_E/\rho g = 30$ mwc, $H_E = 90$ msl and the available friction loss along the route C-B-E is $h_{f,CBE} = 110-90 = 20$ mwc, cumulated from two pipes connected in series. This friction loss will be reached for the flow of $Q_{CBE} = 68.63$ l/s. The straightforward result can be obtained from Spreadsheet Lesson 2–4. The equivalent diameter for this route, $D_{CBE} = 195$ mm (from Spreadsheet Lesson 2–5), rounded off to 200 mm.

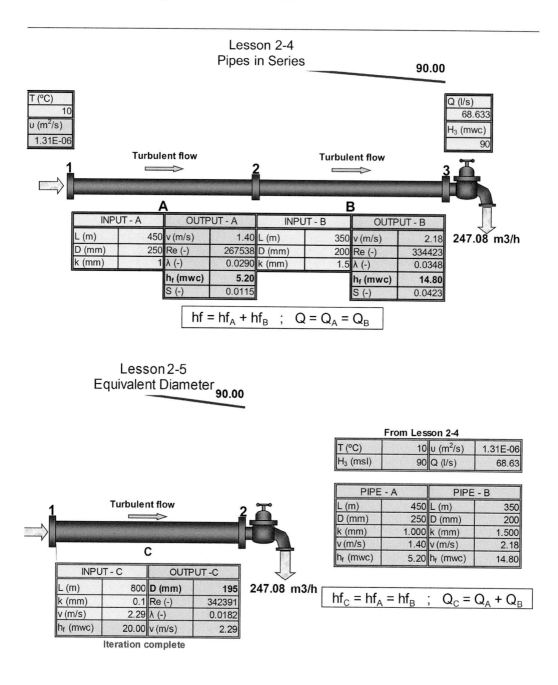

Problem 1.6.14

Areas B, C and D shown in Figure 1.70 receive water by pumping from the reservoirs in A and E. Both of the pumps operate according to the curve shown in Figure 1.71 ($H_d = 40$ mwc, $Q_d = 150$ l/s). The information about the connecting routes is as follows:

Section	L (m)	D (mm)	k (mm)
A – B	400	300	0.5
B – C	600	200	1.5
C – D	500	150	1.0
D – E	700	350	0.5

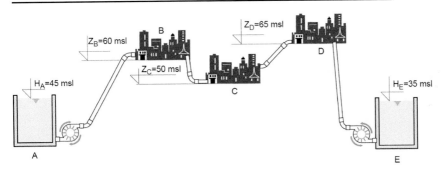

Figure 1.70 Problem 1.6.14

Questions:
a) At a certain moment in time, the pump in A operates at 150 l/s, while the one in E operates at 110 l/s. For these flow conditions, the observed pressure in area C is $p_c/\rho g = 15$ mwc. Determine the demand in all three areas, and the pressure in areas B and D, at the same moment in time.
b) During the maintenance of pump A, the reservoir in E will have to supply the total demand of 260 l/s alone (the section A-B is closed). Two additional pumps of the same type as in Figure 1.71 will be added in parallel to the existing one in E (total three pumps available), and in addition the pipe route along E-D-C-B is to be enlarged for this purpose. What will be the new pipe diameters D_{ED}, D_{DC} and D_{CB} that can provide the same demands in areas B, C and D as under question a), if the pressure of 15 mwc is to be maintained in areas B and D, while the pressure in area C should be 27 mwc? For the diameters of new pipes, use the k value of 0.2 mm, and exactly calculated diameter values, without rounding off the results.
c) For the newly renovated network in question b), calculate the total maximum flow that both reservoirs can supply if the pressure in areas B and D is still to be maintained at 15 mwc (pumping station in A operates with one pump unit, while the one in E operates with all three pumps). What will be the demand of each of the three areas if the pressure in area C is measured at 25 mwc?

Marking matrix:

Question	a	b	c
Max. points	15	15	20

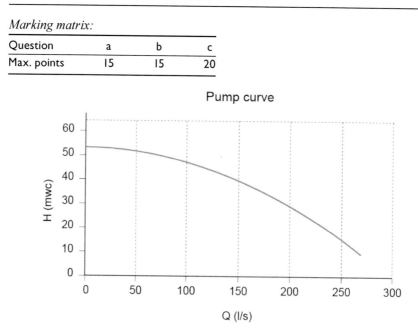

Figure 1.71 Problem 1.6.14 – pump curve for single unit (H_d = 40 mwc, Q_d = 150 l/s)

Answers:

a) For $p_C/\rho g$ = 15 mwc, the head H_C = 65 msl. For the pumping flows $Q_{p,A}$ = 150 l/s and $Q_{p,E}$ = 110 l/s, the corresponding pumping heads are $H_{p,A}$ = 40 msl and $H_{p,E}$ = 46.16 msl. Moreover, the friction losses along A-B and E-D are $h_{f,AB}$ = 7.00 mwc, and $h_{f,ED}$ = 2.98 mwc (both Spreadsheet Lesson 1–2).

INPUT		OUTPUT	
L (m)	400	v (m/s)	2.12
D (mm)	300	υ (m²/s)	1.31E-06
k (mm)	0.5	Re (-)	487262
Q (l/s)	150	λ (-)	0.0229
T (°C)	10	h_f (mwc)	7.00
H₂ (msl)	-	S (-)	0.0175

INPUT		OUTPUT	
L (m)	700	v (m/s)	1.14
D (mm)	350	υ (m²/s)	1.31E-06
k (mm)	0.5	Re (-)	306279
Q (l/s)	110	λ (-)	0.0223
T (°C)	10	h_f (mwc)	2.98
H₂ (msl)	-	S (-)	0.0043

Consequently, the head H_B = 45 + 40–7 = 78 msl, and pressure $p_B/\rho g$ = 18 mwc. Equally, H_D = 35 + 46.16–2.98 = 78.18 msl, and pressure $p_D/\rho g$ = 13.18 mwc.

Based on these heads, the available friction loss along B-C is $h_{f,BC} = 78-65$ = 13 mwc, and the corresponding hydraulic gradient is then $S_{BC} = 13/600$ = 0.02167. At the same time, the available friction loss along D-C is $h_{f,DC} =$ 78.18–65 = 13.18 mwc, leading to $S_{DC} = 13.18/500 = 0.02636$. In the next step, $Q_{BC} = 49.04$ l/s and $Q_{DC} = 26.77$ l/s (both from Spreadsheet Lesson 1–4).

INPUT		OUTPUT	
L (m)	600	h_f (mwc)	13.00
D (mm)	200	u (m²/s)	1.31E-06
k (mm)	1.5	Re (-)	238802
S (-)	0.021667	λ (-)	0.0349
T (°C)	10	v (m/s)	1.56
H₂ (msl)	-	Q (l/s)	49.04

INPUT		OUTPUT	
L (m)	500	h_f (mwc)	13.18
D (mm)	150	u (m²/s)	1.31E-06
k (mm)	1	Re (-)	173361
S (-)	0.02636	λ (-)	0.0338
T (°C)	10	v (m/s)	1.51
H₂ (msl)	-	Q (l/s)	26.77

Finally, the demands in the supplied areas are $Q_B = Q_{AB} - Q_{BC} = 150-49.04$ = 100.96 l/s, $Q_D = Q_{ED} - Q_{DC} = 110-26.77 = 83.23$ l/s and $Q_C = Q_{BC} + Q_{DC} =$ 49.04 + 26.77 = 75.81 l/s.

b) With three equal pumps in parallel in source E, each one is supplying $Q_p = 260/3$ = 86.67 l/s at the head $H_p = 48.88$ mwc. From the pressure requirements in the demand areas, the heads, $H_B = 60 + 15 = 75$ msl, $H_C = 50 + 27 = 78$ msl, and H_D = 65 + 15 = 80 msl. Based on these heads, the available friction loss along E-D is $h_{f,ED} = 35 + 48.88–80 = 3.88$ mwc, and the corresponding hydraulic gradient is then $S_{ED} = 3.88/700 = 0.005543$. Furthermore, the available friction loss along D-C is $h_{f,DC} = 80–77 = 3$ mwc, leading to $S_{DC} = 3/500 = 0.006$. Finally, $h_{f,CB} = 77–70$ = 2 mwc, leading to $S_{CB} = 2/600 = 0.003333$. From the answers in question a), the flows $Q_{ED} = 260$ l/s, $Q_{DC} = 260–83.23 = 176.77$ l/s, and $Q_{CB} = 176.77–75.81$ = 100.96 l/s. Finally, the diameters $D_{ED} = 445$ mm, $D_{DC} = 379$ mm, and $D_{CB} =$ 343 mm (all calculated from Spreadsheet Lesson 1–5).

INPUT		OUTPUT	
L (m)	700	h_f (mwc)	3.88
k (mm)	0.2	u (m²/s)	1.31E-06
Q (l/s)	260	D (mm)	445
S (-)	0.005543	Re (-)	569092
T (°C)	10	λ (-)	0.0173
H₂ (msl)	-	v (m/s)	1.67

INPUT		OUTPUT	
L (m)	500	h_f (mwc)	3.00
k (mm)	0.2	υ (m²/s)	1.31E-06
Q (l/s)	176.77	D (mm)	379
S (-)	0.006	Re (-)	454979
T (°C)	10	λ (-)	0.0180
H₂ (msl)	-	v (m/s)	1.57

INPUT		OUTPUT	
L (m)	600	h_f (mwc)	2.00
k (mm)	0.2	υ (m²/s)	1.31E-06
Q (l/s)	100.96	D (mm)	343
S (-)	0.003333	Re (-)	286501
T (°C)	10	λ (-)	0.0189
H₂ (msl)	-	v (m/s)	1.09

c)   For $H_B = 60 + 15 = 75$ msl, and $D_B = 343$ mm, the pump flow $Q_{p,A} = 180.05$ l/s (from Spreadsheet Lesson 6–2).

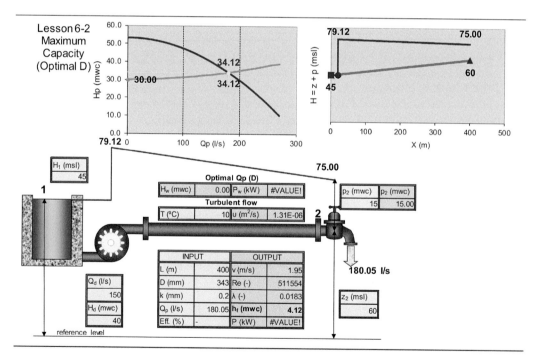

Because of the same pressure in D, the three pumps in E will still pump the flow $Q_{p,E} = 260$ l/s; hence, the total pumping from the two reservoirs will be 440.05 l/s. Based on the heads in B, C and D, the available friction loss along B-C is $h_{f,BC} = 75–75 = 0$ mwc. Hence, $Q_{BC} = 0$ l/s meaning that the area in B is entirely supplied from the reservoir in A. At the same time, the available friction loss along D-C is $h_{f,DC}$ = 80–75 = 5 mwc, leading to S_{DC} = 5/500 = 0.01. In the next step, Q_{DC} = 16.41 l/s (both from Spreadsheet Lesson 1–4).

INPUT		OUTPUT	
L (m)	500	h_f (mwc)	5.00
D (mm)	150	u (m²/s)	1.31E-06
k (mm)	1	Re (-)	106772
S (-)	0.01	λ (-)	0.0341
T (°C)	10	v (m/s)	0.93
H₂ (msl)	-	Q (l/s)	**16.41**

Finally, the demands in the supplied areas are $Q_B = Q_{AB}$ = 180.05 l/s, Q_D = $Q_{ED} - Q_{DC}$ = 260–16.41 = 243.59 l/s and Q_C = 16.41 l/s.

Problem 1.6.15

Area C shown in Figure 1.72 receives water combined by gravity from the reservoir in B, and by the pumping station from the reservoir in A. The pumping station is also supplying reservoir B and comprises two equal pump units connected in series, each of them operating according to the curve shown in Figure 1.73 (H_d = 20 mwc, Q_d = 100 l/s). The information about the connecting routes is as follows:

Section	L (m)	D (mm)	k (mm)
A – B	500	300	0.6
B – C	400	100	1.0
A – C	900	200	0.3

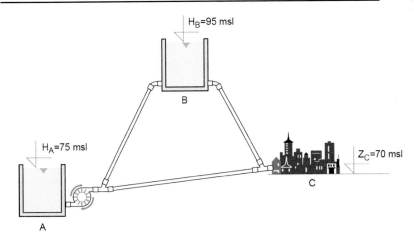

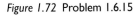

Figure 1.72 Problem 1.6.15

Questions:
a) At a certain point in time, the observed flow in area C is Q_C = 25 l/s supplied at the pressure $p_c/\rho g$ = 24 mwc. Calculate the flows in all three pipes, A-B, B-C and A-C, for this scenario if one pump unit is in the operation.

b)   At another point in time, the observed flow in area C grows to $Q_C = 40$ l/s supplied at the pressure $p_c/\rho g = 23$ mwc. Calculate the flows in all three pipes,
A-B, B-C and A-C, for this scenario if two pump units are in operation.

c)   In the event of the closure of section A-C, what would need to be the diameter
of section B-C to enable the supply of the same demand Q_C and the pressure
$p_c/\rho g$ as in question b), entirely from the reservoir in B? How much flow
would the pumping station then supply to this reservoir with two pumps in
operation? The absolute roughness of the new pipe can be assumed at $k_{BC} =$
0.2 mm.

Marking matrix:

Question	a	b	c
Max. points	20	15	15

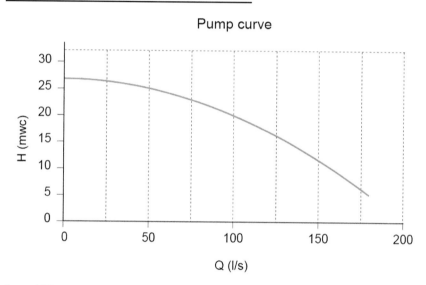

Pump curve

Figure 1.73 Problem 1.6.15 – pump curve for single unit ($H_d = 20$ mwc, $Q_d = 100$ l/s)

Answers:

a)   For $p_c/\rho g = 24$ mwc, the available friction loss $h_{f,BC} = 95 - (70 + 24) = 1$
mwc, leading to $S_{BC} = 1/400 = 0.0025$. In the next step, $Q_{BC} = 2.73$ l/s (from
Spreadsheet Lesson 1–4).

INPUT		OUTPUT	
L (m)	400	h_f (mwc)	1.00
D (mm)	100	u (m²/s)	1.31E-06
k (mm)	1	Re (-)	26789
S (-)	0.0025	λ (-)	0.0405
T (°C)	10	v (m/s)	0.35
H₂ (msl)	-	Q (l/s)	2.73

Furthermore, $Q_C = Q_{AC} + Q_{BC}$ and $Q_p = Q_{AB} + Q_{AC}$. The pump is pumping both of these flows with the same head H_p. The integrity of the hydraulic grade line along A-B-C is therefore to be ensured by $H_p = H_B - H_A + h_{f,AB} = H_C + h_{f,AC} - H_A$, while respecting the flow continuity and the relation between the pumping flow and pumping head. In this scenario, $Q_{AC} = 25–2.73 = 22.27$ l/s and the corresponding friction loss $h_{f,AC} = 2.74$ mwc (from Spreadsheet Lesson 1–2), leading to $H_p = 70 + 24 + 2.74–75 = 21.74$ mwc.

INPUT		OUTPUT	
L (m)	900	v (m/s)	0.71
D (mm)	200	u (m²/s)	1.31E-06
k (mm)	0.3	Re (-)	108513
Q (l/s)	22.27	λ (-)	0.0238
T (°C)	10	h$_f$ (mwc)	2.74
H₂ (msl)	-	S (-)	0.0030

At this head, the pump delivers $Q_p = 86$ l/s, leading to $Q_{AB} = 86–22.27 = 63.73$ l/s. As a check, for this flow the friction loss $h_{f,AB} = 1.69$ mwc (from Spreadsheet Lesson 1–2), leading to $H_p = 95–75 + 1.69 = 21.69$ mwc ≈ 21.74 mwc.

INPUT		OUTPUT	
L (m)	500	v (m/s)	0.90
D (mm)	300	u (m²/s)	1.31E-06
k (mm)	0.6	Re (-)	207021
Q (l/s)	63.73	λ (-)	0.0244
T (°C)	10	h$_f$ (mwc)	1.69
H₂ (msl)	-	S (-)	0.0034

b) The same process applies as in question a), except that two pumps operate in series with the composite pump curve described using $H_d = 40$ mwc, and $Q_d = 100$ l/s. For $p_c/\rho g = 23$ mwc, the available friction loss $h_{f,BC} = 2$ mwc, leading to $S_{BC} = 0.005$, leading to $Q_{BC} = 3.90$ l/s (Spreadsheet Lesson 1–4).

INPUT		OUTPUT	
L (m)	400	h$_f$ (mwc)	2.00
D (mm)	100	u (m²/s)	1.31E-06
k (mm)	1	Re (-)	38269
S (-)	0.005	λ (-)	0.0398
T (°C)	10	v (m/s)	0.50
H₂ (msl)	-	Q (l/s)	3.90

Furthermore, Q_{AC} = 40–3.90 = 36.10 l/s and the corresponding friction loss $h_{f,AC}$ = 7 mwc (from Spreadsheet Lesson 1–2), leading to H_p = 25 mwc.

INPUT		OUTPUT	
L (m)	900	v (m/s)	1.15
D (mm)	200	u (m²/s)	1.31E-06
k (mm)	0.3	Re (-)	175902
Q (l/s)	36.1	λ (-)	0.0231
T (°C)	10	hf (mwc)	7.00
H₂ (msl)	-	S (-)	0.0078

At this head, the pumps deliver Q_p = 145.78 l/s, leading to Q_{AB} = 109.68 l/s. As a control, for this flow the friction loss $h_{f,AB}$ = 4.92 mwc (from Spreadsheet Lesson 1–2), leading to H_p = 24.92 mwc ≈ 25 mwc.

INPUT		OUTPUT	
L (m)	500	v (m/s)	1.55
D (mm)	300	u (m²/s)	1.31E-06
k (mm)	0.6	Re (-)	356286
Q (l/s)	109.68	λ (-)	0.0241
T (°C)	10	hf (mwc)	4.92
H₂ (msl)	-	S (-)	0.0098

c) In case the same flow and the pressure as in question b) need to be delivered from the reservoir in B, while the section A-C is closed, the pipe diameter along B-C has to be enlarged to D_{BC} = 223 mm (from Spreadsheet Lesson 1–5), rounded off to 250 mm.

INPUT		OUTPUT	
L (m)	400	hf (mwc)	2.00
k (mm)	0.2	u (m²/s)	1.31E-06
Q (l/s)	40	D (mm)	223
S (-)	0.005	Re (-)	174449
T (°C)	10	λ (-)	0.0210
H₂ (msl)	-	v (m/s)	1.02

At the same time, the supply of the reservoir in B with two pumps in series will be $Q_{AB} = 138.39$ l/s (from Spreadsheet Lesson 6–2).

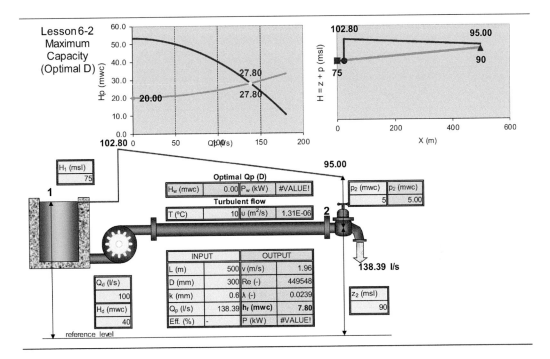

1.7 Examination true-false tests

The following selection of 15 True-False tests was given, slightly modified, as an introductory part of the open book examinations run at IHE Delft in the period 2004–2016. The attached drawings show the particular position of the hydraulic grade line over a pipe section, each with ten statements that are to be judged as True or False. Each correct judgement earns two points (total 20 points). The available time to complete this part of the exam is approximately 30 minutes. The correct answers are given at the end of this paragraph.

In all the problems (unless specified differently):

1. The position of the hydraulic grade line is described by the elevations of the piezometric head.
2. Any change of the flow conditions and/or the pipe properties influences this position as described correctly (or incorrectly) by the given statement.
3. Fixed (i.e. unchanged) values suggested in the statements are those shown in the drawing.
4. The conditions changed in one statement do not automatically apply to the following statements.

Test 1.7.1

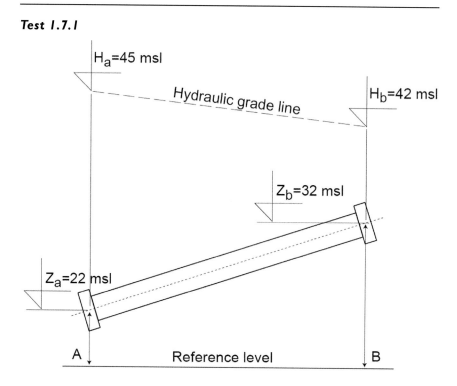

1	The pressures in cross sections A and B are equal.	True	False
2	The flow rate in the pipe equals 0 l/s.	True	False
3	The head in cross section A is lower than the head in cross section B.	True	False
4	The pressure in cross section B, $p_b/\rho g$ = 10 mwc.	True	False
5	The friction loss along pipe section A-B is 3 mwc.	True	False
6	For fixed H_a = 45 msl, the velocity along pipe section A-B will reduce if H_b drops to 37 msl.	True	False
7	For fixed H_a = 45 msl, the pressure along pipe section A-B becomes partly negative if H_b drops to 22 msl.	True	False
8	If the elevations $Z_a > Z_b$, the flow direction will reverse.	True	False
9	If the inner pipe diameters $D_a < D_b$, the flow velocity $v_a < v_b$.	True	False
10	If the pipe diameters $D_a = D_b$ and the heads $H_a > H_b$, the pipe flow rates $Q_a > Q_b$.	True	False

Test 1.7.2

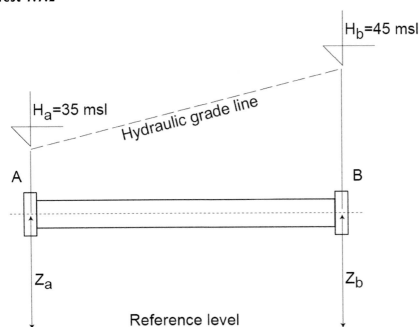

1	The flow direction in the pipe is from B to A.	**True**	**False**
2	The flow velocity in the pipe equals 0 m/s.	**True**	**False**
3	The head in cross section A is higher than the one in cross section B.	**True**	**False**
4	The flow rate in the pipe is constant alongside its length.	**True**	**False**
5	The pressure in cross section A equals $p_a/\rho g$ = 35 mwc.	**True**	**False**
6	For fixed H_a and the pipe length, the drop of the head in cross section B will reduce the hydraulic gradient.	**True**	**False**
7	The uniform increase of the pipe diameter will increase the flow velocity.	**True**	**False**
8	The increase of water temperature will reduce the friction loss.	**True**	**False**
9	For fixed H_a and the pipe length, the increase of flow velocity will increase the hydraulic gradient.	**True**	**False**
10	If the inner pipe diameters $D_a = D_b$, the kinetic energies in cross sections A and B are equal.	**True**	**False**

Test 1.7.3

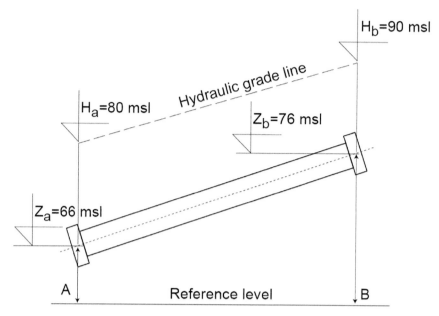

I	The water flows from cross section A to cross section B.	**True**	False
2	The pressure drop along the pipe is 10 mwc.	**True**	False
3	The pressure in cross section A equals the pressure in cross section B.	**True**	False
4	The total head loss along the pipe equals 14 mwc.	**True**	False
5	The potential energies in cross sections A and B are equal.	**True**	False
6	If the inner pipe diameters $D_a = D_b$, the velocity in cross section A equals the velocity in cross section B	**True**	False
7	If the inner pipe diameters $D_a = D_b$, the increase of velocity v_b will cause the increase of velocity v_a.	**True**	False
8	For the constant flow rate, the uniform increase of pipe diameter will reduce the kinetic energy.	**True**	False
9	The increase of head H_b will always cause the increase of head H_a.	**True**	False
10	For fixed pipe diameter and length, if the flow rate doubles, the value of the hydraulic gradient will then also double.	**True**	False

Test 1.7.4

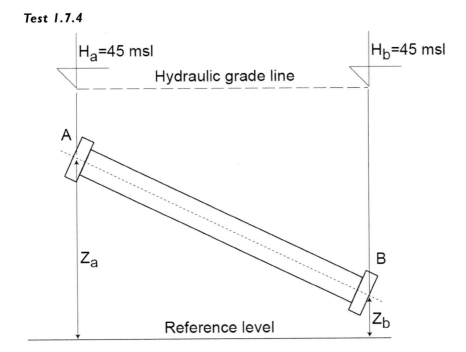

1	The flow direction in the pipe is from A to B.	**True**	False
2	The flow rate in the pipe is constant alongside its length.	**True**	False
3	The flow velocity in the pipe equals 0 m/s.	**True**	False
4	The kinetic energies in cross sections A and B are equal.	**True**	False
5	The piezometric head in cross section A is lower than the one in cross section B.	**True**	False
6	For Z_b = 10 msl, the pressure in cross section B equals $p_b/\rho g$ = 35 mwc.	**True**	False
7	For fixed values of $H_a = H_b$, a smaller pipe diameter would increase the flow velocity.	**True**	False
8	If $H_a - H_b = Z_a - Z_b$, the pressure alongside the section A-B is constant.	**True**	False
9	If Z_b becomes bigger than Z_a, the flow direction in the pipe will reverse.	**True**	False
10	If H_a and H_b grow to 55 msl, the kinetic energy increases for 10 mwc.	**True**	False

Test 1.7.5

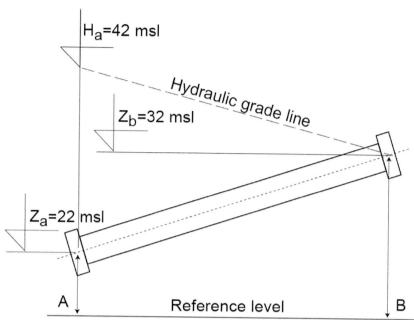

1	The water along pipe section A-B flows from B to A.	**True**	False
2	The potential energy in cross section B equals 0 msl.	**True**	False
3	The pipe is partly filled with water.	**True**	False
4	The pressure in cross section B equals $p_b/\rho g$ = 32 mwc.	**True**	False
5	For constant inner pipe diameter, the velocity along pipe section A-B is constant.	**True**	False
6	The pressure in cross section A is greater than the pressure in cross section B.	**True**	False
7	For fixed flow rate, the diameter increase of pipe section A-B will reduce the friction loss.	**True**	False
8	If H_a drops to 32 msl, for fixed H_b, the flow along rate A-B becomes 0 l/s.	**True**	False
9	If H_a drops to 22 msl, for fixed H_b, the flow velocity along A-B becomes 0 m/s.	**True**	False
10	In case Z_a = 54 msl, the pressure along A-B will be negative.	**True**	False

Test 1.7.6

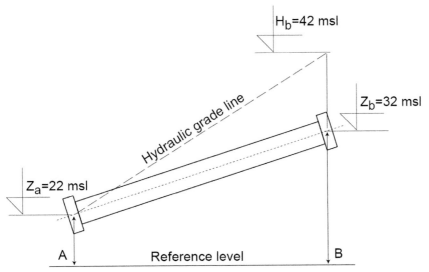

1	The water along pipe section A-B flows from B to A.	**True**	False
2	The potential energy in cross section B equals 42 msl.	**True**	False
3	The pipe is partly filled with water.	**True**	False
4	The pressure in cross section B equals $p_b/\rho g$ = 32 mwc.	**True**	False
5	For constant inner pipe diameter, the Reynolds number along pipe section A-B is constant.	**True**	False
6	The pressure in cross section A is lower than the pressure in cross section B.	**True**	False
7	For constant inner diameter, the kinetic energy in cross section A is lower than the one in cross section B.	**True**	False
8	If H_a grows to 42 msl, for fixed H_b, the flow rate along A-B becomes 0 l/s.	**True**	False
9	If H_a drops below 22 msl, for fixed H_b, the pressure along A-B becomes partly negative.	**True**	False
10	If Z_b drops below 22 msl, the flow direction will reverse.	**True**	False

Test 1.7.7

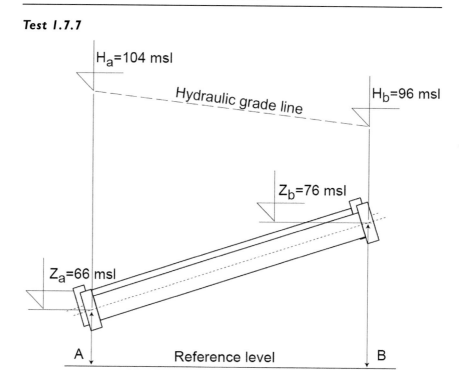

For the two pipes above, connected in parallel:

1	The flow direction in both pipes is from B to A.	**True**	**False**
2	If both pipes have the same inner diameter and the same length, their flow rate is equal.	**True**	**False**
3	The friction loss along pipe section A-B is h_f = 4 mwc in each pipe (total 8 mwc).	**True**	**False**
4	The friction loss along section A-B is always equal in both pipes.	**True**	**False**
5	The hydraulic gradient along section A-B is always equal in both pipes.	**True**	**False**
6	The pressure in cross section A, $p_a/\rho g$ = 38 mwc.	**True**	**False**
7	The head in the middle of the section A-B equals 100 msl in both pipes.	**True**	**False**
8	If H_a drops below 96 msl, for fixed H_b, the flow along pipe section A-B will reverse.	**True**	**False**
9	If the two pipes are replaced with one of equivalent diameter, this diameter will equal the sum of the two pipe diameters.	**True**	**False**
10	The equivalent diameter pipe will generate less friction loss if it is shorter than any of the two pipes in parallel.	**True**	**False**

Test 1.7.8

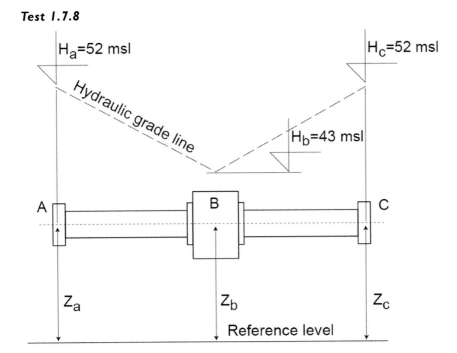

1	The component in cross section B is receiving water.	**True**	False
2	If $Z_a = Z_b = Z_c$, the flow rate along pipe section A-B-C equals 0 l/s.	**True**	False
3	If $Z_b = 33$ msl, the pressure in cross section B equals $p_b/\rho g = 10$ mwc.	**True**	False
4	The water in pipe sections A-B and B-C flows in opposite directions.	**True**	False
5	The friction loss along pipe section B-C equals $h_f = 9$ mwc.	**True**	False
6	For uniform pipe roughness, if the pipe lengths $L_{AB} = L_{BC}$, the diameters D_{AB} and D_{BC} must be equal.	**True**	False
7	If the inner pipe diameter $D_a > D_c$, the kinetic energy in cross section A is higher than the one in cross section C.	**True**	False
8	For fixed values $H_a = H_c$, if the discharge in cross section B becomes equal to 0, $H_b = 52$ msl.	**True**	False
9	For increased values $H_a = H_c = 72$ msl, the discharge in cross section B increases if $H_b = 52$ msl.	**True**	False
10	If $Z_b = 33$ msl and $Z_a = Z_c = 22$ msl, the water will flow from cross section B in the direction of the cross sections A and C.	**True**	False

Test 1.7.9

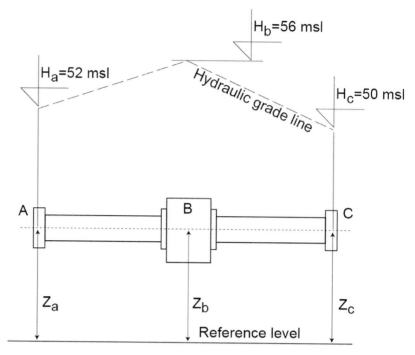

1	The component in cross section B is supplying water.	**True**	False
2	The water in pipe sections A-B-C flows from A to C.	**True**	False
3	For uniform pipe roughness, if the pipe lengths $3L_{AB} = 2L_{BC}$, and diameters $D_{AB} = D_{BC}$, the flows Q_{AB} and Q_{BC} are equal.	**True**	False
4	If $Z_b = 46$ msl, $Z_a = 42$ msl and $Z_c = 40$ msl, the pressure along pipe section A-B-C is constant.	**True**	False
5	If $Z_b = 44$ msl, and $Z_a = Z_c = 48$ msl, the water will flow towards cross section B.	**True**	False
6	For fixed H_b, if the flow rate along pipe section B-C becomes equal to 0 l/s, $H_c = 56$ msl.	**True**	False
7	For fixed H_b, if the flow on pipe section B-C becomes equal to 0, $H_a = 56$ msl, too.	**True**	False
8	For fixed H_a and H_c, if head H_b drops below 50 msl, the flow direction in both pipe sections, A-B and B-C, will be reversed.	**True**	False
9	The increase of flow in pipe section A-B increases the value of H_b.	**True**	False
10	For fixed H_b, if both flows $Q_{AB} = Q_{BC} = 0$ l/s, the potential energy along pipe section A-B-C equals 0 (zero).	**True**	False

Test 1.7.10

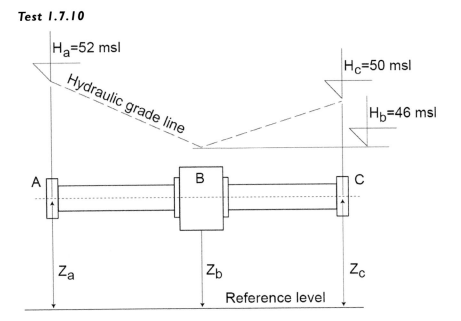

I	The component in cross section B is a pressure-reducing valve.	**True**	**False**
2	The water in pipe sections A-B and B-C flows in opposite directions.	**True**	**False**
3	If $Z_a = Z_b = Z_c$, the water pressure along the pipe section A-B-C is constant.	**True**	**False**
4	If $Z_a > Z_b > Z_c$, the water will flow towards cross section C.	**True**	**False**
5	For uniform pipe roughness, if the pipe lengths $L_{AB} = L_{BC}$, and the inner diameters $D_{AB} = D_{BC}$, the velocities $v_{AB} = v_{BC}$.	**True**	**False**
6	If the flow $Q_{AB} = Q_{BC} = 0$ l/s, the heads $H_a = H_b = H_c$.	**True**	**False**
7	Flows Q_{AB} and Q_{BC} along the section A-B-C are always equal.	**True**	**False**
8	If the hydraulic gradients $S_{AB} = S_{BC}$, the pipe lengths $L_{ab} = L_{bc}$.	**True**	**False**
9	For fixed H_a and H_c, if H_b grows to 53 msl, both flows Q_{AB} and Q_{BC} will reverse their directions.	**True**	**False**
10	For fixed H_a and H_c, if H_b grows to 51 msl, the hydraulic gradients $S_{AB} = S_{BC}$.	**True**	**False**

Test 1.7.11

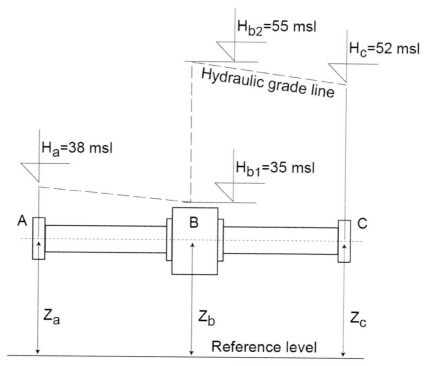

1	The component in cross section B is a valve.	**True**	**False**
2	The water flows from cross section C to cross section A by gravity.	**True**	**False**
3	The pressure drop in cross section B is 20 mwc.	**True**	**False**
4	The friction losses along the pipe sections A-B and B-C are equal.	**True**	**False**
5	The flow rates along the pipe sections A-B and B-C are equal.	**True**	**False**
6	For the same pipe roughness, if the pipe lengths $L_{AB} = L_{BC}$, the diameters D_{AB} and D_{AC} must be equal.	**True**	**False**
7	If the inner pipe diameters $D_{AB} = D_{BC}$, the kinetic energies in cross sections A and C will be equal.	**True**	**False**
8	If the elevations $Z_a = Z_c$, the pressure difference between the cross sections A and C is 14 mwc.	**True**	**False**
9	For fixed H_a, the flow rate increase along the section A-B will increase the head H_{b2}.	**True**	**False**
10	For fixed H_a, the flow rate increase along the section B-C will reduce the head H_{b1}.	**True**	**False**

Test 1.7.12

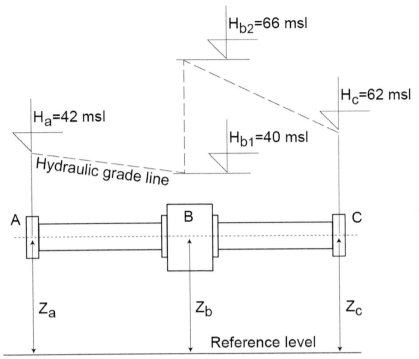

Reference level

I	The component in cross section B is a pump.	**True**	False
2	The water in pipe sections A-B-C flows from A to C.	**True**	False
3	If $Z_a = Z_b = Z_c = 38$ msl, the flow $Q_{AB} = Q_{BC} = 0$.	**True**	False
4	For uniform roughness along the pipe section A-B-C, if the pipe length $L_{AB} = L_{BC}$, also the diameter $D_{AB} = D_{BC}$.	**True**	False
5	The pressure in cross section B, $p_b/\rho g = 26$ mwc.	**True**	False
6	For fixed H_a, the increase of flow rate along pipe section B-C decreases the value of H_{b2}.	**True**	False
7	If the flow rate along pipe section B-C becomes equal to 0, $H_c = 66$ msl.	**True**	False
8	If the pump is switched-off, the head $H_{b1} = H_a$.	**True**	False
9	If the pump is switched-off, the head $H_{b1} = H_{b2}$.	**True**	False
10	The head H_c can never be greater than the head H_{b2}.	**True**	False

Test 1.7.13

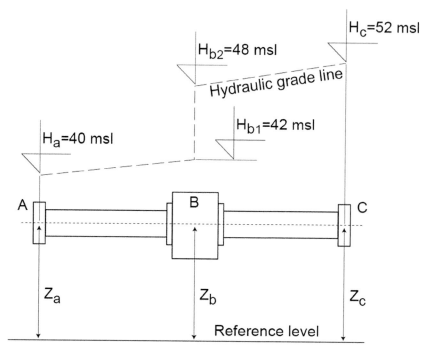

1	The component in cross section B is a pump.	**True**	False
2	The water flows from cross section A to cross section C.	**True**	False
3	The pressure drop in cross section B is 6 mwc.	**True**	False
4	The total head loss along the route A-B-C equals 12 mwc.	**True**	False
5	If the pipe lengths $2L_{AB} = L_{BC}$, the values of hydraulic gradients S_{AB} and S_{BC} are equal.	**True**	False
6	The value of head H_a can never grow above 42 msl.	**True**	False
7	For the same pipe roughness, if the pipe diameters $D_a = D_c$, the kinetic energies in cross sections A and C are equal.	**True**	False
8	If the flow rate along the route A-B-C doubles, the value of hydraulic gradients S_{AB} and S_{BC} will also double.	**True**	False
9	For fixed H_a and H_c, the increase of head H_{b1} will cause the increase of head H_{b2}.	**True**	False
10	For $H_{b2} > H_{b1}$, if the values $H_a = H_{b1}$, then the values $H_{b2} = H_c$.	**True**	False

Test 1.7.14

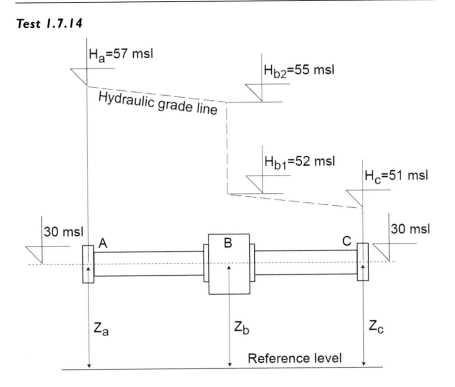

For the pressure reducing valve (PRV) above:

1	The PRV is fully opened.	**True**	False
2	If $Z_a = Z_b = Z_c = 30$ mwc, the PRV setting is 22 mwc.	**True**	False
3	The flow rates $Q_{AB} = Q_{BC}$.	**True**	False
4	The energy drop in cross section B is 3 msl.	**True**	False
5	The pressure in cross section A, $p_a/\rho g = 27$ mwc.	**True**	False
6	If the pipe lengths $2L_{AB} = L_{BC}$, the values of hydraulic gradients S_{AB} and S_{BC} are equal.	**True**	False
7	The head H_a can drop below 51 msl.	**True**	False
8	If $H_c > 52$ msl, the flow direction will reverse.	**True**	False
9	If $H_a < 55$ msl, the flow direction will reverse.	**True**	False
10	If $H_a < 52$ msl, the PRV will be closed.	**True**	False

Test 1.7.15

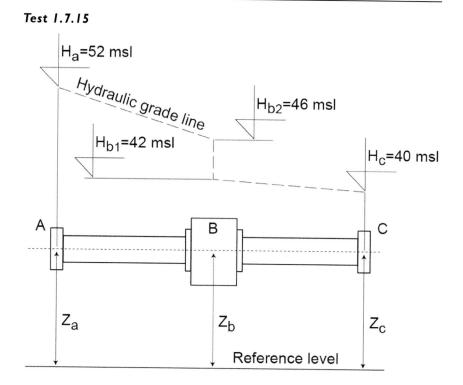

For the butterfly valve above:

1	The valve is partly closed.	True	False
2	The water along the pipe route A-B-C flows from A to C.	True	False
3	If Z_c = 39 msl, and Z_a = Z_b = 36 mwc, the water flow will reverse.	True	False
4	For uniform pipe roughness, if the pipe lengths L_{AB} = L_{BC}, the diameters D_{AB} = D_{BC}.	True	False
5	The pressure drop in cross section B is 42 mwc.	True	False
6	For the same valve opening, if H_c grows to 43 msl, the flow along the pipe section B-C becomes equal to 0 l/s.	True	False
7	For the same valve opening, if H_a grows to 54 msl, the head H_b will also grow to 44 msl.	True	False
8	For the same valve opening, if H_a drops to 42 msl, the pressure drop in cross section B is then 0 mwc.	True	False
9	For the same valve opening, if head H_{b1} drops to 40 msl, H_b = 46 msl remains unchanged.	True	False
10	If both flows Q_{AB} = Q_{BC} = 0, the pressure drop in cross section B must be 0 mwc.	True	False

Answers

Test 1.7.1

1		False	$p_a/\rho g$ = 23 mwc; $p_b/\rho g$ = 10 mwc.
2		False	The hydraulic grade line is not horizontal.
3		False	$H_a > H_b$.
4	True		$H_b - Z_b$ = 10 mwc.
5	True		$H_a - H_b$ = 3 mwc.
6		False	The hydraulic gradient will increase and so will the velocity, too.
7	True		$H_b - Z_b < 0$ (-10 mwc).
8		False	Flow rate in pressurised pipes does not depend on their slopes.
9		False	The opposite is true for constant flow rate along A-B.
10		False	$Q_a = Q_b$ regardless the diameter and head values.

Test 1.7.2

1	True		$H_b > H_a$.
2		False	The hydraulic grade line is not horizontal.
3		False	The opposite is true ($H_b > H_a$).
4	True		Based on the Law of Continuity.
5		False	H_a = 35 msl; $p_a/\rho g < 35$ mwc.
6	True		Based on the friction loss reduction.
7		False	The opposite is true, based on the constant flow rate.
8	True		A higher water temperature reduces the kinematic viscosity.
9	True		The friction loss grows with the velocity increase.
10	True		Based on the equal flow velocities ($v_a = v_b$).

Test 1.7.3

1		False	$H_b > H_a$.
2		False	$H_b - H_a$ = 10 mwc but no pressure drop.
3	True		$p_a/\rho g = p_b/\rho g$ = 14 mwc.
4		False	ΔH = 10 mwc.
5		False	$H_b > H_a$.
6	True		Based on the Law of Continuity.
7	True		Based on the Law of Continuity.
8	True		Based on the reduction of the flow velocity.
9		False	That will depend on the pipe flow rate, too.
10		False	The theoretical relation between pipe flow rates and hydraulic gradients is (nearly) quadratic.

Test 1.7.4

1		False	There is no flow in the pipe.
2	True		The flow rate is 0 l/s.
3	True		There is no flow (friction loss) in the pipe.
4	True		Both kinetic energies in A and B equal 0 mwc.
5		False	$H_a = H_b$.
6	True		$p_b/\rho g = H_b - Z_b$.
7		False	The flow velocity equals 0 m/s in all cases when $H_a = H_b$.
8	True		$p_a/\rho g = H_a - Z_a$ and $p_b/\rho g = H_b - Z_b$.
9		False	The flow rate stays at 0 l/s regardless the position of the pipe.
10	True	False	The kinetic energy equals 0 mwc in all cases when $H_a = H_b$.

Test 1.7.5

1		False	The flow direction is opposite ($H_a > H_b$).
2		False	$H_b = Z_b$.
3		False	The flow is pressurised ($H_a > Z_a$).
4		False	$p_b/\rho g = 0$ mwc.
5	True		Based on the flow uniformity.
6	True		$p_a/\rho g = 20$ mwc; $p_b/\rho g = 0$ mwc.
7	True		The flow velocity will reduce.
8	True		$H_a = H_b$.
9		False	The flow direction will reverse ($H_b > H_a$).
10	True		$Z_a > H_a$; $Z_b = H_b$.

Test 1.7.6

1	True		$H_b > H_a$.
2	True		$H_a = 42$ msl.
3		False	The flow is pressurised ($H_b > Z_b$).
4		False	$Z_b = 32$ mwc.
5	True		The flow velocity and the water temperature are also constant.
6	True		$p_a/\rho g = 0$ mwc; $p_b/\rho g = 10$ mwc.
7		False	The flow velocity is constant and so is the kinetic energy, too.
8	True		$H_a = H_b$.
9	True		$Z_a > H_a$.
10		False	Flow direction in pressurised pipes does not depend on their slopes.

Test 1.7.7

1		False	$H_a > H_b$.
2	True		Based on identical hydraulic conditions.
3		False	The friction loss $h_f = 8$ mwc in each pipe.
4	True		Based on the connectivity of pipes in parallel.
5		False	Based on possibly different lengths of each pipe.
6	True		$H_a - Z_a = 38$ mwc.
7	True		Based on the proportional length of the location.
8	True		$H_b > H_a$.
9		False	The hydraulic calculation is to be conducted based on equal friction loss in both pipes.
10		False	The hydraulic gradient will change (increase), not the friction loss.

Test 1.7.8

1	True		$H_a > H_b$ and $H_c > H_b$.
2		False	Flow rate in pressurised pipes does not depend on their slopes.
3	True		$p_b/\rho g = H_b - Z_b$.
4	True		B is supplied from A and C.
5	True		$h_f = H_c - H_b$.
6	True		Based on the friction loss equations.
7		False	$v_a < v_c$, based on the Law of Continuity.
8	True		$H_a = H_b = H_c = 52$ msl. Consequently, $h_f = 0$ mwc.
9	True		$h_f = H_c - H_b = H_c - H_a = 20$ mwc.
10		False	Flow direction in pressurised pipes does not depend on their slopes.

Test 1.7.9

1	True		$H_b > H_a$ and $H_b > H_c$.
2		False	The water flows from B towards A and C.
3	True		Based on the friction loss equations. L_{AB} is proportionally smaller than L_{BC}; pipe length has a linear relation to the friction loss.
4	True		The pressure along pipe section A-B-C is 10 mwc.
5		False	Flow direction in pressurised pipes does not depend on their slopes.
6	True		$H_b = H_c = 56$ msl. Consequently, $h_f = 0$ mwc.
7		False	B is a supply point. H_a is not dependant on $H_b - H_c$.
8	True		$H_a > H_b$ and $H_c > H_b$.
9		False	B is a supply point.
10		False	If $Q_{AB} = Q_{BC} = 0$ l/s, $H_a = H_b = H_c = 56$ msl > 0 msl.

Test 1.7.10

1		False	B is receiving water from A and C.
2	True		$H_a > H_b$ and $H_c > H_b$.
3		False	$p_a/\rho g > p_c/\rho g > p_b/\rho g$.
4		False	Flow direction in pressurised pipes does not depend on their slopes.
5		False	Based on different friction losses along A-B and B-C indicating different flows Q_{AB} and Q_{BC}.
6	True		The friction loss along A-B-C $h_f = 0$ mwc.
7		False	The Law of Continuity does not necessarily apply along the section A-B-C.
8		False	Based on different friction losses along A-B and B-C.
9	True		$H_b > H_a$ and $H_b > H_c$.
10		False	The friction losses along A-B and B-C become equal ($h_f = 1$ mwc) but not necessarily the hydraulic gradients.

Test 1.7.11

1		False	The component in cross section B is a pump.
2		False	The water is pumped in the opposite direction.
3		False	The pumping head (lift) in cross section B is 20 mwc.
4	True		$h_f = 3$ mwc along both cross sections.
5	True		Based on the Law of Continuity.
6	True		Based on the same friction loss and continuous flow rate.
7	True		$v_a = v_c$, based on the Law of Continuity.
8	True		$H_c - H_a = 14$ mwc.
9		False	The flow rate increase reduces the pumping head.
10		False	The flow rate increase increases the friction loss.

Test 1.7.12

1	True		Based on the shape of the hydraulic grade line.
2	True		Based on the slope of the hydraulic grade line along A-B-C.
3		False	Flow rate in pressurised pipes does not depend on their slopes.
4		False	Based on the different friction losses along A-B and B-C.
5		False	The pumping head (lift) in cross section B, $H_p = 26$ mwc.
6	True		The flow rate increase reduces the pumping head.
7		False	The pump is switched off in this case.
8	True		$Q_{AB} = 0$. The friction loss along A-B also $h_f = 0$ mwc
9		False	The head H_{b1} depends on the upstream hydraulic conditions, while the head H_{b2} depends on the downstream conditions.
10	True		$H_c > H_{b2}$ implies reversed flow direction, from C to B, which is not allowed in pumping stations.

Test 1.7.13

1		False	The component in B causes a loss of head (could be a valve).
2		False	Based on the slope of the hydraulic grade line along A-B-C.
3	True		$H_{b2} - H_{b1} = 6$ mwc.
4	True		$H_c - H_a = 12$ mwc.
5	True		Based on the two times smaller friction loss along A-B, compared to the one along B-C.
6		False	The reverse flow is impossible only in case of specific valves.
7	True		$v_a = v_c$, based on the Law of Continuity.
8		False	The theoretical relation between pipe flow rates and hydraulic gradients is (nearly) quadratic.
9		False	The increase of H_{b1} in this situation indicates the flow rate increase, i.e. a reduction of pressure drop (an opened valve).
10	True		$Q_{AB} = Q_{BC} = 0$ l/s (a closed valve).

Test 1.7.14

1		False	Based on the shape of the hydraulic grade line.
2	True		$H_{b1} - Z_b = 22$ mwc.
3	True		Based on the Law of Continuity.
4	True		$H_{b2} - H_{b1} = 3$ mwc.
5	True		$H_a - Z_a = 27$ mwc.
6		False	Based on the twice bigger friction loss along the twice shorter pipe length along A-B.
7	True		Based on the upstream hydraulic conditions.
8		False	PRVs do not allow reversed flows.
9		False	The PRV condition $H_a < H_{b2}$ is impossible in theory.
10		False	If $H_a < 52$ msl, the PRV will be fully opened.

Test 1.7.15

1	True		The fully open valve would likely generate smaller head loss.
2	True		Based on the slope of the hydraulic grade line.
3		False	Flow rate in pressurised pipes does not depend on their slopes.
4		False	Based on the different friction loss along A-B and B-C.
5		False	The pressure drop in cross section B is $H_{b2} - H_{b1} = 4$ mwc.
6		False	The reversed flow is possible in the case of butterfly valves.
7		False	H_{b2} also depends on the downstream hydraulic conditions.
8		False	The pressure drop in cross section B also depends on the downstream hydraulic conditions.
9		False	The pressure drop in cross section B depends on the change in flow rate, next to the (same) valve opening.
10	True		No head loss is generated for the no flow conditions.

Network modelling workshop

2.1 Learning objectives and set-up

The objective of this assignment is the analysis of the hydraulic performance of a simple water distribution network. The main goal of the workshop is to understand the basic functionality of standard demand-driven network computer modelling software and the implications resulting from modifications of the system configuration and/or its operational modes. When given as a lecture, the workshop contact time consists of three sessions of two hours, with the following set-up:

INTRODUCTION (Session 1)
- Case description
- Introduction to the software
- Start-up, model preparation

PART 1: DIRECT PUMPING (Session 2)
- Pump selection
- Pump operation
- System extension
- Future demand
- System failure

PART 2: BALANCING STORAGE (Session 3)
- Balancing volume
- Pressure optimisation
- Future demand
- Water tower
- Fire demand

The above hydraulic aspects classified for two typical distribution schemes are analysed by using EPANET[1] pressurised network modelling software. Brief

---

1 EPANET (Version 2) is software for hydraulic and water quality modelling of water distribution networks, developed by the United States Environmental Protection Agency (US EPA)

information about the installation and use of this software is given in Appendix 8. The original 200-page manual can be downloaded as a PDF file, together with a programme that is available on the Internet in the public domain.

Ten problem areas classified into two groups will be analysed, and a total of 100 partly filled answers available at the end of each question will be completed. Before starting work, two separate computer files should be prepared:

- *CWSDP.NET*, the network file of the model with the direct pumping scheme that is to be used for PART 1 of the assignment,
- *CWSBS.NET*, the network file of the model with the balancing storage scheme that is to be used for PART 2 of the assignment.

These files can be easily created from the available *CWS.NET* file that contains raw information about the network. The procedure to upgrade this file is explained in Section 2.3 – Preparing the models. The text file format showing the contents of the *CWS.NET* is given in this appendix.

The necessary instructions on all aspects of the assignment and the software are to be given during the three sessions listed above, which are also meant as discussion hours. The work on the assignment is to be completed at home as a self-study.

Limited modification of the network models is required while answering the questions. Instructions and guidance on how this should be done are given for each question, namely:

- how to modify the input file before running the simulation,
- some useful hints to better understand the problem and approach its solution,
- partly completed answers related to the question.

All the adaptations to the model within the same distribution scheme are adopted as the work progresses; *the modified model of the previous question is to be used as the input for the following question.* Furthermore, selected nodes/junctions and pipes/links in some questions are determined from an individual number (the values between 120 and 150), or from a value between 1 and 5 reflecting the serial number of five topographical data sets available in the assignment (see Table 2.1), in order to create a unique combination of data for each student. Various combinations of these numbers impact the magnitude of the problem differently.

The answers can be completed directly in the text of this appendix, or in the available MS Word file (*the answer sheets*). If the suggested intervention in the network is not necessary, the corresponding space should still be filled: with 'x', 'invalid', 'not applicable', 'n/a', 'not needed', etc. *Leaving any space empty will be regarded as an unanswered question; only a fully completed answer will be assessed.* After completing the assignment, the hard or electronic copy of the answer sheets should be submitted to the teacher prior to the given deadline.

In principle, each correct answer earns one point, with a maximum of 100 for the entire assignment. Alternatively, a pattern of weighting the answers/questions can be applied by the lecturer.

Based on the average student performance, the estimated study load for this exercise at IHE Delft is 28 hours, or one credit of the European Credit Transfer

System (ECTS), which includes the contact hours. The solved example for individual number 150 and topographical data set number 1 is given at the end of this appendix.

2.2 Case introduction

A town with a population of approximately *90,000* is located in a hilly area with ground elevations ranging between *A* and *F msl*, as shown in Table 2.1 (only one set to be selected/assigned!).

Figure 2.1 shows the configuration of the terrain together with the layout of the network secondary mains. The node numbers and pipe diameters are indicated for each pipe in the system.

Table 2.1 Topography, data sets (altitudes of isometric lines in *msl*)

Data set number	A	B	C	D	E	F
1	10	15	20	25	30	35
2	10	20	30	40	30	20
3	30	20	10	20	30	40
4	10	12	11	13	12	10
5	50	45	40	35	30	25

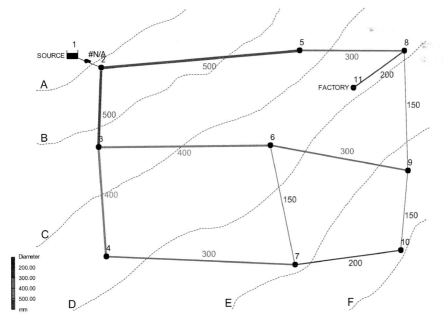

Figure 2.1 Node numbers and pipe diameters (mm)

The source of supply, located at *Node 1*, is groundwater stored after simple treatment in a groundwater reservoir from where it is pumped into the system. A rapid expansion is expected in the coming years because of regional development activity. Consequently, a factory for production of agricultural machinery is going to be built. Due to the demand growth, the existing water distribution network appears to be of insufficient capacity. Its reconstruction, which includes a general overhaul of the pumping station at the source, should provide sufficient amounts of water and *minimum pressure of 20 mwc* during regular operation at all times. In addition, an uninterrupted supply of water to the new factory is required in case of breakdowns elsewhere in the system.

The water consumption in the area is mainly of a domestic type, except for the new factory that is supplied from the system at an average capacity of *1080 m³/d.* The factory works at a constant (average) capacity of *12 hours a day*, between *7:00 and 19:00 hours* (two shifts). Based on the population distribution, the nodal base demands have been determined in *l/s* as shown in Figure 2.2 (together with the pipe lengths in *m*).

The typical diurnal domestic demand pattern is shown in Figure 2.3, with the peak factor values given in Table 2.2. These values include physical water losses. The seasonal demand variations are represented by the factor that equals *individual number/100*, which describes the ratio between the maximum consumption day and the average consumption day. Neither the seasonal variations nor the water losses apply in the case of the factory.

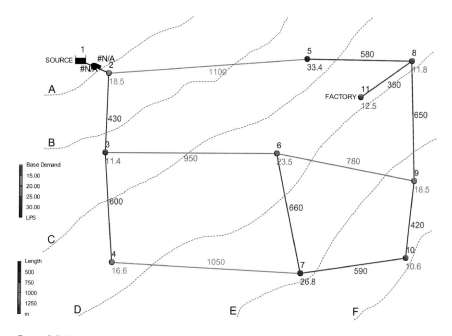

Figure 2.2 Average nodal demands (l/s) and pipe lengths (m)

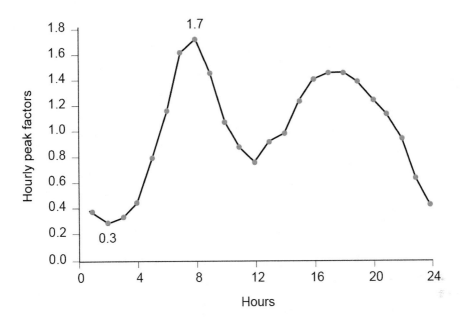

Figure 2.3 Diurnal domestic pattern (the value range 0.3–1.7)

Table 2.2 Diurnal peak factors – Figure 2.3.

Hour	1	2	3	4	5	6	7	8	9	10	11	12
pf_h	0.38	0.30	0.34	0.45	0.78	1.15	1.70	1.60	1.44	1.06	0.87	0.76
Hour	13	14	15	16	17	18	19	20	21	22	23	24
pf_h	0.91	0.98	1.23	1.40	1.45	1.45	1.38	1.24	1.13	0.94	0.64	0.42

Table 2.3 Efficiency curve.

Q (l/s)	$0.25Q_d$	$0.50Q_d$	$1.00Q_d$	$1.50Q_d$	$1.75Q_d$
Efficiency (%)	20	60	75	65	30

PVC is chosen as the pipe material for the network. Due to this choice, it is assumed that the roughness of the pipe remains low. The accepted k-value of *0.5 mm* includes impacts of local (minor) losses in the network.

Finally, the existing pumping station is old and of insufficient capacity. Due to frequent failures in the past, it is operating at the moment with only one unit, out of the three initially installed in parallel. The duty head and duty flow of this unit are $H_d = 40$ *mwc* and $Q_d = 200$ *l/s*, respectively. Table 2.3 describes how the pump efficiency curve depends on the value of the duty flow.

The energy costs for pumping are charged at a flat rate of *EUR 0.15 per kWh*.

2.3 Preparing the models

After studying the contents of Appendix 8 and having installed the EPANET programme together with the network case folder, execute the following ten steps:

2.3.1 Modifying and loading the backdrop map

1. Start MS PowerPoint and open the file *CWS.ppt* from the electronic materials available with this exercise. Replace the letters A-F on the topography slide with the data set selected/assigned from Table 2.1.
2. Save the modification as a Windows Metafile (extension WMF) under an arbitrary name (MS PowerPoint menu command **File>>Save as . . .**); answer the question in PowerPoint that comes after selecting the WMF format as 'No'. Exit the programme.
3. Start EPANET and open the modified file *CWS.NET*. Load the backdrop map (menu command **View>>Backdrop>>Load . . .** ; browse within the sub-directory and load the created WMF file).

2.3.2 Adjusting nodal elevations

4. Based on the given isometric lines, assign appropriate elevations to the nodes (from the graph, search for the shortest distance connecting two consecutive lines through the node in between them).
5. Modify this information in the *CWS.NET* file (double-click on each node and adjust the **Elevation** initially set to 0 in the property editor; click on the following node; close the editor after the elevations in all the nodes have been modified). To check the input for all the nodes, show it on the map (browser option **Map>>Nodes >>Elevation**).
6. The source node (1) is an EPANET component called **Reservoir** and the information to be modified there is **Total Head**. This indicates the fixed piezometric head i.e. the water level on the suction side of the pumping station. This is a reference-surface level from which the pumping takes place, its variation being neglected in this case. As an approximation, take the same value as that for isometric line A for the total head of the reservoir.

2.3.3 Adjusting the seasonal variation factor and the nodal demand of the factory

7. Select menu option **Project>>Analysis Options . . .** and replace the **Demand Multiplier** value of 1.0 in the property editor by the value generated from the selected/assigned individual number.
8. Double-click on the factory node (11) and decrease its **Base Demand** proportionally, i.e. divide it by the demand multiplier. This is necessary in order to cancel the effect of the seasonal variations on the factory demand. Close the property editor.

Table 2.4 Nodal (junction) elevations for each topographical data set (msl).

Topography number	1	2	3	4	5
Node 1	10.0	10.0	30.0	10.0	50.0
Node 2	10.2	10.4	29.6	10.1	49.8
Node 3	16.5	23.0	17.0	11.7	43.5
Node 4	23.1	36.2	16.2	12.3	36.9
Node 5	18.3	26.6	13.4	11.4	41.7
Node 6	22.7	35.4	15.4	11.9	37.3
Node 7	29.5	31.0	20.0	12.1	30.5
Node 8	24.4	38.8	18.8	12.9	35.6
Node 9	31.7	26.6	33.4	11.4	28.3
Node 10	34.7	20.6	39.4	10.2	25.3
Node 11	24.0	38.0	18.0	12.8	36.0

2.3.4   Saving the information

9. Save the modifications as a new project file under the name *CWSDPxyyy. NET* (recommended, not compulsory) by using the menu command **File>>Save as . . . .** Index '*x*' in the file name denotes the number of the used topography data set from Table 2.1 (values 1–5), while '*yyy*' stands for the applied individual number (values 120–150). This model will be used for Part 1 (Direct Pumping) in the workshop.

10. Run the menu command **File>>Save as . . .** once more and save the current NET file under the name *CWSBSxyyy.NET* (recommended, not compulsory). This model will be used for Part 2 (Balancing Storage) of the workshop.

All the other input information, referred to in the case introduction, already exists in the above models. The last step, to create a tank in the CWSBS model, will be explained in Section 2.5.

NOTE 1: Nodal elevations given in Table 2.4 can be used as assistance in Step 4.

NOTE 2: To proceed faster with the preparation of the EPANET model files, the above steps 1 to 6 can be alternatively omitted by loading the already prepared *CWS topX.NET* EPANET file, and the corresponding *topX* WMF-topography file (where '*X*' is the number of selected topographical data sets).

2.4   Questions part 1 – direct pumping

2.4.1   Pump selection

Question 1.1.1

Using the initial model, analyse the pressures in the system. Identify three nodes with the longest duration of low pressure (below the threshold of 20 mwc) and register the hours when this pressure occurs. What is the absolute minimum pressure in the system and at what time?

Model modification:
- Load the model *CWSDPxyyy.NET*.
- Crosscheck the individual data modified by the procedure in Section 2.3 (the **Browser** option **Map>>Nodes>>Elevation** and **Map>>Nodes>>Base Demand**).

Hints:
- Demand multiplier: an example of the individual number that equals 135 means a seasonal factor of 1.35.
- During simulation runs, EPANET multiplies each base demand by the demand multiplier, with no exception. For nodes where this multiplication is to be neutralised, the base demand should be manually divided by the demand multiplier.
- In this case, the seasonal demand variations do not apply for the factory. Consequently, for the same seasonal factor of 1.35, the corresponding base demand of node 11 would be 12.5/1.35 = 9.26 *l/s*.
- During extended period simulations (EPS), EPANET further multiplies each base demand by the corresponding peak factor stored in the network model component called **Pattern** (available in the **Data** tab of the **Browser**). When the duration of EPS is longer than the duration of the pattern, EPANET will continue repeating the pattern until the end of the EPS.
- *CWS.NET* is prepared for 24-hour EPS (at 1-hour time steps) and the **Pattern** *Domestic* contains 24 diurnal peak factors shown in Figure 2.3/Table 2.2. In this, a commonly applied EPS format, the results are always shown from 00:00–24:00 hours i.e. for 25 time steps while using 24 diurnal peak factors.
- It may be therefore slightly confusing that EPANET takes the first value indicated in the pattern editor as *Time Period 1* to calculate the first time step at 00:00 hours, which is midnight. This will result in a shift of the maximum demand hour from 7:00 hours to 6:00 hours, in this case.
- This is not necessarily a significant deviation, especially if it is not clear whether the peak factor values represent a specific moment, or an average for a particular hour. If however a full synchronisation with the data as shown in Table 2.2 is needed, the **Pattern** *Domestic* should start with the value indicated for a period of 24 hours (0.42) that will be used twice: for the first and last snapshot calculations. In this exercise, the maximum demand hour has been kept at 6:00 hours although the diurnal demand pattern shows it to be at 7:00 hours; the actual meaning of this value in EPANET is the demand representation between 6:00 and 7:00 hours.
- On the same note, the *Factory* demand pattern has been adjusted to reflect the exact operation of the factory between 7:00 and 19:00 hours. The peak factor values are consequently introduced for the time periods 8 through 19. The value 2.0 is used to indicate that the average hourly demand needs to be twice as high to generate the actual daily demand because the factory works for only half a day. The value of 1.0 would also be correct provided the base demand is doubled.
- Looking at the snapshot results for the pressures, identify the most critical nodes (bear in mind the topography!). Analyse these nodes further by displaying them in a time series graph (the main menu command **Report>>Graph**

. . . >>**Time Series**) or as a table (the main menu command **Report>>Table** . . . >>**Time series for node**); specific results can be displayed by choosing parameters in the tab **Columns**, and filtered in the tab **Filters**.

Answers:
Within a 24-hour period:
1. Node _____ has a pressure below 20 *mwc* during the following times: _____ hours.
2. Node _____ has a pressure below 20 *mwc* during the following times: _____ hours.
3. Node _____ has a pressure below 20 *mwc* during the following times: _____ hours.
4. The lowest pressure of _____ *mwc* occurs at node _____ at _____ hours.

Question 1.1.2

Calculate the overall duty head and flow of the pumping station at the source that can provide a minimum pressure of 20 *mwc* throughout the network. Determine the daily energy consumption and the cost of pumping, assuming 24-hour operation of the pump.

Model modification:
*   Modify the pump curve of the existing pump (the **Browser** option **Data>>Curves**, double-click on PST-HQ).
*   Adjust the pump efficiency curve proportionately to the change in the duty flow, by using the pattern from Table 2.2 (the **Browser** option **Data>>Curves**, double-click on PST-EFF).

Hints:
*   Based on the specified duty head and flow, EPANET generates a synthetic pump curve, using the formula: $h_p = c - aQ_p^2$. A curve like this has a theoretical maximum flow (for $h_p = 0$) of twice the specified duty flow, and a maximum head (for $Q_p = 0$) that equals 4/3 of the specified duty head. E.g. for the existing pump from the initial file, the curve is defined by the following three h_p (*mwc*)/Q_p (*l/s*) points: (1)-53.33/0, (2)-40/200 and (3)-0/400, respectively.
*   The duty head and flow represent a composite pump curve if it is assumed that several units are in operation. Increasing only the duty head has the effect of arranging the pumps in series. Increasing only the duty flow has the effect of arranging the pumps in parallel. Increasing both the duty head and flow has the effect of installing bigger pumps.
*   The selection of the pump duty head needs to take into consideration the topography (i.e. elevation difference between the source and the most critical point in the system), the network resistance, and the minimum pressures required for the given demand scenario. Answer 4 suggests the deficit of the pressure in the system. This deficit results from insufficient pumping and/ or high friction losses. The latter can be checked by analysing the hydraulic gradients (the **Browser** option **Map>>Links>>Unit Headloss**).

- To deliver the pressure of 20 *mwc* in the most critical node of the network, the pump needs to overcome the elevation difference between the source and this node, and the head loss along the distance connecting the two points. The latter can be determined as the piezometric head difference between the pressure side of the pump and the most critical node (the **Browser** option **Map>>Nodes>> Head**).
- The selection of the pump duty flow takes into consideration the range of diurnal demand in the network. Compare this range in the initial network model with the value of the current duty flow. A sensible starting point is to assume the duty flow is equal to the average flow in the network. This average can be obtained by presenting the pump flow in tabular form (the main menu command **Report>>Table . . . >>Time series for link**). Clicking in the upper left corner (cell *Time Hours*) marks the entire table for copying to the clipboard, and pasting it into MS Excel.
- Because of the 25 snapshot results displayed in EPANET, any further processing of the time series results in tabular form should take into consideration that the first or the last result need to be eliminated from the analysis, because they are identical.
- The efficiency curve can be adjusted only once, at the end when the final values for the duty flow have been determined. The last simulation with the revised curve will then provide the accurate energy consumption to be filled in Answer 10.
- For energy consideration, take the average *hourly* pump power into consideration (menu option **Report>>Energy>>Average Kwatts**)

Answers:

5. The required duty head and flow of pump _____ are _____ *mwc* and _____ *l/s*, respectively.
6. The highest demand of _____ *l/s* occurs in the system at _____ hours.
7. The lowest pressure of _____ *mwc* occurs at node _____ at _____ hours.
8. The lowest demand of _____ *l/s* occurs in the system at _____ hours.
9. The highest pressure of _____ *mwc* occurs at node _____ at _____ hours.
10. The daily energy use and the cost of pumping are _____ *kWh* and _____ *EUR*, respectively.

2.4.2   Pump operation

Question 1.2.1

For the above pumping station an arrangement of *three pump units in parallel arrangement* has been planned. Determine the duty head and duty flow of each pump that can simulate the same operation as in Question 1.1.2.

Model modification:

- Add two extra pump units next to the existing one. The procedure for doing this is explained after running the menu options **Help>>Help Topics>>Index>>Adding Objects>>Links**.

- To avoid an overlap between the pumps and their corresponding results on the map, use the Vertex feature (i.e. curving of the links). The procedure for this is explained under **Help>>Help Topics>>Index>>Curved Links**.
- Click on the right mouse button at the original pump on the map and select the **Copy** option; click on the right mouse button at the two new pumps and select the **Paste** option. This action will copy the properties of the first pump to the other two. For convenience, change the pump ID in the property editor of each pump by double-clicking on it (suggested ID: 101, 102 and 103, respectively).
- Modify the existing pump curve (the **Browser** option **Data>>Curves**, double-click on PST-HQ). For the same pump types the same curve (name) can be used. Alternatively a new name(s) can be created for each pump; the new ID of the curve should be assigned in the pump property editor.

Hints:
- All three pumps pump the water from node 1 to node 2; using the same node ID with a different link ID represents their parallel arrangement.
- If the duty heads and duty flows are properly selected, the pumping station should generate exactly the same pressures in the system (without any control of the pumps!) as in Question 1.1.2.
- For the sake of simplicity, it is advisable to use the same pump types while connected in a parallel arrangement. In this scenario, each pump will have the same duty head and one third of the duty flow defined in Question 1.1.2.

Answers:
11. The required duty head and flow of pump _____ are _____ *mwc* and _____ *l/s*, respectively.
12. The required duty head and flow of pump _____ are _____ *mwc* and _____ *l/s*, respectively.
13. The required duty head and flow of pump _____ are _____ *mwc* and _____ *l/s*, respectively.
14. The lowest pressure of _____ *mwc* occurs at node _____ at _____ hours.

Question 1.2.2

For the above selection of pumps, determine the optimum schedule for manual operation that can reduce the pressure variation in the system during a period of 24 hours.

Model modification:
- Adjust the pump efficiency curve proportionally to the change in the duty flow, by using the pattern from Table 2.3 (the **Browser** option **Data>>Curves**, double-click on PST-EFF). If different duty flows have been selected, create additional efficiency curves and assign them to the corresponding pump units.
- All the pumps in Question 1.2.1 are operating 24 hours a day, unless stated differently by the control commands that have to be written manually. This is

done under the **Browser** option **Data>>Controls** (double-click on **Simple**). The **Help** button in the opened window shows the syntaxes of all the available control commands and gives a few examples. Use the second format for this question (pump status **OPEN** or **CLOSED AT TIME** ...).

- The **Initial Status** (i.e. at the beginning of the simulation) can be changed from **OPEN** to **CLOSED** in the property editor of each pump. This status is valid until changed by the control commands.

Hints:

- Change the **Initial Status** of the two pumps to **CLOSED** and start the simulation with one pump only; check the pressures in the system hour by hour. For an hour when the pressure in any node drops below the minimum of 20 *mwc*, add a new control line that switches an additional pump on at that hour; re-run the simulation. In contrast, for an hour when the pressure in any node jumps substantially, write a new control line that switches one of the pumps off at that hour.
- In the maximum peak hour, all the pumps are in operation and the pumping station should generate exactly the same pressures in the system as in Question 1.1.2.
- The syntax of the control commands is to be strictly followed.
- There is no limitation to the number of command lines. The last setting listed for each pumping unit is valid until the end of the simulation. Example:

```
LINK  101  OPEN    AT TIME 1
LINK  102  OPEN    AT TIME 1
LINK  101  CLOSED  AT TIME 10
LINK  102  CLOSED  AT TIME 12
LINK  101  OPEN    AT TIME 16
```

means that the first pump unit operates for the whole day except in the period from 10:00 to 16:00 hours, while the second unit works until 12:00 hours only.

- For energy consideration, the pump power expressed through the *Average Kwatts* in the energy report needs to take into consideration the *Percent Utilisation* of the pumps.

Answers:
Within a 24-hour period:

15. Pump ____ is manually operated during the following times: _____
    _____ hours.
16. Pump ____ is manually operated during the following times: _____
    _____ hours.
17. Pump ____ is manually operated during the following times: _____
    _____ hours.
18. The daily energy use of pump ____ is _____ *kWh* at a cost of _____
    EUR.

19. The daily energy use of pump ____ is _____ *kWh* at a cost of _____
    EUR.
20. The daily energy use of pump ____ is _____ *kWh* at a cost of _____
    EUR.
21. The lowest pressure of _____ *mwc* occurs at node ____ at _____ hours.
22. The highest pressure of _____ *mwc* occurs at node ____ at _____ hours.

2.4.3 System extension

Question 1.3.1

A new residential area for *6000 inhabitants* has been planned in the vicinity of
node number **10 – Topography Data Set Number**. The specific demand of **Individual Number** *l/c/d* is selected as a design parameter assuming the same diurnal
pattern from Figure 2.3. What will be the maximum peak demand of the node after
the residential area is completed? Does the pump selection from problem area1.2
allow the supply of such demand without a pressure drop in the system (below the
threshold)? Test the conditions during the maximum consumption hour. Identify
the node in the system with the most critical pressure as a result of the extension.
If required, calculate the additional pumping capacity by modifying the existing
and/or adding a maximum of two new pump units. In addition, consider new
diameters for maximum two pipes to restore the normal pressure. Assume k-value
of *0.1 mm* for the new pipes.

Model modification:
*   Calculate the additional base demand and add it to the node in question by
    activating its property editor; run the simulation and check the pressures in
    the system during the maximum consumption hour.
*   The duty flow and head of the existing pump units can be modified as
    explained in Question 1.1.2. If additional pumps are to be added, the proce-
    dure as in problem area 1.2 can be applied.
*   The diameters of the selected pipes are modified through their property editor.

Hints:
*   The control of the pumps is not relevant in this question as only the peak
    supply conditions are considered; it is logical that all the available pumps are
    in operation during the maximum consumption hour. However, adding new
    pump(s) will require a change in the pump schedules whereby a situation
    may occur that the lowest pressure appears outside the maximum consump-
    tion hour due to some units being switched off.
*   Increasing the capacity of the existing pumps or adding new pumps does not
    exclude alternative measures for changing the pipe diameters, and vice versa.
    A combined solution is often applicable.
*   Answers 25 and 26: if the existing pump ID is specified in the answer(s), it is
    assumed that the existing pump(s) will be replaced with another (bigger) unit.
    If a new pump ID is specified, it will simulate additional pumping unit(s)
    connected in parallel with the existing pumps.

- For maintenance purposes, it is advisable to use the same pump types while connected in parallel arrangement, provided excessive energy consumption can be reduced.
- Answers 27 and 28: if the existing pipe ID is specified in the answer(s), it is assumed that the existing pipe diameter will be replaced with another (larger) diameter. If a new pipe ID is specified with already connected nodes, it will simulate an additional pipe connected in parallel.
- Introducing pipes between the nodes that are not initially connected is possible in theory but is not advised in this exercise as the routes of such pipes normally follow the streets in the area.

Answers:

As a result of the system extension:

23. The highest demand of _____ *l/s* occurs at node _____ at _____ hours.
24. The lowest pressure of _____ *mwc* occurs at node _____ at _____ hours.

(Some of) the following measures are proposed to restore the pressure in the system:

25. The required duty head and flow of pump _____ are _____ *mwc* and _____ *l/s*, respectively.
26. The required duty head and flow of pump _____ are _____ *mwc* and _____ *l/s*, respectively.
27. Pipe _____ that connects nodes _____ and _____ has a _____ *mm* diameter.
28. Pipe _____ that connects nodes _____ and _____ has a _____ *mm* diameter.

As a result of these measures:

29. The lowest pressure of _____ *mwc* occurs at node _____ at _____ hours.
30. The highest hydraulic gradient of _____ *m/km* occurs at pipe _____ at _____ hours.

2.4.4   Future demand

Question 1.4.1

Annual domestic demand growth of *3 %* is assumed for the next *10 years*. In the same period, the water demand of the factory will increase by *30 %*. Analyse the consequences for the water distribution system if the existing demand patterns were to apply throughout this period. Identify the node with the lowest pressure. Test the conditions during the maximum consumption hour. Propose a reconstruction of the system that could restore the normal pressure. Establish additional pumping capacity by modifying the existing and/or adding maximum two pump units and/or new diameters of maximum four pipes. Assume k-value of *0.1 mm* for the new pipes.

Model modification:
- Calculate the demand increase after ten years and modify the general multiplier of nodal demands accordingly; increase the average demand of the factory. The same procedure applies as explained in Section 2.3, steps 7 and 8.
- Other suggestions are the same as in the previous question.

Hints:
* Integrate the seasonal factor and the demand growth factor. Use the exponential model for the demand growth. Note that this model does not affect the factory demand that has its own pattern of increase, largely dependent on the production capacity.
* The same hints for the system reconstruction apply as in the case of the previous problem.
* Higher pressures are often achieved by enlarging pipes along the same route (in series), rather than changing the diameters of scattered pipes that do not have a direct hydraulic connection.

Answers:
As a result of the demand growth, after ten years:

31. The highest demand of _____ *l/s* occurs in the system at _____ hours.
32. The lowest pressure of _____ *mwc* occurs at node ____ at _____ hours.

(Some of) the following measures are proposed to restore the pressure in the system:

33. The required duty head and flow of pump _____ are _____ *mwc* and _____ *l/s*, respectively.
34. The required duty head and flow of pump _____ are _____ *mwc* and _____ *l/s*, respectively.
35. Pipe _____ that connects nodes _____ and _____ has a _____ *mm* diameter.
36. Pipe _____ that connects nodes _____ and _____ has a _____ *mm* diameter.
37. Pipe _____ that connects nodes _____ and _____ has a _____ *mm* diameter.
38. Pipe _____ that connects nodes _____ and _____ has a _____ *mm* diameter.

As a result of these measures:

39. The lowest pressure of _____ *mwc* occurs at node ____ at _____ hours.
40. The highest hydraulic gradient of _____ *m/km* occurs at pipe ____ at _____ hours.

2.4.5 System failure

Question 1.5.1

For the future system from problem area 1.4, analyse the effect if failure of the pipe number **Topography Data Set Number + 1** occurs during the maximum consumption hour. Identify the most critical parts of the system. Propose a reconstruction that can restore the normal pressure. Establish the required duty head and flow with a maximum of two stand-by pump units and/or new diameters of maximum three pipes. Assume a k-value of *0.1 mm* for the new pipes.

Model modification:
* To simulate the pipe burst, change the **Initial Status** in the property editor of the pipe in question from **OPEN** to **CLOSED**; run the simulation and analyse the pressures during the maximum consumption hour.

Hints:

- Opting for stand-by pump units of the same type and size as the pumps used for regular operation gives an advantage in operation and maintenance (O&M).
- The burst of a pipe affects the pressure in the system depending on the pipe conveying capacity. Larger pipes may cause severe pressure drops that cannot be mitigated by increasing the pumping capacity.
- Pipe replacement for the sake of increased system reliability should be applied selectively. Bear in mind that such pipes may create permanent problems with low velocities during regular operation.

Answers:

As a result of the pipe burst:

41. The lowest pressure of _____ *mwc* occurs at node _____ at _____ hours.
42. The highest hydraulic gradient of _____ *m/km* occurs at pipe _____ at _____ hours.

(Some of) the following measures are proposed to restore the pressure in the system:

43. The required duty head and flow of pump _____ are _____ *mwc* and _____ *l/s*, respectively.
44. The required duty head and flow of pump _____ are _____ *mwc* and _____ *l/s*, respectively.
45. Pipe _____ that connects nodes _____ and _____ has a _____ *mm* diameter.
46. Pipe _____ that connects nodes _____ and _____ has a _____ *mm* diameter.
47. Pipe _____ that connects nodes _____ and _____ has a _____ *mm* diameter.

As a result of these measures:

48. The lowest pressure of _____ *mwc* occurs at node _____ at _____ hours.
49. The highest hydraulic gradient of _____ *m/km* occurs at pipe _____ at _____ hours.
50. The highest velocity of _____ *m/s* occurs at pipe _____ at _____ hours.

2.5 Questions part 2 – balancing storage

An alternative design of the network has been considered with an elevated tank at the highest point in the system. The tank bottom is proposed at *25 m* above ground level and the cross-section diameter is *20 m*. The minimum and maximum water depth in the tank have been set at *2* and *6 meters*, respectively, leading to an available balancing volume of approximately *1260 m³*. The tank is connected to the system with a pipe of *200 mm* diameter and the length of **Individual Number** *m*. The roughness factor of the pipe is *0.5 mm*. It is assumed that the tank is approximately *50 %* full around midnight and it should balance the demand in the system over a period of *24 hours*.

2.5.1 Tank location

Question 2.1.1

Using the initial model, identify the nodes with absolute minimum and maximum pressure in the system and record the hours when these pressures occur. At what period of the day is the balancing volume exhausted (i.e. the tank is at the minimum depth) or at maximum (the tank is full)?

Model modification:
- Load the model *CWSBSxyyy.NET*. Check if the same layout, demand multiplier and nodal elevations are used as in Part 1 of the exercise (Question 1.1.1).
- Add the tank into the system by clicking on the tank icon (not the reservoir icon!) from the map toolbar (see also **Help>>Help Topics>>Index>> Adding Objects>>Nodes**). Position the tank on the highest isometric line, in the vicinity of the node that is nearest to that isometric line.
- Choose the **Select Object** tool (i.e. the black arrow icon) from the map toolbar to open the tank property editor; adjust its contents with the above-mentioned characteristics of the tank: elevation, initial, minimum and maximum depths, and diameter of the cross-section area.
- Connect the tank to the rest of the system in the node with the highest elevation.
- Open the property editor of the connection pipe and adjust its length; keep the default pipe diameter of 200 mm and the roughness of 0.5 mm unchanged.

Hints:
- Tank **Elevation** in the property editor assumes an absolute altitude of the bottom level i.e. the ground elevation plus the tank height. E.g. for a tank with an elevation of 25 *m* positioned on the isometric line of 35 *msl*, the tank elevation in EPANET is going to be 60 msl. This is done for the sake of presenting the tank depth variation by using the same pressure parameter as for ordinary demand nodes.
- For monitoring the water level variation in the tank choose the **Graph** tool from the map toolbar (see also **Report>>Graph**). In the new window, select **Time Series** for **Nodes**, choose parameter **Pressure** and **Add** the tank to the list.
- EPANET signals a potential problem when the tank water level (i.e. the balancing volume) is at the minimum. The tank will be disconnected from the system although the reserve volume is still available. If this volume is to be used (e.g. in the analysis of emergency scenarios, the minimum depth should be reduced to zero).
- Low pressures in the system do not exclusively occur during the maximum consumption hour but can also be a consequence of an empty tank.

Answers:
The tank is connected to the system in the following way:

51. Pipe _____ that connects nodes _____ and _____ has a length of _____ *m*.
52. The bottom level and initial depth of tank _____ are _____ *msl* and _____ *m*, respectively.

As a result of such a position of the tank, within a 24-hour period:

53. The lowest pressure of _____ *mwc* occurs at node _____ at _____ hours.
54. Tank _____ is at the minimum depth during the following times _____ _____ hours.

Question 2.1.2

Assuming constant operation of the pumping station during a period of 24 hours, adjust the duty head and flow, and corresponding tank volume (diameter and/or maximum/minimum depth) to enable the balancing function of the tank. Keep the bottom level of the tank fixed at 25 m above ground. Analyse the pressures in the system.

Model modification:
- The pump and efficiency curves are modified in the same way as in Question 1.1.2.
- The tank dimensions are modified in the corresponding property editor: the data on **Initial Level, Minimum Level, Maximum Level** and **Diameter**.

Hints:
- No particular control of the pumps is needed. With the balancing tank in the system, the pumping station will usually operate at constant (i.e. average) flow throughout the day, when the aim is to obtain savings in the energy costs. As a consequence, one (composite) pumping curve will be sufficient to reflect the operation of the entire pumping station.
- In the sequence of trial and error simulations, the first run can be executed by assuming an arbitrary initial water depth in the tank. A well-balanced demand is achieved if a similar depth as the initial one can be met at the end of the simulation i.e. every 24 hours.
- If the final depth is lower, the tank loses its volume over a period of 24 hours. Depending on the pressures in the system, the remedy is to increase the pumping and/or reduce the tank elevation. The first action alone will boost the pressures in the system while the second one will reduce them.
- An opposite action is needed if the final depth is higher that the initial depth, meaning that the tank receives too much water over a period of 24 hours. To solve this problem, the pumping should be reduced and/or the tank elevation should be increased. The first action alone will lower the pressures in the system while the second one will boost them.
- The tank volume is insufficient if the balancing is performed too quickly: the tank successively overflows and empties during a period of 24 hours.
- Introducing the balancing volume into the system should result in smaller pumps than in the case of the direct supply option (in Part 1 of the exercise).
- The pumps are most efficient if they operate close to the selected duty head and flow. Adjust the efficiency curve only once, at the end when the final value for the duty flow has been determined. The last simulation with the revised curve will then provide the accurate energy consumption required in Answer 62.

Answers:
To establish the balancing function of the tank, the required system modifications are:

55. The required duty head and flow of pump _____ are _____ *mwc* and _____ *l/s*, respectively.
56. The diameter and initial level of tank _____ are _____ *m* and _____ *m*, respectively.

57. The minimum and maximum depth of tank _____ are _____ *m* and _____ *m*, respectively.

During the operation of the modified system:

58. The lowest pressure of _____ *mwc* occurs at node _____ at _____ hours.
59. The highest pressure of _____ *mwc* occurs at node _____ at _____ hours.
60. The lowest water depth of _____ *m* occurs in tank _____ at _____ hours.
61. The highest water depth of _____ *m* occurs in tank _____ at _____ hours.
62. The daily energy use of pump _____ is _____ *kWh* at a cost of _____ EUR.

2.5.2   Pressure optimisation

Question 2.2.1

Run the simulation of the system designed in Question 2.1.2 by reducing the height of the tank to *15 m* above ground level. Check the tank operation. Identify the node (other than the tank itself) with the longest pressure duration below *20 mwc*. Establish the lowest pressure in the system.

Model modification:
• Adjust the tank **Elevation** figure in the tank property editor.

Hints:
• The reduced height will probably result in the tank overflowing during periods of low demand.
• EPANET also signals a potential problem when the tank water level is at the maximum; further filling of the tank will be stopped, assuming that the tank is equipped with a float valve for overflow prevention. This effectively disconnects the tank from the network until the piezometric head in the connecting node becomes lower than the one in the tank.

Answers:
Within a 24-hour period:

63. Tank _____ is at the maximum depth during the following times _____ _____ hours.
64. Node _____ has a pressure below 20 *mwc* during the following times _____ _____ hours.
65. The lowest pressure of _____ *mwc* occurs at node _____ at _____ hours.

Question 2.2.2

Assuming constant pump operation during a period of 24 hours, establish the optimal parameters of the pumping station and the tank (including its height) that can provide both the balancing function and minimum pressures in the system close to the threshold of *20 mwc*.

Model modification:
- The same as in Question 2.1.2.

Hints:
- Through a continuing trial and error procedure try to finetune the parameters of the pumping station and the tank until the pressure in the most critical node is approximately 20 *mwc* at the most critical moment.
- More refined determination of the balancing volume and the initial tank depth can be obtained, based on the system flow variation, by applying the approach explained in the book, Section 4.2.3 on storage design, and using the Spreadsheet Lesson 8–10.

Answers:
To establish optimal pressures in the system, the required system modifications are:

66. The required duty head and flow of pump _____ are _____ *mwc* and _____ *l/s*, respectively.
67. The height and diameter of tank _____ are _____ *m* and _____ *m*, respectively.
68. The minimum and maximum depth of tank _____ are _____ *m* and _____ *m*, respectively.

During the operation of the modified system:

69. The lowest pressure of _____ *mwc* occurs at node _____ at _____ hours.
70. The lowest water depth of _____ *m* occurs in tank _____ at _____ hours.
71. The highest water depth of _____ *m* occurs in tank _____ at _____ hours.
72. The initial and final water depth of tank _____ are _____ *m* and _____ *m*, respectively.
73. The daily energy use of pump _____ is _____ *kWh* at a cost of _____ *EUR*.

2.5.3   Future demand

Question 2.3.1

For the overall demand increase of *30 %* (also in the factory) analyse the consequences for the water distribution system from Question 2.2.2. Identify the node with the lowest pressure. Propose a reconstruction of the system that can restore the balancing function of the tank and normal pressures in the system. The measures could include an increase in pumping capacity, tank diameter, tank height and/or new diameters for maximum three pipes. Assume a k-value of *0.1 mm* for the new pipes.

Model modification:
- Increase the general demand multiplier by 30 % (as in Section 2.3, step 7).

Hint:
* A change in the tank height is allowed but not a preferred measure if other listed measures can solve the problem, as it actually means constructing another elevated tank next to the existing one.
* In reality, a larger tank volume can be constructed now to cater for the future demand. This will result in a larger emergency volume than necessary at present, which will gradually convert into a balancing volume as a result of the demand increase.

Answers:
As a result of the demand growth, within a 24-hour period:

74. The lowest pressure of _____ *mwc* occurs at node ____ at _____ hours.
75. Tank ____ is at the minimum depth during the following times _____ _____ hours.

(Some of) the following measures are proposed to restore the pressure in the system:

76. The required duty head and flow of pump ____ are _____ *mwc* and _____ *l/s*, respectively.
77. The height and diameter of tank ____ are _____ *m* and _____ *m*, respectively.
78. Pipe ____ that connects nodes ____ and ____ has a _____ *mm* diameter.
79. Pipe ____ that connects nodes ____ and ____ has a _____ *mm* diameter.
80. Pipe ____ that connects nodes ____ and ____ has a _____ *mm* diameter.

As a result of these measures:

81. The lowest pressure of _____ *mwc* occurs at node ____ at _____ hours.
82. The initial and final water depth of tank ____ are _____ *m* and _____ *m*, respectively.
83. The daily energy use of pump ____ is _____ *kWh* at a cost of _____ EUR.

2.5.4 Water tower

Question 2.4.1

In the system from Question 2.3.1 reduce the tank volume to approximately **4 x Individual Number** m^3. Alter the diameter and the minimum/maximum depths to achieve this. The tank operates as a water tower i.e. without an emergency volume. Establish the optimum duty head and flow of *two units arranged in parallel* in the pumping station. The first pump operates continuously over a period of 24 hours while the second pump is used to cover the morning and afternoon peaks, and switches on/off automatically based on the water level in the tank. Calculate the 'on' and 'off' levels of the second pump that help to maintain sufficient pressures in the system. What is the total energy consumption in this scenario?

Model modification:
- Add an additional pumping unit and adjust the pump and efficiency curves according to the procedure discussed in Question 1.1.2.
- To introduce automatic operation of the second pump, follow a similar procedure for the manual operation as explained in Question 1.2.2. An example of the command lines written for an automatic mode of operation is listed below:

LINK   102   OPEN   IF NODE 12 BELOW 1.5
LINK   102   CLOSED   IF NODE 12 ABOVE 4.0

  referring to the water levels in the tank at node 12. According to this regime, pump 102 is switched on when the water depth in the tank is between the levels of 1.5 to 4.0 meters.
- The *Initial Status* of pump 101 is 'Open' and of pump 102 is 'Closed'.

Hints:
- Water towers have the prior purpose to maintain stable pressures in the system and not to contribute with a buffer volume in emergency situations. Therefore, the minimum depth can be set at 0 *m*, or just above the tank bottom level.
- By reducing the volume, the elevated tank has almost entirely lost its demand balancing function. Consequently, the pumps operate increasingly to fit the diurnal demand pattern. In this question, the schedule of the second pump is determined by the water level variation in the tank. The switching levels should be determined throughout a number of trial and error simulations, until stable operation has been achieved.
- Because the tank volume has been reduced, more pumping energy will be needed to meet the diurnal demand at acceptable pressures than in the case of larger tanks in combination with pumping stations. However, minimal capacity still exists to optimise the energy consumption in the case of water towers, by appropriate selection of the switching levels.
- The selection of the pump switching levels is dependent on the selected simulation time step. A one-hour time step in this exercise may be rather coarse, resulting in too fast a level variation between two consecutive snap-shoot calculations. It is not strictly required in this exercise, but possible shortening of the time step, for more refined analyses, would also require expansion of the series of diurnal peak factors for two available demand categories (*Domestic* and *Factory*) in line with the corresponding hint in Question 1.1.1. The change in the time step can be done under the **Browser** option **Data>>Options** (double-click on **Times** and modify the value of the **Hydraulic Time Step** in the property editor).

Answers:
84. The diameter and initial level of tank _____ are _____ *m* and _____ *m*, respectively.
85. The minimum and maximum depth of tank _____ are _____ *m* and _____ *m*, respectively.

86. The required duty head and flow of pump ____ are _____ *mwc* and _____ *l/s*, respectively.
87. The required duty head and flow of pump ____ are _____ *mwc* and _____ *l/s*, respectively.
88. The switch on and off levels for pump ____ are _____ *m* and _____ *m*, respectively.

As a result of the automatic pump operation, within a 24-hour period:

89. Pump ____ is switched on during the following times _____ _____ hours.
90. The daily energy use of pump ____ is _____ *kWh* at a cost of _____ *EUR*
91. The daily energy use of pump ____ is _____ *kWh* at a cost of _____ *EUR*.
92. The lowest water depth of _____ *m* occurs in tank ____ at _____ hours.
93. The highest water depth of _____ *m* occurs in tank ____ at _____ hours.
94. The lowest pressure of _____ *mwc* occurs at node ____ at _____ hours.

2.5.5   Fire demand

Question 2.5.1

For the system in Question 2.4.1, analyse the consequences of a fire occurring in the factory. On top of the regular factory demand, the additional quantity of water of *30 l/s* is required at a minimum pressure of *30 mwc*. Assume that the fire event occurs at the worst moment, during the morning peak demand period at 6:00, 7:00 and 8:00 hours (a *total of three hours*). Identify the node with the most critical pressure as a result of the fire. Establish the required duty head and flow of a third (stand-by) unit that is sufficient to provide additional flow and the required pressure in the factory.

Model modification:
• Add the third pumping unit and write the control line(s) for manual operation during the three critical hours.
• The factory demand during the fire can be simply simulated by increasing the diurnal peak factors during the three critical hours (pattern **Factory** under the browser option **Data>>Patterns**).

Answers:
As a result of the fire event:

95. The factory demands are _____ *l/s* at 6:00, 7:00 and 8:00 hours, respectively.
96. The factory pressures are _____ *mwc* at 6:00, 7:00 and 8:00 hours, respectively.
97. The absolute lowest pressure of _____ *mwc* occurs in the network at node ____ at _____ hours.

To provide sufficient pressure for the firefighting:

98. The required duty head and flow of pump ____ are _____ *mwc* and _____ *l/s*, respectively.

As a result of this measure:

99. The factory pressures are _____ *mwc* at 6:00, 7:00 and 8:00 hours, respectively.
100. The absolute lowest pressure of _____ *mwc* occurs in the network at node ____ at _____ hours.

2.6 Epanet raw case network model – cws.net (INP-format)

```
[JUNCTIONS]
ID          Elev          Demand          Pattern
 2           0            18.5            Domestic
 3           0            11.4            Domestic
 4           0            16.6            Domestic
 5           0            33.4            Domestic
 6           0            23.5            Domestic
 7           0            26.8            Domestic
 8           0            11.8            Domestic
 9           0            18.5            Domestic
10           0            10.6            Domestic
11           0            12.5            Factory
```

```
[RESERVOIRS]
ID    Head
 1     0
```

```
[PIPES]
ID  Node1 Node2 Length Diameter Roughness Minor Loss Status
 2  2      3     430    500      .5        0          Open
 3  3      4     600    400      .5        0          Open
 4  2      5     1100   500      .5        0          Open
 5  3      6     950    400      .5        0          Open
 6  4      7     1050   300      .5        0          Open
 7  5      8     580    300      .5        0          Open
 8  6      9     780    300      .5        0          Open
 9  7      10    590    200      .5        0          Open
10  6      7     660    150      .5        0          Open
11  8      9     650    150      .5        0          Open
12  9      10    420    150      .5        0          Open
13  8      11    350    200      .5        0          Open
```

```
[PUMPS]
ID         Node1            Node2            Parameters
 1          1                2               HEAD PST-HQ;


[TAGS]
NODE     11               Industry


[PATTERNS]
ID                  Multipliers
Domestic Demand Pattern
 Domestic  .38      .3       .34      .45      .78      1.15
 Domestic  1.7      1.6      1.44     1.06     .87      .76
 Domestic  .91      .98      1.23     1.4      1.45     1.45
 Domestic  1.38     1.24     1.13     .94      .64      .42
Factory Pattern
 Factory   0        0        0        0        0        0
 Factory   0        2        2        2        2        2
 Factory   2        2        2        2        2        2
 Factory   2        0        0        0        0        0


[CURVES]
ID                X-Value        Y-Value
PUMP: Pump Curve
 PST-HQ         200               40
EFFICIENCY: PST Efficiency Curve
 PST-EFF         50               20
 PST-EFF        100               60
 PST-EFF        200               75
 PST-EFF        300               65
 PST-EFF        350               30


[ENERGY]
 Global Efficiency   75
 Global   Price      .1
 Demand Charge       0
 Pump       1               Efficiency       PST-EFF
 Pump       1               Price            .15


[REACTIONS]
 Order Bulk            1
 Order Tank            1
 Order Wall            1
 Global Bulk           0
 Global Wall           0
 Limiting Potential    0
 Roughness Correlation 0
```

```
[TIMES]
 Duration                      24
 Hydraulic Timestep            1:00
 Quality Timestep              0:05
 Pattern Timestep              1:00
 Pattern Start                 0:00
 Report Timestep               1:00
 Report Start                  0:00
 Start ClockTime               12 am
 Statistic                     None

[REPORT]
 Status                        No
 Summary                       No
 Page                          0

[OPTIONS]
 Units                         LPS
 Headloss                      D-W
 Specific Gravity              1
 Viscosity                     1
 Trials                        40
 Accuracy                      0.001
 Unbalanced                    Continue 10
 Pattern                       Domestic
 Demand Multiplier             1.0
 Emitter Exponent              0.5
 Quality                       None mg/L
 Diffusivity                   1
 Tolerance                     0.01

[COORDINATES]
;Node         X-Coord              Y-Coord
 2            334.85               1314.75
 3            319.29               920.58
 4            355.59               378.61
 5            1338.41              1392.54
 6            1188.01              923.18
 7            1309.89              331.93
 8            1870.02              1387.36
 9            1885.58              793.52
 10           1846.68              399.35
 11           1610.70              1205.83
 1            187.03               1374.39
```

```
[LABELS]
;X-Coord          Y-Coord        Label & Anchor Node
  8.10            1395.1         "SOURCE"
  1431.77         1226.58        "FACTORY"

[BACKDROP]
  DIMENSIONS      0.00           0.00        2000.00      1600.00
  UNITS           Meters
  FILE            CWStopography.wmf
  OFFSET          0.00           0.00

[END]
```

2.7   Answer sheets – solutions 1/150

The following answers (in red font) are given for the combination of topographical data set number 1 and the individual number of 150. This combination resembles a scenario in which the source is at the lowest point in the network. The most critical node is the furthest away from the source, and at the highest altitude. The highest individual number available means the highest demand. Hence, to provide sufficient pressures, relatively big pumps are needed, which makes this scenario the most extreme of all those available in this exercise. The solutions offered are not necessarily unique, or the best; they serve as <u>one example</u> of how to mitigate the pressure problems. The EPANET files are available for each question, to view the results and possibly search for alternative solutions. Specific comments on the results are given at the end.

PART 1 – Direct Pumping

Question 1.1.1

Answers:
Within a 24-hour period:

1. Node 10 has a pressure below 20 *mwc* between 4:00 and 22:00 hours.
2. Node 9 has a pressure below 20 *mwc* between 4:00 and 21:00 hours.
3. Node 7 has a pressure below 20 *mwc* between 4:00 and 21:00 hours.
4. The lowest pressure of -42.48 *mwc* occurs at node 10 at 06:00 hours.

Question 1.1.2

Answers:
5. The required duty head and flow of pump 1 are 85 *mwc* and 300 *l/s*, respectively.
6. The highest demand of 436.30 *l/s* occurs in the system at 06:00 hours.
7. The lowest pressure of 21.04 *mwc* occurs at node 10 at 06:00 hours.

8. The lowest demand of 77.00 *l/s* occurs in the system at 01:00 hours.
9. The highest pressure of 111.27 *mwc* occurs at node 2 at 01:00 hours.
10. The daily energy use and the cost of pumping are 8107.44 *kWh* and 1216.11 *EUR*, respectively.

Question 1.2.1

Answers:

11. The required duty head and flow of pump 101 are 85 *mwc* and 100 *l/s*, respectively.
12. The required duty head and flow of pump 102 are 85 *mwc* and 100 *l/s*, respectively.
13. The required duty head and flow of pump 103 are 85 *mwc* and 100 *l/s*, respectively.
14. The lowest pressure of 21.04 *mwc* occurs at node 10 at 06:00 hours.

Question 1.2.2

Answers:
Within a 24-hour period:

15. Pump 101 is manually operated during all 24 hours.
16. Pump 102 is manually operated between 4:00 and 22:00 hours.
17. Pump 103 is manually operated between 6:00 and 8:00, and 14:00 and 19:00 hours.
18. The daily energy use of pump 101 is 2792.16 *kWh* at a cost of 418.84 *EUR*.
19. The daily energy use of pump 102 is 2229.17 *kWh* at a cost of 334.37 *EUR*.
20. The daily energy use of pump 103 is 1068.03 *kWh* at a cost of 160.20 *EUR*.
21. The lowest pressure of 21.04 *mwc* occurs at node 10 at 06:00 hours.
22. The highest pressure of 96.34 *mwc* occurs at node 2 at 01:00 hours.

Question 1.3.1

Answers:
As a result of the system extension:

23. The highest demand of 85.17 *l/s* occurs at node 5 at 06:00 hours.
24. The lowest pressure of 12.21 *mwc* occurs at node 10 at 06:00 hours.

(Some of) the following measures are proposed to restore the pressure in the system:

25. The required duty head and flow of pump 104 are 85 *mwc* and 100 *l/s*, respectively.
26. The required duty head and flow of pump n/a are n/a *mwc* and n/a *l/s*, respectively.
27. Pipe 11a that connects nodes 8 and 9 has a 200 *mm* diameter.
28. Pipe 12a that connects nodes 9 and 10 has a 200 *mm* diameter.

As a result of these measures:

29. The lowest pressure of 25.72 *mwc* occurs at node 10 at 13:00 hours.
30. The highest hydraulic gradient of 5.27 *m/km* occurs at pipe 7 at 07:00 hours.

Question 1.4.1

Answers:
As a result of the demand growth, after ten years:

31. The highest demand of 622.11 *l/s* occurs in the system at 06:00 hours.
32. The lowest pressure of -24.31 *mwc* occurs at node 10 at 13:00 hours.

(Some of) the following measures are proposed to restore the pressure in the system:

33. The required duty head and flow of pump 105 are 85 *mwc* and 100 *l/s*, respectively.
34. The required duty head and flow of pump n/a are n/a *mwc* and n/a *l/s*, respectively.
35. Pipe 7a that connects nodes 5 and 8 has a 300 *mm* diameter.
36. Pipe 13a that connects nodes 8 and 11 has a 200 *mm* diameter.
37. Pipe 8a that connects nodes 7 and 9 has a 300 *mm* diameter.
38. Pipe n/a that connects nodes n/a and n/a has a n/a *mm* diameter.

As a result of these measures:

39. The lowest pressure of 21.04 *mwc* occurs at node 10 at 23:00 hours.
40. The highest hydraulic gradient of 7.62 *m/km* occurs at pipe 11 at 06:00 hours.

Question 1.5.1

Answers:
As a result of the pipe burst:

41. The lowest pressure of -241.14 *mwc* occurs at node 7 at 06:00 hours.
42. The highest hydraulic gradient of 373.65 *m/km* occurs at pipe 11 at 06:00 hours.

(Some of) the following measures are proposed to restore the pressure in the system:

43. The required duty head and flow of pump n/a are n/a *mwc* and n/a *l/s*, respectively.
44. The required duty head and flow of pump n/a are n/a *mwc* and n/a *l/s*, respectively.
45. Pipe 2a that connects nodes 2 and 3 has a 500 *mm* diameter.
46. Pipe n/a that connects nodes n/a and n/a has a n/a *mm* diameter.
47. Pipe n/a that connects nodes n/a and n/a has a n/a *mm* diameter.

As a result of these measures:

48. The lowest pressure of 21.07 *mwc* occurs at node 10 at 23:00 hours.
49. The highest hydraulic gradient of 6.97 *m/km* occurs at pipe 11 at 06:00 hours.
50. The highest velocity of 1.79 *m/s* occurs at pipe 2a at 06:00 hours.

PART 2 – Balancing Storage

Question 2.1.1

Answers:
The tank is connected to the system in the following way:

51. Pipe 1 that connects nodes 10 and 12 has a length of 150 *m*.
52. The bottom level and initial depth of tank 12 are 60.00 *msl* and 3.00 *m*, respectively.

As a result of such a position of the tank, within a 24-hour period:

53. The lowest pressure of -42.48 *mwc* occurs at node 10 at 06:00 hours.
54. Tank 12 is at the reserve volume (minimum depth) between 4 and 24 hours.

Question 2.1.2

Answers:
To establish the balancing function of the tank, the required system modifications are:

55. The required duty head and flow of pump 101 are 63 *mwc* and 255 *l/s*, respectively.
56. The diameter and initial level of tank 12 are 26 *m* and 3.89 *m*, respectively.
57. The minimum and maximum depth of tank 12 are 2.00 *m* and 6.00 *m*, respectively.

During the operation of the modified system:

58. The lowest pressure of 20.47 *mwc* occurs at node 9 at 07:00 hours.
59. The highest pressure of 75.63 *mwc* occurs at node 2 at 01:00 hours.
60. The lowest water depth of 2.75 *m* occurs in tank 12 at 20:00 hours.
61. The highest water depth of 6.00 *m* occurs in tank 12 at 05:00 hours.
62. The daily energy use of pump 101 is 5179.68 *kWh* at a cost of 776.94 *EUR*.

Question 2.2.1

Answers:
Within a 24-hour period:

63. Tank 12 is disconnected (at maximum depth) during 4:00, 5:00, 6:00, 14:00, and 24:00 hours.
64. Node 10 has the pressure below 20 mwc between 6:00 and 8:00, and between 15:00 and 18:00 hours.
65. The lowest pressure of 16.14 *mwc* occurs at node 9 at 07:00 hours.

Question 2.2.2

Answers:
To establish optimal pressures in the system, the required system modifications are:

66. The required duty head and flow of pump 101 are 62 *mwc* and 255 *l/s*, respectively.

67. The height and diameter of tank 12 are 25.00 *m* and 25 *m*, respectively.
68. The minimum and maximum depth of tank 12 are 2.20 *m* and 6.00 *m*, respectively.

During the operation of the modified system:

69. The lowest pressure of 20.02 *mwc* occurs at node 9 at 07:00 hours.
70. The lowest water depth of 2.30 *m* occurs in tank 12 at 20:00 hours.
71. The highest water depth of 6.00 *m* occurs in tank 12 at 05:00 hours.
72. The initial and final water depth of tank 12 are 3.50 *m* and 3.49 *m*, respectively.
73. The daily energy use of pump 101 is 5093.76 *kWh* at a cost of 764.07 *EUR*.

Question 2.3.1

Answers:
As a result of the demand growth:

74. The lowest pressure of -36.33 *mwc* occurs at node 10 at 16:00 hours.
75. Tank 12 is at the reserve volume (minimum depth) between 10:00 and 11:00 and between 13:00 and 22:00 hours.

(Some of) the following measures are proposed to restore the pressure in the system:

76. The required duty head and flow of pump 101 are 62 *mwc* and 345 *l/s*, respectively.
77. The height and diameter of tank 12 are 25.00 *m* and 25 *m*, respectively.
78. Pipe 12 that connects nodes 9 and 10 has a 200 *mm* diameter.
79. Pipe n/a that connects nodes n/a and n/a has a n/a *mm* diameter.
80. Pipe n/a that connects nodes n/a and n/a has a n/a *mm* diameter.

As a result of these measures:

81. The lowest pressure of 20.18 *mwc* occurs at node 9 at 07:00 hours.
82. The initial and final water depth of tank 12 are 3.70 *m* and 3.76 *m*, respectively.
83. The daily energy use of pump 101 is 6796.08 *kWh* at a cost of 1019.42 *EUR*.

Question 2.4.1

Answers:
84. The diameter and initial level of tank 12 are 16 *m* and 3.00 *m*, respectively.
85. The minimum and maximum depth of tank 12 are 0.00 *m* and 4.00 *m*, respectively.
86. The required duty head and flow of pump 101 are 62 *mwc* and 220 *l/s*, respectively.
87. The required duty head and flow of pump 102 are 62 *mwc* and 220 *l/s*, respectively.
88. The switch-on and off levels for pump 102 are 2.0 *m* and 3.5 *m*, respectively.

As a result of the automatic pump operation, within a 24-hour period:

89. Pump 102 is switched on between 6:00 and 11:00 and between 13:00 and 21:00 hours.
90. The daily energy use of pump 101 is 4306.80 *kWh* at a cost of 646.02 *EUR*

91. The daily energy use of pump 102 is 2592.60 *kWh* at a cost of 388.89 *EUR*.
92. The lowest water depth of 0.00 *m* occurs in tank 12 at 08:00 hours.
93. The highest water depth of 4.00 *m* occurs in tank 12 at 01:00 hours.
94. The lowest pressure of 21.68 *mwc* occurs at node 10 at 08:00 hours.

Question 2.5.1

Answers:
As a result of the fire event:

95. The factory demands are 30.00, 62.49, and 62.49 *l/s* at 6:00, 7:00 and 8:00 hours, respectively.
96. The factory pressures are 31.95, 22.83, and 21.26 *mwc* at 6:00, 7:00 and 8:00 hours, respectively.
97. The absolute lowest pressure of 17.88 *mwc* occurs in the network at node 10 at 08:00 hours.

To provide sufficient flow and pressure for the firefighting:

98. The required duty head and flow of stand-by pump 103 are 62 *mwc* and 220 *l/s*, respectively.

As a result of this measure:

99. The factory pressures are 43.60, 34.02, and 36.53 *mwc* at 6:00, 7:00 and 8:00 hours, respectively.
100. The absolute lowest pressure of 20.86 *mwc* occurs in the network at node 10 at 09:00 hours.

Specific observations on the results

Question 1.1.1

1. Nodes, 7, 9 and 10 are the most critical because they are located the furthest away from the source, but also at the highest elevations. Thus, the combined effect of friction losses and elevation difference makes their pressures negative (as well as in other parts of the network), which means that the specified nodal demand cannot be delivered.
2. EPANET shows the minimum pressure occurring in node 10 at 6:00 hours. Based on the input and the real meaning of the diurnal peak factors, this could also be interpreted as the minimum pressure at 7:00 hours, or the average minimum pressure between 6:00 and 7:00 hours.

Question 1.1.2

3. Answer 4 suggests the maximum pressure deficit in the system is in the order of 62.48 mwc at 6:00 hours (in node 10). At the same moment in time, the system flow rate (Answer 6) is more than twice as high as the pump duty flow. The current pump is therefore obviously far too small for the preferred

demand scenario; it is seemingly the biggest contributor to the negative pressures in the network. A check of the pipe hydraulic gradients (*Unit Headloss* in EPANET) throughout the network would confirm this conclusion; it shows that the pipes are of sufficiently large diameters to convey the flows.

4. In the current layout, the water is pumped from the reservoir elevation of 10 *msl* towards the most distant and the highest elevated node (number 10), located at 34.7 *msl*, where the pressure needs to be maintained at minimum 20 *mwc*. The (piezometric) head difference is then 44.7 *mwc*. In addition, the pump needs to recover the system head loss at 6:00 hours, which can be estimated from the *Head* difference in nodes 2 and 10. In the model in Question 1.1.1 these values are −0.12 *msl* and −7.78 *msl*, respectively, indicating a head loss of 7.66 *mwc*.

5. Thus, the task of the pump at 6:00 hours is to deliver a maximum flow of 436.30 *l/s* with a head of minimum 44.7 + 7.66 = 52.36 *mwc*. Consequently, the pumping curve equation used in EPANET would show that $52.36 = c - a436.30^2$, where $c = 4H_d/3$ and $a = H_d/(3Q_d^2)$. Using these relations, the duty head H_d can be calculated for the selected duty flow Q_d as $H_d = 3H_p^{min}/(4 - (Q_p^{max}/Q_d)^2)$, where Q_p^{max} is the maximum pumping flow to be delivered at the minimum pumping head H_p^{min}.

6. In the first assumption, the duty flow Q_d is made equal to the average flow in the network. In the initial demand scenario, this average is 269.14 *l/s*. Applying the above equation, this results in the duty head H_d of 114.65 *mwc*. With some rounding of values i.e. including a minor safety factor, the first assumption of the duty flow and duty head can be based on values 270 *l/s* and 115 *mwc*, respectively.

7. The hydraulic simulation with these values (and the efficiency curve adapted accordingly) yields a pressure in node 10 of 20.87 *mwc* at 6:00 hours, and a total pump energy consumption of 9882.96 *kWh* at a cost of *EUR* 1482.44. While the minimum pressure condition is satisfied, the energy consumption would not necessarily be optimal because much depends on the shape of the diurnal pattern, which could partly eliminate the preliminary choice of Q_d. It is therefore advisable to reiterate with slightly higher/lower values of Q_d/H_d than those calculated, which will eventually lead to the results as presented in the answers. For the $Q_d = 300$ *l/s* the above equations yields $H_d = 83.33$ *mwc*, rounded to 85 *mwc* in Answer 5.

Question 1.2.1

8. Three pumps in parallel arrangement offer identical hydraulic performance, provided their duty flow is one third of the value of the big pump of equivalent capacity, and the duty head is the same as for the big pump used in the simulation run in the previous question.

9. The efficiency curve needs to be adapted accordingly, but the total energy consumption will still be the same if all three pumps are switched on during the entire period of EPS. This can be seen by comparing the energy report of the simulation run in this question with the one in Question 1.1.2.

Question 1.2.2

10. Using EPANET syntax, the modified pump scheduling looks as follows:

    LINK   102   OPEN AT TIME 4
    LINK   103   OPEN AT TIME 6
    LINK   103   CLOSED AT TIME 9
    LINK   103   OPEN AT TIME 14
    LINK   103   CLOSED AT TIME 20
    LINK   102   CLOSED AT TIME 23

11. Manual operation of the pumps clearly helps to reduce the energy costs. For the pump operation given in answers 15–17, the total energy consumption is 6089.36 *kWh* at a cost of 913.41 *EUR*, which is a saving of approximately 25 % compared to the operation of one big pump of equivalent capacity used in the simulation run in Question 1.1.2. The improved performance can also be seen from the improved efficiency of the pumps (*Average Efficiency*) and the lower pumping costs per m^3 of supplied water (*Kw-hr/m3*) both presented in the energy report.

12. Furthermore, the maximum pressure of 111.27 *mwc* indicated in Answer 9 has been reduced to 96.34 *mwc*, given in Answer 22 (or by approximately 14 %). This reduction is more moderate than in the case of the energy, because the duty head of the pumps is rather high. This high pressure can only be further mitigated by increasing the number of pumps and/or mixing/reducing their size.

Question 1.3.1

13. In this data set, the residential area is planned in the vicinity of node 9 (10–1, which is the topography data set number). 6000 inhabitants should receive on average 150 *l/c/d*. This additional consumption converts into a surplus of *Base Demand* in node 9 of 10.42 *l/s*, increasing it to 28.92 *l/s*.

14. Three pumps in operation as in the previous question result in a minimum pressure below 20 *mwc* at 6:00 hours (Answer 24).

15. The proposed solution is to add an additional pump with the same capacity as the existing three, and two pipes in parallel. The modified pump scheduling looks as follows:

    LINK   102   OPEN AT TIME 4
    LINK   103   OPEN AT TIME 5
    LINK   104   OPEN AT TIME 6
    LINK   104   CLOSED AT TIME 8
    LINK   103   CLOSED AT TIME 10
    LINK   103   OPEN AT TIME 14
    LINK   103   CLOSED AT TIME 21
    LINK   102   CLOSED AT TIME 23

16. For this scenario, the pressure in node 10 at 6:00 hours is 44.19 *mwc*, which is rather high. However, with pump 104 switched off this pressure would drop to 14.68 *mwc*, which is below the required 20 *mwc*. Consequently, the minimum pressure appears to be at 13:00 hours, when only two pumps are in operation, yet well above 20 *mwc*, as shown in Answer 29.

Question 1.4.1

17. The annual demand increase of 3 % over 10 years means an exponential demand growth of 34.4 % (1.03^{10} = 1.344). Consequently, the initial demand multiplier in this case grows from 1.5 to 2.016.

18. Furthermore, the base demand of the factory needs to be adapted to reflect a demand growth of 30 %, which will result in the value of 12.5/2.016 multiplied by 1.3, which is 8.06 *l/s*.

19. Four pumps in operation as in the previous question each deliver 155.53 l/s, which makes a total of 622.11 *l/s* at 6:00 hours, in the future demand scenario (Answer 31). The minimum pressure in node 10 at the same point in time is 8.48 *mwc*, thus below 20 *mwc*. However, this is not the absolute minimum pressure because the lowest appears at 13:00 hours (Answer 32) due to only two pumps being in operation at this point in time, according to the schedule in the previous question.

20. The proposed solution is to add an additional pump with the same capacity as the existing four, and an additional three pipes in parallel. The modified pump scheduling will then look as follows:

LINK   102   OPEN AT TIME 3
LINK   103   OPEN AT TIME 5
LINK   104   OPEN AT TIME 6
LINK   105   OPEN AT TIME 6
LINK   105   CLOSED AT TIME 8
LINK   104   CLOSED AT TIME 9
LINK   104   OPEN AT TIME 14
LINK   104   CLOSED AT TIME 20
LINK   103   CLOSED AT TIME 22
LINK   102   CLOSED AT TIME 23

21. For this scenario, the pressure in node 10 at 6:00 hours is 36.33 *mwc*, which is relatively high for the same reason as in Question 1.3.1. The minimum pressure appears to be at 23:00 hours, when only one pump is in operation, yet above 20 *mwc*, as shown in Answer 39.

Question 1.5.1

22. The burst of pipe 2 is analysed (1, which is the topography data set number, + 1). This pipe is connected to the source and conveys a significant flow to the network. Its failure is therefore considered to be a major problem, which is illustrated by the extreme values of the pressure and hydraulic gradient shown in answers 41 and 42, respectively. Also, the disturbed flow distribution moves the most critical point from node 10 to node 7, and also the parallel pipes 11 and 11a are clearly a bottleneck for they are not able to convey the surplus of flow resulting from the burst of pipe 2.

23. The magnitude of the problem is therefore so high that the available remedies of additional pumping and/or the increase of other pipe diameters become questionable, being very expensive and thus ineffective. The only

reasonable intervention is to lay a parallel pipe next to pipe 2, as suggested in Answer 45. This makes the hydraulic operation look similar to Question 1.4.1; the minor improvements seen in answers 48 and 49, compared to 39 and 40, originate from the lower roughness value of the added parallel pipe.

Question 2.1.1

24. The length of the pipe connecting the tank with the network has been selected based on the assigned individual number.
25. The tank bottom level includes the ground elevation and the height of the tank. It is actually the altitude of the tank bottom, expressed in *msl*.
26. The tank loses its volume very quickly because the pump is too weak to fill it. The minimum depth is therefore already reached at 4:00 hours when the tank is disconnected from the network. From there, the system operates as the direct pumping system for the rest of the day i.e. exactly as in Question 1.1.1. That is what makes the result in Answer 53 identical to the one in Answer 5.

Question 2.1.2

27. The duty head and flow, as well as the initial tank depth, have been obtained by running a number of trial and error simulations in order to reach a 'proper' demand balancing of the tank. 'Proper' here means: (1) a similar water depth at the beginning and the end of the day, (2) a full tank before the morning peak demand hour occurs, and (3) the tank depth ideally at the minimum after the afternoon peak demand has been supplied.
28. The latter condition has not been satisfied (Answer 60), which leaves room for further optimisation of the tank design, yet the current result means more volume in emergency situations i.e. additional safety.
29. Because the tank is connected to the highest elevated node (10), this node is no longer the most critical one, as long as the tank water depth is above the minimum. The pressure in node 10 is mostly influenced by the height of the tank (25 *m* above the ground), which means that node 9 is the most critical. The lowest pressure, shown in Answer 58, appears in this node one hour after the maximum consumption hour, which results from the actual water depth in the tank at that moment.
30. Having the balancing tank clearly helps to reduce the pump energy costs. Compared to the pump operation given in answers 18–20, the results in Answer 60 suggest further energy savings of 15 % compared to the operation of three pumps in the simulation run in Question 1.2.2. The improved performance can also be seen from the slightly improved efficiency of the pump (*Average Efficiency*) and further reduced cost of pumping per m^3 of supplied water (*Kw-hr/m3*). In addition, this distribution scheme has more favourable consequences for the maximum pressures in the network (75.63 *mwc* in Answer 59, against 96.34 *mwc* in Answer 22; nearly 22 %).

Question 2.2.1

31. The tank volume is positioned too low and more water fills the tank than needed, which disturbs the balancing function. For part of the day, the tank is full and therefore disconnected from the system.
32. The lowest pressure still occurs in node 9, but this is not the node with the longest duration of pressures below 20 *mwc*.
33. Node 10 is actually the most affected by the reduced tank height because of being directly connected to the tank.
34. The pressure drop is not significant because the tank is still able to supply pressures above 17 *mwc*. When the tank is disconnected, the demand becomes lower and therefore the pump can still deliver the pressures above 20 *mwc*.

Question 2.2.2

35. The hydraulic simulation of the network gives the system/pump flow in *l/s* which can be viewed in tabular form and shows the following values:

Table 2.5 Diurnal system/pump flows, Question 2.2.2

Hour	1	2	3	4	5	6	7	8
Q_p(l/s)	159.09	165.91	185.77	240.86	282.08	349.10	350.38	328.25
Hour	9	10	11	12	13	14	15	16
Q_p(l/s)	290.27	272.47	257.28	276.10	282.79	301.77	323.17	330.79
Hour	17	18	19	20	21	22	23	24
Q_p(l/s)	331.68	323.35	299.29	292.18	271.05	224.27	183.95	175.48

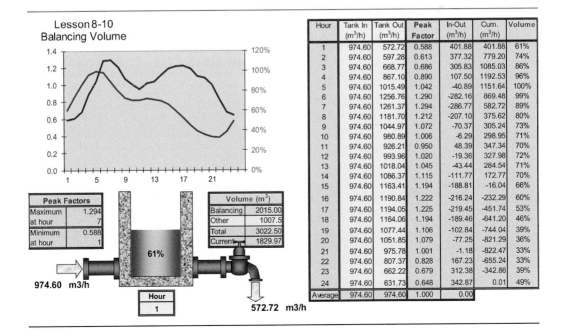

Hour	Tank In (m³/h)	Tank Out (m³/h)	Peak Factor	In-Out (m³/h)	Cum. (m³/h)	Volume
1	974.60	572.72	0.588	401.88	401.88	61%
2	974.60	597.28	0.613	377.32	779.20	74%
3	974.60	668.77	0.686	305.83	1085.03	86%
4	974.60	867.10	0.890	107.50	1192.53	96%
5	974.60	1015.49	1.042	-40.89	1151.64	100%
6	974.60	1256.76	1.290	-282.16	869.48	99%
7	974.60	1261.37	1.294	-286.77	582.72	89%
8	974.60	1181.70	1.212	-207.10	375.62	80%
9	974.60	1044.97	1.072	-70.37	305.24	73%
10	974.60	980.89	1.006	-6.29	298.95	71%
11	974.60	926.21	0.950	48.39	347.34	70%
12	974.60	993.96	1.020	-19.36	327.98	72%
13	974.60	1018.04	1.045	-43.44	284.54	71%
14	974.60	1086.37	1.115	-111.77	172.77	70%
15	974.60	1163.41	1.194	-188.81	-16.04	66%
16	974.60	1190.84	1.222	-216.24	-232.29	60%
17	974.60	1194.05	1.225	-219.45	-451.74	53%
18	974.60	1164.06	1.194	-189.46	-641.20	46%
19	974.60	1077.44	1.106	-102.84	-744.04	39%
20	974.60	1051.85	1.079	-77.25	-821.29	36%
21	974.60	975.78	1.001	-1.18	-822.47	33%
22	974.60	807.37	0.828	167.23	-655.24	33%
23	974.60	662.22	0.679	312.38	-342.86	39%
24	974.60	631.73	0.648	342.87	0.01	49%
Average	974.60	974.60	1.000	0.00		

Lesson 8-10
Balancing Volume

Peak Factors	
Maximum	1.294
at hour	7
Minimum	0.588
at hour	1

Volume (m³)	
Balancing	2015.00
Other	1007.5
Total	3022.50
Current	1829.97

61%

974.60 m3/h

Hour
1

572.72 m3/h

The average hourly flow for this scenario is 270.72 *l/s* (or 974.60 *m³/h*). Transposing these values in the Spreadsheet Lesson 8–10 (see Appendix 7) make it possible to calculate the tank balancing volume straight forwardly, using the principles explained in Section 4.2.3. The balancing volume obtained in this way is 2015 *m³*. Assuming the minimum and maximum tank levels remain at 2 and 6 m, respectively, makes the emergency volume half of the balancing volume, or 1007.5 *m³*.

36. In the above calculations, the tank which has a volume of 3022.5 *m³* needs to be 61% full at the beginning of the day, suggesting an initial level of 3.66 *m* and a tank diameter of 25.36 *m*. Not surprisingly, these parameters are relatively close to those selected in Question 2.1.2.

37. The tank is fed and supplies the demand through the same pipe, which makes these patterns different to those shown in the above spreadsheet table. Moreover, the pump does not feed the tank directly but through the network where it follows the nodal demand patterns; it can therefore not operate at the average flow. However, the above tank parameters will yield a smooth balancing curve with the exception that the minimum tank level is 2.60 m; thus, higher than needed. This can be reduced by further elevating the tank, which will unnecessarily increase the cost and pressures in the network. An alternative option is to reduce the pumping, which may have the opposite effect on the pressures (the pipe hydraulic gradients should be checked in this case to possibly mitigate this problem).

38. After minor fine-tuning, the scenario presented in answers 66–73 results in the tank balancing curve as shown in Figure 2.4.

39. The question gives no possibility to modify the pipe diameters. Relatively high hydraulic gradients in pipes 1, 9 and 12 give the impression that enlarging these diameters would create a significant positive effect on the surrounding pressures, leaving the option open to further reduce the pumping and the tank height. This would require additional fine-tuning of the demand balancing curve, which is left outside the scope of this exercise but can be tested as self-study.

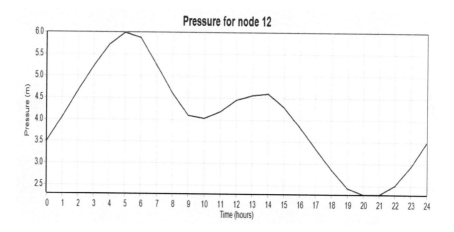

Figure 2.4 Variation of the tank depth (answers 67–72)

Question 2.3.1

40. The demand increase of 30 % cannot be met without an increase in the tank and/or the pump capacity. The drop of pressure in Answer 74 is a combined consequence of the pump that is unable to feed the tank with the required volume, and the tank that is losing the volume too fast due to increased demand. The negative pressure appears at the moment the tank is empty (i.e. at the minimum volume = disconnected) and the pump alone cannot deliver the demand required in the network.

41. The scenario described in answers 76–83 assumes the increase of the pump capacity alone, as a cheaper investment option. The implication is that the energy costs will increase because the surplus of the demand is entirely supplied by direct pumping. Comparing the answers 73 and 83, the demand increase of 30 % has been mitigated by the increase in pumping costs of nearly 34%. Lastly, the diameter increase of pipe 12 is needed to provide the required minimum pressure in the most critical node.

42. The selected pump of the duty head/flow of 62 *mwc*/345 *l/s* is the equivalent of adding a smaller pump of 62 *mwc*/90 *l/s* to the existing one of 62 *mwc*/255 *l/s*. With the same size of tank, and slightly increased initial depth (to 3.70 *m*), the balancing curve for the future demand looks as shown in Figure 2.5.

43. The option with the 30 % larger diameter of the tank has also been tested. Making a bigger tank volume however does not work without increasing the pump capacity to be able to feed the tank. In the scenario of the tank diameter increase from 25 to 32.5 *m*, in combination with the pump of 62 *mwc*/320 *l/s*, the minimum pressure in node 9 at 7:00 hours is 22.12 *mwc* and the pumping energy cost is 971.10 EUR; thus, somewhat lower than in the case of Answer 83. The tank balancing curve for this option will look similar to the one above, which is shown in Figure 2.6. Further refinement would possibly improve the energy savings at the cost of the tank and pipe enlargement (as another self-study problem).

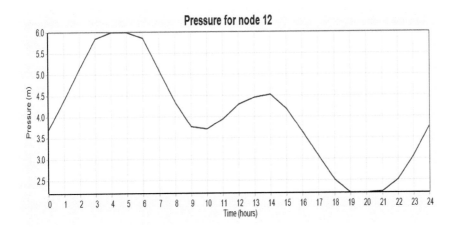

Figure 2.5 Variation of the tank depth (answers 77 and 82)

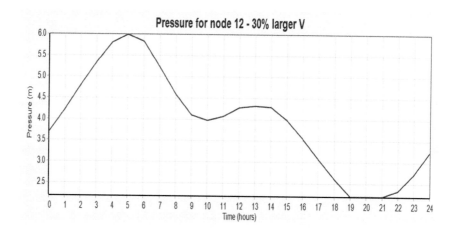

Figure 2.6 Variation of the tank depth (30 % larger volume)

Question 2.4.1

44. The elevated tank is insufficiently high and too small to deliver the pressure in the whole network alone; therefore, one of the pumps must always be switched on.
45. In selecting the switch-off level for the second pump, care needs to be taken not to choose a level that would switch the pump off prematurely during the peak demand hours, which would result in an instant drop in pressure.
46. The pumping costs increase as a result of the tank volume reduction. The smaller the tank volume, the more closely the distribution scheme becomes to the direct pumping. The opposite extreme is where the pumps operate on average flow, in combination with maximised tank volume. A range of solutions between these two extremes can also be considered, based on economic grounds and reliability of supply.
47. Due to the significantly reduced volume of the tank, the selected switching levels for the second pump (102) result in a rather erratic variation between the maximum depth (for three hours) and the minimum depth (for one hour). Having the tank disconnected in those periods however does not affect the network pressures in any negative way because the entire demand can be supplied by the pumps alone.

Question 2.5.1

48. The base demand in the factory in Question 2.4.1 is 8.33 l/s. The actual hourly demand is calculated by multiplying this demand by the diurnal peak factor of 2.0 (specified in the factory demand pattern for time intervals 8:00–19:00), and the general demand multiplier of 1.95. Hence, by the time it is working, the factory water demand is 32.49 l/s = 116.95 m^3/h, which is 1403.44 m^3/d (over 12 hours per day).

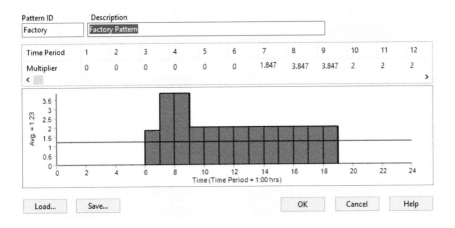

Figure 2.7 Modelling of the fire demand in the pattern editor

49. The additional 30 l/s needs to be provided for the fire extinguishing purposes in the period 6:00–8:00 hours, by adapting the diurnal peak factors for the time intervals 7:00–9:00. At 6:00 hours the factory is still closed, and the peak factor to generate the demand of 30 l/s will be 30/1.95/8.33 = 1.847, while the peak factor at 7:00 and 8:00 hours will be (32.49 + 30)/1.95/8.33 = 3.847 (equal to 1.847 of the fire demand + 2.0 of the regular factory demand). The *Factory* pattern will consequently look in the EPANET as shown in Figure 2.7.

50. Obviously, in reality the regular diurnal pattern may also change during an irregular event (in the case of the factory, the start of morning production is delayed due to the fire). Keeping the regular pattern unchanged in this case will probably add additional safety to the selected option.

51. The modified pattern does not create a significant drop in pressures, as can be seen in answers 96 and 97. At 6:00 hours, the pressure is already above the required 30 *mwc* because the regular factory demand is 0 *l/s* at that point in time (it only opens at 7:00 hours). Adding a third pump unit with the same size as the existing two solves the pressure problem in the factory at 7:00 and 8:00 hours relatively easily.

52. The following adapted pump scheduling has been applied in the shown scenario:

```
LINK  102  OPEN     IF NODE 12 BELOW 2.5
LINK  102  CLOSED   IF NODE 12 ABOVE 3.5
LINK  102  OPEN     AT TIME 6
LINK  102  CLOSED   AT TIME 9
LINK  103  OPEN     AT TIME 6
LINK  103  CLOSED   AT TIME 9
```

Pumps 102 and 103 are closed at the beginning of the simulation and additional time-based controls are needed between 6:00 and 9:00 to ensure that all three pumps work during the fire, regardless of the water level in the tank.

53. Normal automatic operation of pump 102 will be restored after the emergency period is over. Nevertheless, the pump switching schedule may be disturbed by the fire event and correction of the switching levels is needed to avoid low pressures because the pump is off. This is why in the above controls the switch-on level is increased to 2.5 m, from 2.0 m used in Question 2.4.1. Equally, the initial tank level may also be corrected in order to synergize the pump operation with the demand variation (2.6 m in this case, compared to 2.5 m in Question 2.4.1).

Appendix 3

Network design exercise

3.1   Learning objectives and set-up

This assignment concerns the design of a new water distribution network. The main objective of this exercise is to set appropriate pipe diameters and analyse the hydraulic performance of looped network configurations in two possible supply schemes:

- Alternative A – direct pumping, and
- Alternative B – pumping combined with balancing storage.

This should also include the appropriate combination of the required pump capacity and the storage volume, sufficient to maintain the required service level expressed by the minimum pressure to be guaranteed over the entire design period; thus, maintaining the anticipated demand growth. The selected designs are then to be tested in failure scenarios by using a simplified demand-driven approach, and finally evaluated financially using the approach explained in Section 4.1.2. When lectured, the design exercise contact time consists of six sessions of three hours, with the following set-up:

INTRODUCTION (session 1)
- Case description
- Software introduction
- Start-up, model preparation

PART 1: HYDRAULIC DESIGN (sessions 2 and 3)
- Preliminary design concept
- Determination of nodal consumptions
- Selection of pipe diameters
- Required pumping heads and flows
- Design of storage volume

PART 2: NETWORK OPERATION (sessions 4, 5 and 6)
- Optimisation of network operation
- Pipe burst, fire requirements
- Reliability assessment
- Choice of final alternatives
- Cost comparisons and final conclusions

The network design is developed based on the availability of two water sources: an initial one with limited capacity, and a future one connected when the capacity of the current source has been exceeded. To provide a variety of solutions, each student works with an individual data set combining the two sources in different locations with a number (Nr), which they select at random between 150 and 200, and which is used in the calculation of particular data.

The design alternatives are analysed with the help of a computer modelling tool. As in the computer workshop elaborated in Appendix 2, EPANET software (Version 2) is used throughout the tutorial. For this purpose, a model of the case network has been prepared, which consists of a group of EPANET input files describing the various steps of the analysis. All file names follow uniform coding: 'Sxyz.NET', where:

- S stands for Safi (the name of the hypothetical town used in this example),
- x = 6 or 30, indicates respectively the first or the last year of the design period,
- y = A or B, indicates the two alternatives requested by the assignment, and
- z = the serial number of the particular network layout explored.

The results of the calculations are presented in the tables and figures. These may be accompanied by:

- *Clarification* that explains some (calculation) details from the table/figure,
- *Conclusions* that suggest further steps based on the calculation results,
- *Comments* that elaborate points in a wider context of the problem, and/or
- *Reading* that refers to the chapters of the main text in Volume 1 that are related to the problem.

Although the explanations in the tutorial should be understood without necessarily using the computer programme, more will be gained if the case network is studied in parallel on a PC. For this purpose, the full version of the EPANET software and the case network files mentioned in the text are available with the descriptions. General information relevant for this exercise, which includes the programme installation and digested instructions for use, is given in Appendix 8.

The output of the work is to be described in a design exercise report that is to be submitted for evaluation; an oral exam can be organized for this purpose. Based on the average student performance, the estimated study load for this exercise is 56 hours, or two credits of the European Credit Transfer System (ECTS), which includes the contact time. The solved example presented in this appendix contains partly modified data where the network is laid on a different topography; again, for the sake of preventing the copying of the solutions that are, in fact, not necessarily the best ones.

3.2   Case introduction – the town of Safi

Safi is a secondary town located in a moderately hilly area. Over the coming years, a rapid expansion of the area around the town is expected because of good regional development activities. A factory for production of agricultural machinery is going to be built near the town itself.

The existing water distribution network, which presently covers just a half of the area of the town, is in very poor condition and the construction of a completely new system has been proposed. The pumping station at the source is also rather old and of insufficient capacity. Its general overhaul has also been planned for during the project implementation. An international loan for both purposes was requested and has been approved.

A looped-type network is planned for the construction. The planned design period of the new distribution network is *25 years*. Within that time, the system should provide sufficient amounts of water and pressure variations of between *20–60 mwc* during regular operation. Based on the evaluation of the present situation and the estimates for future growth, a technically and economically acceptable solution for construction of the network needs to be proposed.

3.2.1 Topography

The town is located in a hilly area with ground elevations ranging between *6 and 26 msl*. Figure 3.1 shows the configuration of the terrain together with the layout of the main streets in the urban area (blue dotted lines). The grey dotted border line indicates the boundary of the town area supplied by the distribution network.

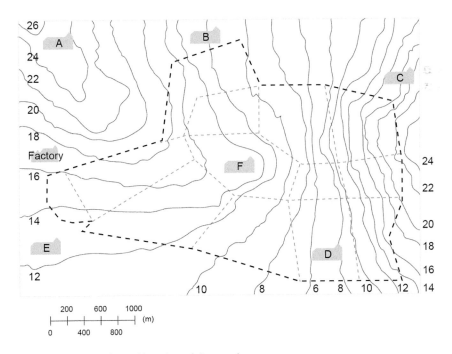

Figure 3.1 Topography and location of the supply sources

3.2.2   Supplying source

The source of supply located at **[A][B][C][D][E][F]**[1] is groundwater stored after simple treatment in a ground reservoir from where it is pumped into the system. The new pumping station will need to be built at the source, together with the clear water reservoir at the suction node. Centrifugal, fixed-speed pumps are going to be installed in parallel arrangement. The number and size of the pumps should be sufficient to provide smooth operation of the system, which includes a number of anticipated irregular supply scenarios.

The maximum production capacity of the source is **10 Nr[2]** *m³/h*. The additional groundwater source located at **[A][B][C][D][E][F]**[3] can be put into operation once the indicated capacity has been exceeded.

3.2.3   Population distribution and future growth

The town is divided into six districts (A-F, in Figure 3.2), each of a homogenous population density. The total surface areas of the districts and sub-districts are given in Table 3.1.

The present population density for each area, expressed as the number of inhabitants per hectare, is shown in Table 3.2.

In the absence of more reliable data, the indicated population growth is assumed to be constant throughout the entire design period.

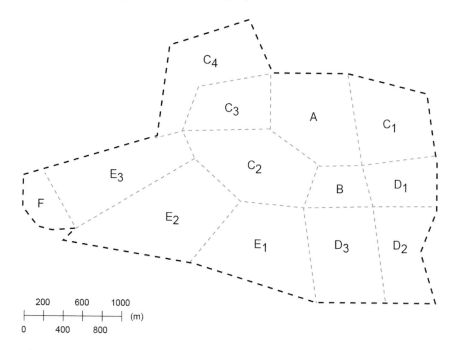

Figure 3.2 Areas of different population density

---

1 One supply point should be chosen as Source 1, used in the current demand scenario.
2 Students each select an individual number at random between 150 and 200.
3 Another supply point should be chosen as Source 2, to be used in the future.

Table 3.1 The town of Safi – surface area of the districts and sub-districts

District	Area (ha)	District	Area (ha)	District	Area (ha)
A	83.2	C_3	50.0	D_3	78.0
B	30.8	C_4	84.4	E_1	97.2
C_1	73.6	D_1	37.2	E_2	117.2
C_2	80.8	D_2	55.6	E_3	83.2
				F	29.6

Table 3.2 The town of Safi – population density and growth

District	Present (inhab./ha)	Annual growth (%)
A	0.6 Nr*	2.0
B	0.7 Nr	1.0
$C_{1,2,3,4}$	0.4 Nr	2.4
$D_{1,2,3,4}$	0.5 Nr	2.8
$E_{1,2,3}$	0.3 Nr	3.5
F	0.2 Nr	3.5

* - Number selected by individual students

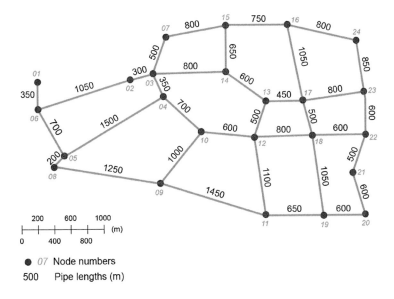

Figure 3.3 Node numbers and pipe routes

3.2.4   Distribution system

The distribution pipes will be laid alongside the main streets of Safi. The network layout is shown in Figure 3.3, with the lengths of the pipe routes. The intersections (nodes) are numbered. Connection of the factory (Node 01) to the system is planned to be in *Node 06*.

Regarding the local situation (soil conditions, local manufacturing), PVC has been chosen as the pipe material for the network. Because of this choice, it is assumed that the roughness of the pipe will remain low throughout the design period. The accepted k value of *0.5 mm* includes the impact of local losses in the network.

3.2.5 Water demand and leakage

The water consumption in the area is mainly domestic, except for the new factory that will also be supplied from the distribution system. In the new system, the domestic consumption per capita is planned to be *150 l/d*. In addition, the projected water demand of the new factory is *1080 m³/d*. The water loss is not expected to exceed *10 %* of the water production after the commissioning of the system. This is assumed to increase by a further *10 %* by the end of the design period, i.e. would grow to *20 %* of the water production.

The typical diurnal domestic consumption pattern is shown in Figure 3.4 (the peak factor values given in Table 3.3), which does not include water losses. Variation of the consumption during the week is within the range of *0.95* and *1.10* of

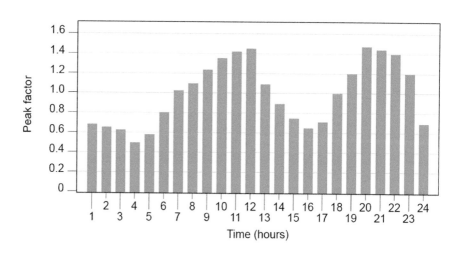

Figure 3.4. Consumption pattern

Table 3.3 Diurnal peak factors – Figure 3.4

Hour	1	2	3	4	5	6	7	8	9	10	11	12
pf_h	0.70	0.68	0.65	0.62	0.50	0.58	0.80	1.08	1.10	1.22	1.35	1.42

Hour	13	14	15	16	17	18	19	20	21	22	23	24
pf_h	1.45	1.10	0.90	0.75	0.65	0.73	1.00	1.20	1.47	1.45	1.40	1.20

the average daily figure. In addition, the monthly (seasonal) variations are in the annual range of between *90* and *115 %* of the average consumption. The factory will be working at a constant (average) capacity for *12 hours a day*, between *7 am* and *7 pm* (two shifts). Neither seasonal variations nor water loss percentages apply in this case.

3.2.6   Financial elements

An international loan has been requested and approved with a payback period of *30 years* and an interest rate of *8 %*. The loan is intended only for construction of the distribution part of the network, which does not include the tertiary network or service connections.

The monthly instalments are to be financed entirely through the sale of water. It is therefore agreed that the repayments will start only after the system has been commissioned, i.e. five years from now (in *year 6*).

Procurement of the equipment will be according to the prices listed in Table 3.4. If considering the possibility of laying of pipes in parallel, the price indicated under 1.1–1.8 should be doubled.

Operation and maintenance costs (O&M) are determined as a percentage of the investment costs, which is shown in Table 3.5:

As all cost calculations are to be made in EUR, the effect of local inflation can be ignored.

Table 3.4 Investment costs of the equipment

No.	Component	EUR	Per
1.1	Laying pipe D = 80 mm	60.00	m
1.2	Laying pipe D = 100 mm	70.00	m
1.3	Laying pipe D = 150 mm	90.00	m
1.4	Laying pipe D = 200 mm	130.00	m
1.5	Laying pipe D = 300 mm	180.00	m
1.6	Laying pipe D = 400 mm	260.00	m
1.7	Laying pipe D = 500 mm	310.00	m
1.8	Laying pipe D = 600 mm	360.00	m
2.	Pumping station	$0.5 \times 10^4 \times Q^{0.8}$	$Q = Q_{max}$ m^3/h
3.	Reservoir	$35 \times 10^4 + 150 \times V$	$V = V_{tot}$ m^3
4.	Support structure for H m-elevated tank	$3 \times H \times V$	$V = V_{tot}$ m^3

Table 3.5 Annual operation and maintenance costs (O&M)

No.	Component	% of investment
1.	Distribution pipes	0.5
2.	Pumping station	2.0
3.	Storage	0.8

3.3 Questions

3.3.1 Hydraulic design

Preliminary concept

1.1 Calculate the demand increase throughout the design period.
Two possible alternatives for the network design have to be analysed:

A) Supply by direct pumping (source pumping stations), and
B) Supply by pumping and by gravity (balancing tank or water tower).

For both alternatives:

1.2 Regarding the supply points:

  - determine if and when the second source will need to be put into opera-
    tion and calculate its maximum required capacity by the end of the
    design period, and
  - decide in which nodes the sources need to be connected to the network,
    and find the lengths of the connecting pipes.

1.3 Develop a preliminary supply strategy: suggest possible phases in the devel-
    opment of the system, function of the pumping stations, reservoirs, etc.

Nodal consumption

1.4 Calculate the average consumption for nodes 01–24:

  - at the beginning of the design period, and
  - at the end of the design period.

Network layout

1.5 Size the pipe diameters in the system at the beginning and the end of the
    design period.

Pumping stations

1.6 For the source pumping stations, determine:

  - the required duty head and duty flow, and
  - the provisional number and arrangement of the pumping units, both at
    the beginning and the end of the design period.

1.7 Determine the location, duty head and flow of the booster station(s), if needed
    in the system.

Storage

1.8 Determine the volume of the clear water reservoir at the suction side of the
    source pumping stations, at the beginning and the end of the design period
    (assume ground-level tanks).

1.9 For Alternative B only: determine the provisional location, volume and dimensions of the balancing storage in the network at both the beginning and the end of the design period.

Summary

1.10 Draw conclusions based on the hydraulic performance of the system throughout the entire design period. Explain the phased development of the system, if applied. Suggest a preference for layout alternative A or B.

3.3.2 System operation

Upgrade the computer model of the selected layout alternative A and B at the end of the design period. Model the pumping station operation by assuming a parallel arrangement of the pump units.

For both alternatives, run the simulation of the maximum consumption day and consider the following steps:

Regular operation

2.1 Propose a plan of operation for both pumping stations during regular supply conditions. Compare the manual pump operation with an automatic operation based on the pressure in the selected critical node (or tank) in the network.

Factory supply under irregular conditions

2.2 A reliable supply and also fire protection have to be provided for the factory. Analyse the preliminary layout under the maximum consumption hour conditions, if:

- a single pipe, either 15–16, 16–24 or 16–17, bursts, and/or
- a requirement for firefighting of **Nr**[4] m^3/h at a pressure of *30 mwc* is needed in Node 01.

If necessary, propose an operation that can provide a supply of the required quantities and pressures.

Network reliability assessment

2.3 Assess the reliability of the distribution network. Determine the consequences of single pipe failure events if a supply of minimum *75 %* of the maximum consumption hour demand has to be maintained in the system.

Choice of final layouts

Summarise all the findings from the exercise and adjust the network layouts for alternatives A and B, where required. For these final layouts:

2.4 Show the network layout, number and size of the pump units and distribution of the storage volume at the beginning of the design period.

---

4 Student's individually selected number.

2.5 For the manual mode of operation on the maximum consumption day, show the range of pressures in the system. For Alternative B only, show the volume variation in the balancing tank.

2.6 Describe the steps for the reconstruction/extension of the system and show its operation at the end of the design period (as required by Question 2.5).

Cost comparisons

2.7 Calculate the investment costs of the network.

2.8 Calculate the operation and maintenance (O&M) costs of the network.

2.9 Determine the average increase in cost per m³ of water, due to the loan repayment and O&M of the system.

If a phased development of the system is planned, explain its impact on the above costs.

Summary

2.10   Summarise all the conclusions and select the final design.

NOTE: *The tutorial that follows is based on a partly modified case!* The changed input data are:

- population densities and growth (for the values, see Table 3.6),
- maximum capacity of the first source selected in location A, which is *2500 m³/h,*
- location of the factory, which is connected in *Node 16,*
- factory demands, which are *1300 m³/d* for regular operation and *180 m³/h* for firefighting,
- nodal ground elevations, which are extracted from the map shown in Figure 3.5.

All other data are taken from the original data set.

In this context, the demands in the tutorial will be generally higher than from the selection of an individual numbers in the range 150–200. As the figure also shows, the future source is at location D, with the highest elevations in the middle of the town area (a hill), while the student's case in Figure 3.1 shows the

Table 3.6 Safi tutorial – population density and growth

District	Present (inhab./ha)	Annual growth (%)
A	225	1.0
B	248	1.2
$C_{1,2,3,4}$	186	2.0
$D_{1,2,3,4}$	202	2.7
$E_{1,2,3}$	144	2.9
F	115	3.3

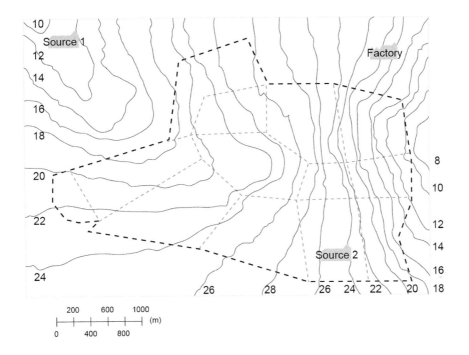

10

Source 1

12

14

16

18

20

22

24

Factory

8

10

12

14

Source 2

16

26   28   26   24   22   20   18

200   600   1000

(m)

0   400   800

Figure 3.5 Safi tutorial – topography and location of supply sources

lowest elevations in the middle of the town (a valley). This difference obviously has implications for the monitoring of the most critical nodes in the network, which will not be in the same area of the town. It also impacts the location of the storage that is to be designed in Alternative B.

3.4 Hydraulic design

3.4.1 Preliminary design concept

By combining the information from tables 3.1 and 3.2, the population and domestic consumption figures are shown in Table 3.7. A spreadsheet calculation is conducted for:

- the present situation (year 1),
- commissioning of the system after five years (year 6), and
- the end of the 25-year design period (year 30).

Clarification:
- An exponential growth model is assumed in the demand calculations.

Table 3.7 Population and domestic consumption growth

Districts	Area (ha)	Density (inh./ha)	Growth (%)	Present	At year 6*	At year 30
A	83.2	225	1.0	18,720	19,872	25,232
B	30.8	248	1.2	7638	8205	10,925
C_1	73.6	186	2.0	13,690	15,417	24,797
C_2	80.8	186	2.0	15,029	16,925	27,223
C_3	50.0	186	2.0	9300	10,473	16,846
C_4	84.4	186	2.0	15,698	17,679	28,435
D_1	37.2	202	2.7	7514	8817	16,711
D_2	55.6	202	2.7	11,231	13,178	24,977
D_3	78.0	202	2.7	15,756	18,487	35,040
E_1	97.2	144	2.9	13,997	16,616	32,998
E_2	117.2	144	2.9	16,877	20,035	39,788
E_3	83.2	144	2.9	11,981	14,223	28,245
F	29.6	115	3.3	3404	4136	9016
			Total	**160,835**	**184,062**	**320,232**
Specific consumption			150	l/c/d		
Total average consumption			m³/d	24,125	27,609	48,035
			m³/h	1005.2	1150.4	2001.5
			l/s	279.23	319.55	555.96

Population (inhabitants)

* The shaded areas in the table indicate the input information used in the calculations.

Conclusions:
- The population i.e. the demand will nearly double in the next 30 years.

Comments:
- The exponential growth model simulates a faster growth of population than the linear growth model does. Local conditions determine which of the two models is more suitable.

Reading:
- Volume 1, Section 2.5: 'Demand forecasting'.

The design alternatives to be compared are:

A) Supply by direct pumping from the source(s), and
B) Supply combined by pumping and by gravity.

The principal difference between these two is:

A) The entire demand in the area is supplied from the source(s) at all times. Hence, the sources must be capable of satisfying the demand during the maximum consumption hour of the maximum consumption day.

B) In peak periods of the day, part of the demand can be supplied from the balancing tank in the system. The source will normally supply the average flow on the maximum consumption day, provided the position and volume of the balancing tank are properly determined.

Consequently, the first source will reach its maximum capacity of 2500 m³/h sooner in the case of Alternative A than in Alternative B. Table 3.8 indicates when this will happen.

Table 3.8 Demand growth

					Seasonal variations		Peak hour at:	
					Max	Min	12	4
	Factory:	1300	m3/d	Factors:	1.265	0.855	1.45	0.5
		Average demands			Peak demands			
	Populat.	Consum.	Leakage	Demand	$Q_{max,day}$	$Q_{min,day}$	$Q_{max,hour}$	$Q_{min,hour}$
	(inh.)	(m³/d)	(%)	(m³/d)	(m³/d)	(m³/d)	(m³/h)	(m3/h)
Present	160,835	24,125	0	24,125	30,518	20,627	1843.8	429.7
year 06	184,062	27,609	10	31,977	40,106	27,529	2452.9	546.4
year 07	188,274	28,241	10	32,679	40,994	28,129	2506.5	558.9
year 08	192,590	28,889	10	33,398	41,904	28,744	2561.5	571.8
year 09	197,013	29,552	10	34,136	42,837	29,374	2617.9	584.9
year 10	201,545	30,232	10	34,891	43,792	30,020	2675.6	598.3
year 11	206,190	30,929	11	36,051	45,260	31,012	2764.3	619.0
year 12	210,951	31,643	11	36,854	46,275	31,698	2825.6	633.3
year 13	215,830	32,375	11	37,676	47,315	32,401	2888.4	647.9
year 14	220,830	33,125	11	38,519	48,381	33,122	2952.8	663.0
year 15	225,955	33,893	11	39,382	49,474	33,860	3018.9	678.3
year 16	231,208	34,681	13	41,163	51,727	35,383	3155.0	710.1
year 17	236,593	35,489	13	42,092	52,902	36,177	3225.9	726.6
year 18	242,112	36,317	13	43,043	54,105	36,991	3298.7	743.6
year 19	247,769	37,165	13	44,019	55,339	37,825	3373.2	760.9
year 20	253,568	38,035	13	45,019	56,604	38,679	3449.6	778.7
year 21	259,513	38,927	16	47,642	59,922	40,922	3650.1	825.5
year 22	265,607	39,841	16	48,730	61,299	41,852	3733.3	844.8
year 23	271,855	40,778	16	49,846	62,710	42,806	3818.5	864.7
year 24	278,259	41,739	16	50,989	64,157	43,784	3905.9	885.1
year 25	284,826	42,724	16	52,162	65,640	44,787	3995.6	906.0
year 26	291,558	43,734	20	55,967	70,454	48,040	4286.4	973.8
year 27	298,460	44,769	20	57,261	72,091	49,147	4385.3	996.8
year 28	305,536	45,830	20	58,588	73,769	50,281	4486.7	1020.4
year 29	312,792	46,919	20	59,949	75,490	51,444	4590.7	1044.7
year 30	320,232	48,035	20	61,344	77,255	52,637	4697.3	1069.5

Clarification:

- The demand calculation in the table is based on Equation 2.1 in Volume 1, Section 2.4
- The growth scenario of the leakage percentage is hypothetical.
- The maximum and minimum consumption day demands are determined based on the seasonal factors $1.10 \times 1.15 = 1.265$ and $0.95 \times 0.90 = 0.855$, respectively.
- The factory is treated as a major user. However, unlike the domestic demand, its demand (1300 m³/d, constant throughout the design period) is included without taking into account any influences of the seasonal variations and leakage.
- The maximum/minimum consumption hour demands are calculated from the peak factors displayed in Figure 3.4/Table 3.3. The factory demand is added based on a 12-hour operation ($1300/12 = 108.3$ m³/h) if the factory was in operation during the maximum/minimum consumption hour.

Conclusions:

- From the domestic diurnal diagram, the maximum hourly peak factor is at 20:00 hours (1.47). Nevertheless, the maximum consumption hour occurs at 12:00 hours due to the factory demand (combined with the peak factor of the domestic demand of 1.45). The minimum consumption hour is at 04:00 hours (peak factor = 0.50 and the factory is closed).
- By applying the direct pumping scheme (Alternative A), the maximum capacity of the first source is going to be reached in year 7 i.e. already by the second year after being commissioned. The additional peak capacity needed in year 30 from the second source is therefore: $4697.3 – 2506.5 \approx 2200$ m³/h (see Table 3.8, column $Q_{max,hour}$).
- By applying the combined system (Alternative B), the capacity of the first source is going to be reached in year 21. If this capacity is supplied 24 hours a day (2500 m³/h = $60{,}000$ m³/d), the additional capacity needed from the second source will be $77{,}255 – 59{,}922 \approx 17{,}300$ m³/d (Table 3.8, column $Q_{max,day}$). The combined supply scheme assumes that the balancing tank will provide/accommodate the difference between any hourly flow and the average flow during 24 hours. This assumption is going to be taken as the starting point in this design alternative.

Comments:

- The calculation procedure applied in Table 3.8 is determined by the origin of the diurnal pattern.
  The crucial questions to be answered are:

  - Does the diurnal diagram represent one or more demand categories?
  - Does it include leakage or not?
  - Which consumption day does it represent (average, maximum, minimum)?

- A distinction should be made between the non-revenue water (NRW) percentage and the leakage percentage (i.e. physical loss). Part of the NRW is water that is in the system but is not paid for, e.g. due to illegal connections or under-reading the water meter. Consequently, including the entire NRW level instead of the leakage level alone reduces the calculated pressures, which works as a kind of safety factor for the design. Moreover, the total NRW is often easier to assess than its contributing components, as this figure is reflected in the loss of revenue.

Reading:
- Volume 1, sections 2.3 and 2.4: 'Water demand patterns' and 'Water demand calculation'.

Initial scenarios

The following initial scenarios are going to be tested while developing the system:

Alternative A
- Both sources 1 and 2 will be connected to the system at the beginning of the design period because the first source alone will become exhausted soon after the system is commissioned. By connecting the second source immediately, the whole construction can be carried out at once.
- Both sources should each supply roughly half of the total demand during regular operation. They are located on the opposite sides of the system which allows for a network of a narrower range of pipe diameters, which is in principle a more convenient layout for future extensions.
- Source 1 will not exceed more than 90% of its maximum capacity during regular operation. The buffer of minimum 10% will be kept as a stand-by for irregular supply (firefighting, pipe bursts, etc.).
- In case of failure or planned maintenance in any of the two sources, the source remaining in operation should have sufficient capacity to supply the factory.

Alternative B
- Source 1 has to supply the average flow on the maximum consumption day. The diurnal demand variation will be satisfied from the balancing tank.
- The balancing tank should also have provision for irregular supply conditions.
- From the year in which it reaches its design capacity, the first source will continue to operate at this capacity until the end of the design period.
- The second source should become operational in year 21, at the latest. If required, this source should also assist with supply during irregular situations.
- It is assumed that the volume of the balancing tank can only be recovered from the two sources (no other source in the vicinity of the tank is available).

In Alternative B, Source 1 is the major source of the system. The constant i.e. average delivery from this source is possible throughout the design period because the peak flows are to be supplied from the balancing tank. This will delay the connection of Source 2, compared to Alternative A where this needs to happen earlier.

In both alternatives:

- A looped structure for the secondary mains will be designed. This is a 'must' in any serious consideration of the system reliability (more reliability = more investment). Question 2.3 stipulates a high reliability in this assignment.
- Pipe and street routes should coincide (no pipe routing underneath buildings).
- The source connections to the system will depend on the selected route for the secondary mains. Moreover:
    - these connections should be directed towards the high demand areas, and
    - their pipe route should be as short as possible.

Reading:
- Volume 1, Section 3.7: 'Hydraulics of storage and pumps',
- Volume 1, sections 4.2.1: 'Design criteria' and 4.2.2: 'Basic design principles'.

3.4.2   Nodal consumption

The average consumption in the nodes has been determined by calculating the specific consumption per metre of pipe in each loop, assuming an even dispersion of house connections throughout the system. A spreadsheet calculation can be carried out based on the procedure described in Volume 1, Section 4.3: 'Computer models as design tools'. The nodal consumption is calculated for the following points of time: at year 6, which is the beginning of the design period, and at year 30, which is the end of the design period. The results are shown in the table below.

Table 3.9 Average nodal consumption

NODE	At year 6		At year 30		Factor of increase
	Q(l/s)	Q(m³/h)	Q(l/s)	Q(m³/h)	
01	30.09	108.3	30.09	108.3	1.000
02	7.15	25.7	13.12	47.2	1.834
03	18.79	67.6	31.00	111.6	1.650
04	18.43	66.3	34.96	125.9	1.897
05	16.91	60.9	34.28	123.4	2.027
06	9.13	32.9	18.83	67.8	2.062
07	16.77	60.4	26.97	97.1	1.608
08	5.42	19.5	10.77	38.8	1.986
09	16.93	60.9	33.62	121.0	1.986
10	17.30	62.3	32.32	116.4	1.869
11	16.66	60.0	32.39	116.6	1.944
12	23.05	83.0	40.59	146.1	1.761
13	12.73	45.8	17.90	64.4	1.405
14	16.75	60.3	24.85	89.5	1.484
15	19.37	69.7	28.81	103.7	1.488
16	15.94	57.4	22.64	81.5	1.420
17	21.45	77.2	32.31	116.3	1.506
18	21.36	76.9	38.17	137.4	1.787
19	13.21	47.6	25.04	90.2	1.895
20	4.10	14.8	7.77	28.0	1.895
21	3.76	13.5	7.12	25.6	1.895
22	7.43	26.7	14.08	50.7	1.895
23	10.59	38.1	18.27	65.8	1.725
24	6.31	22.7	10.15	36.5	1.608
Total	**349.64**	**1258.7**	**586.05**	**2109.8**	**1.676**

Clarification:
- The figures in the table represent the 'baseline' or average consumption. For actual demand at a certain point in time, these figures have to be modified by the hourly peak factors, water losses, and seasonal variations, specified separately in the input file of the computer model.
- As a check, the total demands in years 6 and 30 correspond to the maximum consumption day demands from Table 3.8 (column $Q_{max,day}$). For the water loss of 10 and 20 %, respectively:

$$((1258.7\text{–}108.3) \times 1265/0.90) \times 24 + 1300 \approx 40,106 \text{ m}^3/\text{d}.$$
$$((2109.8\text{–}108.3) \times 1265/0.80) \times 24 + 1300 \approx 77,255 \text{ m}^3/\text{d}.$$

- The increase factor shown in the table indicates the ratio between the demands at the end and the beginning of the design period.

Conclusions:
- Due to different growth percentages in the various districts of the town, the pattern of population (demand) distribution will change throughout the design period. It can be observed from the table that the demands in nodes 02, 04–06, 08–12 and 18–23 grow faster than in the rest of the network, shifting the demand concentration towards the south of the town. This fact should also be taken into consideration while deciding on the network layout.

Comments:
- Analysis of the demand distribution is an important element of the network design. Larger pipe diameters are normally laid in, or on the route to, the areas of higher demand.
- Phased system development should follow the demand development throughout the design period.

Reading:
- Volume 1, Section 4.3: 'Computer models as design tools'.

3.4.3 Network layout

The design of the network layout begins by defining the preliminary routes of the secondary mains. What is important to realise at this stage is the obvious hydraulic correlation between the system parameters:

- the demand distribution throughout the area,
- the location and capacity of the supply points and their connection to the system,
- the pipe diameters,
- the pump units, and their type, number and operating schedule, and
- the position, elevation and volume of the storage.

One single combination of the above parameters results in one unique distribution of the pressures and flows in the system. Modifying even a single nodal demand or pipe diameter will affect this equilibrium, to some degree.

In theory, different network configurations can show a similar hydraulic performance provided that correct judgements have been made while building the model. For a final choice of the network configuration, other criteria such as a reliability assessment, soil conditions, access for maintenance, future extensions, etc. will also be considered.

Alternative A – direct pumping

Based on the preliminary design concept and the demand analysis, the first model has been built for the demand situation after the commissioning of the system (year 6). The given EPANET filename is S6A1.NET and its contents can be studied from the electronic materials available with this tutorial. The main components of the system have been modelled as follows:

*   [JUNCTIONS]: An average nodal consumption as shown in Table 3.9 in year 6 has been introduced in each node.
*   [JUNCTIONS]: Altitudes of nodes 102 and 202 have been assumed to equal the ground elevations taken from Figure 3.2.
*   [DEMANDS]: The multiplier of the nodal demand has been set for the conditions on the maximum consumption day, including the water loss of 10 %; the factor has been calculated as $1.265/(1-10/100) = 1.406$.
*   [RESERVOIRS]: The pump suction heads at the sources are assumed to be fixed at the ground elevation.
*   [PIPES]: Given the fact that the population density is going to increase in the southern part of the town, the connection of the first source has been tested in nodes 06 and 02. The pipe lengths are estimated based on the scale of Figure 3.2: for Pipe 102–06 at 1550 m and for 102–02 at 1700 m.
*   [PIPES]: The factory is connected at Node 16 (pipe length 01–16 = 350 m). To provide good connectivity with Source 2, via the shortest path along route 16–17–18, this source will be connected in Node 18 (pipe length 202–18 = 750 m).
*   [PUMPS]: Equal-size pumping stations have been assumed at the sources, with the following duty head and flow: H_d = 50 mwc; this is an estimate to satisfy the design pressure criteria, and Q_d = 230 l/s is an estimate based on the information from Table 3.8 (at year 6: $Q_{max,day}$ = 40,106 m³/d ≈ 460 l/s). For the duty head and flow specified as a single pair of Q/H points, EPANET generates a synthetic pump curve based on this principle: Q_p (at H_p = 0) equals $2 \times Q_d$, and H_p (at Q_p = 0) equals to $1.33 \times H_d$. As a result, the following three Q/H points define the curve: 0–66.5, 230–50 and 460–0 (in l/s-mwc).

To analyse the demand distribution in the system, the first simulations have been run using uniform pipe diameters of 222 mm; the results are shown in figures 3.6 and 3.7.

Clarification:
*   The graphs show the snapshot at 12:00 hours (the maximum consumption hour).
*   The pipe unit head loss is displayed in m/km.
*   With regard to the pumping stations, the negative figures indicate the actual pumping head in mwc.
*   The reservoir symbols represent the suction nodes of the pumping stations.

Conclusions:

- The source connection to different nodes influences the head-loss (flow) distribution.
- The pipes with a large unit head loss i.e. the hydraulic gradient (greater than 5–10 m/km) imply the routes of the bulk flows, which is a direct consequence of the demand distribution in the nodes. The diameter of these pipes has to be increased whereas the pipes with smaller head loss should be reduced in size.
- Connecting Source 1 to Node 06 suggests that the secondary pipes should pass through the peripheral area of the town; a possible main loop is indicated in Figure 3.6. The logic behind it is to connect both sources and the factory with pipes that can carry additional capacity in case of accidents in the system.
- An alternative connection of Source 1 to Node 02 shifts the resistance towards the central part of the system; the main loop is shown in Figure 3.7. This loop is shorter (= less expensive) but may not provide a satisfactory distribution of water throughout the network. An extra loop or larger lateral pipes should therefore be considered.
- Out of the two equal pumping curves selected for the pumping stations, the one at Source 2 shows a lower head (i.e. higher supply), which is in contradiction to the initial requirement that both sources should supply a similar capacity. The likely cause is in the higher elevation of this pumping station compared to that in Source 1. Consequently, the pumping station at Source 2 can be smaller than the one at Source 1 (to be tested later).

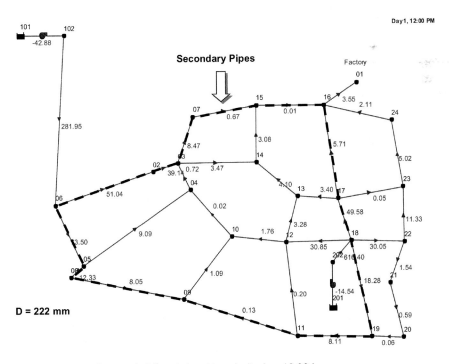

Figure 3.6 S6A1 (Source 1–06): unit head loss (m/km) at 12:00 hours

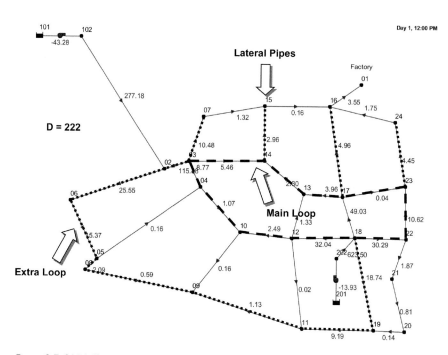

Figure 3.7 S6A1 (Source 1–02): unit head loss (m/km) at 12:00 hours

In order to proceed, the layout from Figure 3.6 has been further developed:

- [PIPES]: The main loop from the figure has been formed from pipes D = 400 mm,
- [PIPES]: The source connections to the system were provided with pipes D = 600 mm,
- [PIPES]: All other pipes in the network have been set at D = 200 mm.

The input file has been named S6A2.NET, with simulation results shown in Figure 3.8.

Conclusions:
- The unit head losses in most of the secondary pipes are low, suggesting that D = 400 mm is too large. Two possible solutions are:
    - to look at each pipe individually and reduce the diameters where required, and/or
    - to choose a smaller loop of secondary pipes

The first approach in the conclusions actually means abandoning the loop concept and switching to a branched structure of the secondary mains, which could potentially raise problems related to the reliability of supply. For these reasons, the approach with a smaller loop has been tried in the S6A3.NET file, created in the same way as S6A2.NET. The proposed layout is shown in Figure 3.9.

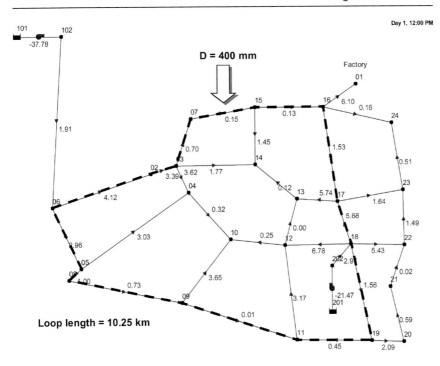

Figure 3.8 S6A2: unit head loss (m/km) at 12:00 hours

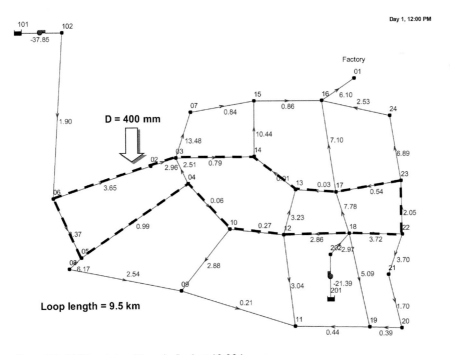

Figure 3.9 S6A3: unit head loss (m/km) at 12:00 hours

Conclusions:

- The length of the loop has been reduced with a slight improvement of the head loss. However, some pipes are still too large (e.g. 04–10 or 13–17).
- The lateral pipes at the northern side (03–07, 14–15, 17–16 and 23–24) show large unit head losses, which is the consequence of making the main loop smaller. These pipes should be enlarged.

The smaller loop of the secondary pipes combined with laterals stretched towards the ends of the system appears to be a promising concept. To explore in addition the effects of the connection between Source 1 and Node 02, file S6A4.NET has been created from S6A3.NET:

- by introducing connection 102–02 instead of 102–06, and
- by reducing further the length of the main loop.

The layout with the results is shown in Figure 3.10.

Conclusions:

- Connection 102–02 allows further reduction of the main loop. As a consequence, the head losses in the lateral pipes will increase.
- The head losses in the main loop remain low, which implies again that D = 400 mm is too large a diameter for the supply conditions at the beginning of the design period.

The network layout is further developed in file S6A5.NET. Through a number of trial and error computer runs, the pipe diameters are adopted to satisfy the design unit head loss of ± 1–5 m/km. The final layout is shown in Figure 3.11.

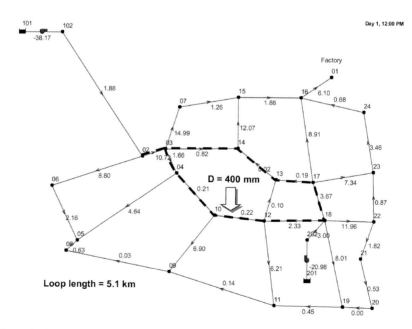

Figure 3.10 S6A4: unit head loss (m/km) at 12:00 hours

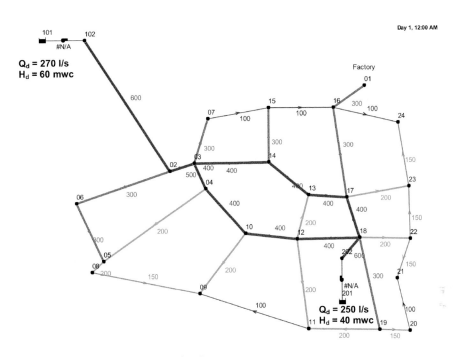

Figure 3.11 S6A5: pipe diameters (mm)

To satisfy the minimum design pressure of 20 mwc while supplying a similar flow from both sources, the pumping characteristics have been adjusted as follows:

- Source 1: $H_{d,1}$ = 60 mwc, $Q_{d,1}$ = 270 l/s, and
- Source 2: $H_{d,2}$ = 40 mwc, $Q_{d,2}$ = 250 l/s.

The results of calculating with these values are shown in figures 3.12 and 3.13.

Conclusions:
- As preferred, the pumping stations now work at an even capacity. The total supply (346 + 336 = 682 l/s = 2455 m³/h) equals approximately the maximum hour on the maximum consumption day at the beginning of the design period (in Table 3.8, $Q_{max,hour}$ = 2452.9 m³/h at year 6).
- The system has been deliberately left with extra capacity at this stage of the analysis. Firstly, lower velocities (head losses) indicate that the pipes possess some spare capacity, which can be of use during irregular supply conditions. Secondly, the larger pipes can deal with demand growth in the longer term, which postpones the system extension. Justification of these assumptions is going to be tested whilst analysing the system operation.
- The minimum pressure in the system is in Node 11 (22.22 mwc). The maximum (night) pressures are to be analysed after the number of pump units, and their arrangement and operation has been decided.
- Assuming the unit costs as given in Table 3.4, the total investment cost for the pipes is calculated at EUR 4,661,500.

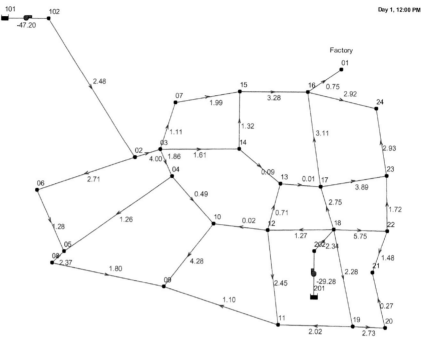

Figure 3.12 S6A5: unit head loss (m/km) at 12:00 hours

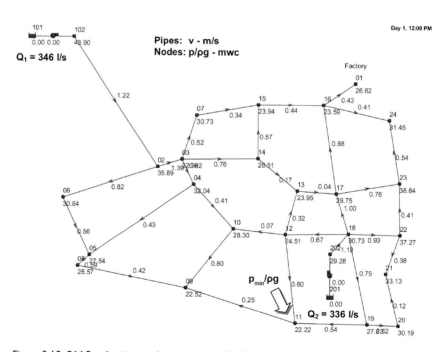

Figure 3.13 S6A5: velocities and pressures at 12:00 hours

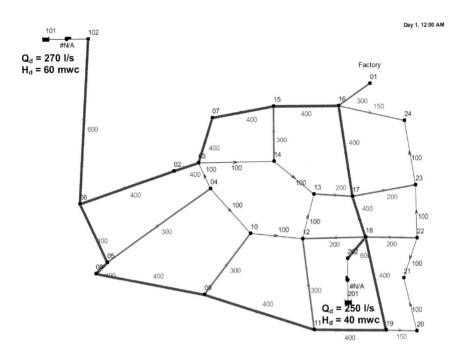

Figure 3.14 S6A6: pipe diameters (mm) for a large loop

For reasons of comparison, two additional alternatives have been developed based on the design hydraulic gradient of ± 1–5 m/km and minimum pressure of 20 mwc:

- An alternative with the large loop of secondary mains, as discussed in Figure 3.6, is developed in file S6A6.NET. This network layout is shown in Figure 3.14. Similar hydraulic performance (Figure 3.15) yields the total cost of the pipes as EUR 5,218,500.
- A branched layout of the secondary mains (file S6A7.NET), shown in figures 3.16 and 3.17. The total cost of the pipes in this case is EUR 4,484,500.

In both cases the same pump characteristics were used as in the S6A5 layout.

Conclusions:
- All three configurations, 5, 6 and 7, provide similar hydraulic performances during the maximum hour on the maximum consumption day. However, given the pattern of the population growth and reliability requirements, the S6A5 layout is anticipated to be the most adequate one. Further development of the direct pumping alternative is going to be based on this layout.
- The S6A6 layout is more expensive than the other two. Nevertheless, this is an acceptable approach in situations where the growth in population is followed by an overall growth of the town's size. As this is not anticipated in this case, additional investment in a large loop of secondary mains does not seem to be justified.
- The S6A7 layout is the cheapest but probably the least reliable. The cost saving compared to the S6A5 alternative does not itself make this more attractive. Its weak points will be illustrated during the reliability assessment of the network.

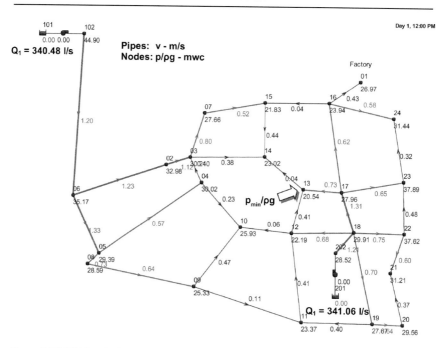

Figure 3.15 S6A6: velocities and pressures at 12:00 hours

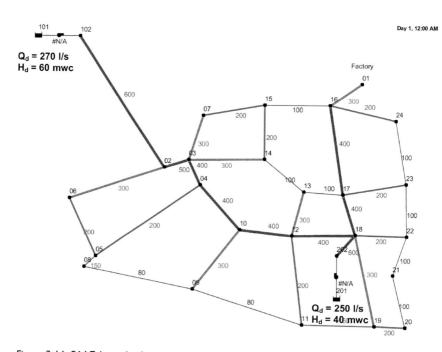

Figure 3.16 S6A7: branched mains – pipe diameters (mm)

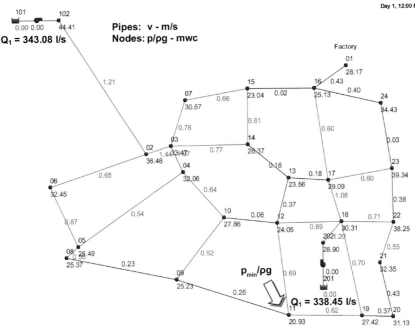

Figure 3.17 S6A7: velocities and pressures at 12:00 hours

Comments:

- While sizing pipes:

  - A pipe with extremely low velocity/head loss is a potential source of water quality problems and should therefore be reduced in diameter. Pipes in which the water stagnates play no role in a proper distribution network. These can be removed without major implications for the overall hydraulic performance of the network, unless needed for service connections or as a reliability provision.
  - Pipes with extremely high velocity/head loss require high energy input for distribution and should therefore be increased in diameter. Often it is necessary to enlarge not just one but a sequence of pipes following the same path/flow direction (i.e. connected in series).

- Choosing appropriate pipe diameters means optimisation of the hydraulic gradients/flow velocities in the system. However, this does not mean:

  - Satisfying the design hydraulic gradient/flow velocity *in each pipe* of the network at a particular point in time,
  - Satisfying the design hydraulic gradient/flow velocity of a single pipe *at any point in time*.

Depending on the demand pattern, one pipe may generate a range of velocities over 24 hours that will be rather wide; this is normal.

- Optimisation of pipe diameters is usually done for the maximum consumption day demand. Thus, a pipe with extremely low velocities on that day will maintain even lower velocities throughout the year. On the other hand, the velocities/head losses tend to increase throughout the design period, as a result of the demand growth, and the same pipe may improve its hydraulic performance within a couple of years.

To finalise the network design of Alternative A, the proposed layout in S6A5 has also been tested for the demand level at the end of the design period, i.e. 30 years from now. File S30A1.NET has been created from S6A5.NET by modifying the following sections:

- [JUNCTIONS]: The nodal consumption in year 30 has been put in (see Table 3.9).
- [DEMANDS]: The multiplier of the nodal consumption represents the maximum consumption day with water loss of 20 %.

The simulation has been run with the same pump characteristics as at the beginning of the design period. The results are shown in Figure 3.18.

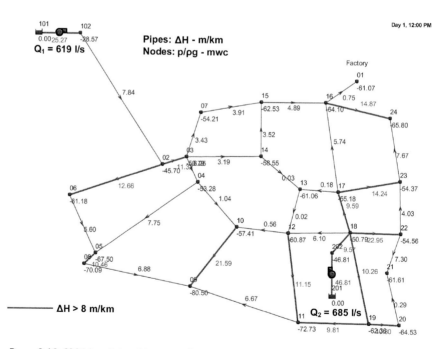

Figure 3.18 S30A1: unit head losses and pressures at 12:00 hours

Conclusions:

- The nodal pressures become negative at a minimum of -80.54 mwc at Node 09. This is a consequence of the demand/population growth that has created two problems:

  - The existing pumps are too weak to deliver the increased flow at sufficient head (the pressures already become negative at the nodes nearest to the supply points),
  - The flow increase causes a substantial increase of the head loss in some pipes ($\Delta H > 8$ m/km).

- The warning message in EPANET will also indicate the negative pressures, which was registered in 12 out of 24 hours; thus, the system has an insufficient supply for at least half of the day.

- Between midnight and 06:00 hours in the morning, the system is able to satisfy demand at the required minimum pressure. The problems occur from 07:00 hours onwards. At that moment the pressure will become negative for the first time, while the head loss in some pipes will start to grow, as Figure 3.18 shows. The total demand of the system at 07:00 hours is 979 l/s ≈ 3526 m³/h, which is the level of demand at the maximum hour of the maximum consumption day in around the year 20–21 (see column $Q_{max,hour}$ in Table 3.8). This is the critical moment but the new investment is likely to be necessary

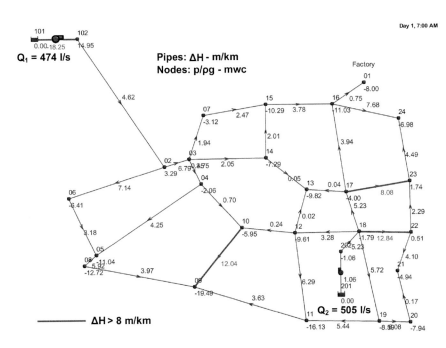

Figure 3.19 S30A1: unit head losses and pressures at 07:00 hours

a few years earlier, to maintain regular supply throughout the entire design period. These include:

- Extension of the pumping stations (the duty flows have to be nearly doubled, based on the calculated demand increase during the design period), and
- Replacement of the critical pipes with larger ones or laying the pipes in parallel.

During the second phase, the pipes already laid will be approximately 10–15 years old. These should normally still be in good condition and their replacement with a larger diameter pipe may not be economically justified, especially if trench widening is required. If reliability is an issue, parallel pipes could be considered instead. In the case of Safi, the combination of both measures has been applied as an illustration for purely educational reasons.

Summary of Alternative A

The layouts for the regular supply conditions have been finalised in files S6A8. NET and S30A2.NET (beginning and end of the design period, respectively). The results are displayed in figures 3.20–3.23.

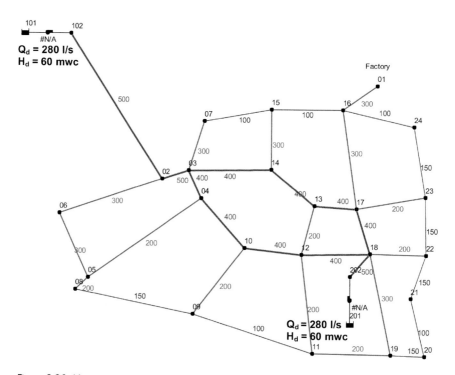

Figure 3.20 Alternative A: pipe diameters (mm) at the beginning of the design period

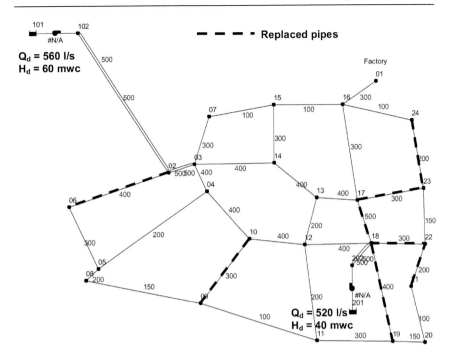

Figure 3.21 Alternative A: pipe diameters (mm) at the end of the design period

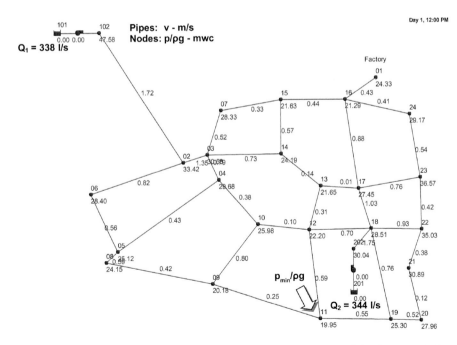

Figure 3.22 Alternative A: operation at 12:00 hours at the beginning of the design period

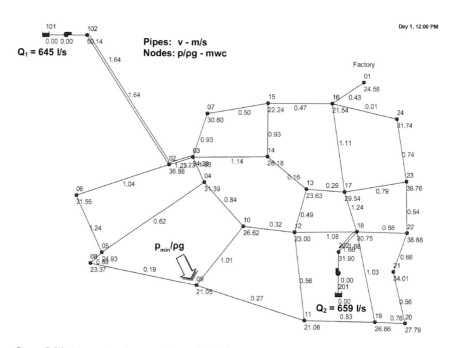

Figure 3.23 Alternative A: operation at 12:00 hours at the end of the design period

Clarification:
*   The bold lines in Figure 3.21 indicate routes where action will be taken in around year 15. Parallel pipes D = 500 mm are planned along routes 102–02, 02–03 and 202–18, while all other indicated pipes will be replaced with the next larger diameter from Table 3.4.
*   At the beginning of the design period, the diameters of pipes 102–02 and 202–18 will be reduced to D = 500 mm instead of 600 mm, as previously determined in S6A5. As a consequence, the pressure will drop by a few mwc compared to the values in Figure 3.13. To prevent it falling below $p_{min}/\rho g \approx 20$ mwc, the pumping capacity has to be slightly increased. In Source 1, $Q_d = 280$ l/s (from 270 l/s), and in Source 2, $Q_d = 260$ l/s (from 250 l/s).

Conclusions:
*   Pipe investment for the network at the beginning of the design period is EUR 4,539,000. The additional cost for the pipes laid in the second phase is EUR 2,330,000.
*   Both network configurations provide stable operation during the regular (maximum) supply conditions. At the end of the design period, the total peak supply of the pumping stations fits the figure from Table 3.8 (645 + 660 =

1305 l/s = 4698 m³/h ≈ $Q_{max, hour}$ in year 30). The first source will have reached 93 % of its maximum capacity (2500 m³/h = 695 l/s).

- For some pipes, there is still room for further reduction of the diameter. Relatively low velocities are registered in pipes 08–09, 09–11, 10–12, 14–13, 13–17, 15–16 and 16–24, some of these indicating a 'hydraulic border' between the part of the network supplied by Source 1, and the one supplied by Source 2. The final decision on this is going to be taken after the reliability aspects have been analysed.

Alternative B – pumping and balancing storage

Based on the preliminary concept and applying the same basic principles, the network layout has been developed for the beginning (file S6B1.NET) and the end of the design period (file S30B1.NET). To utilise the existing topography, the tank has been positioned at the highest altitude in the area (28.0 msl), in the vicinity of Node 13. The layouts and the results of the simulation are shown in figures 3.24–3.29.

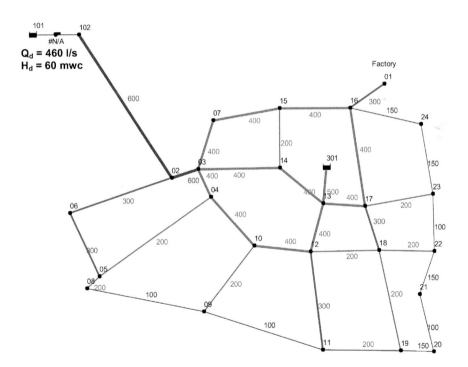

Figure 3.24 Alternative B: pipe diameters (mm) at the beginning of the design period

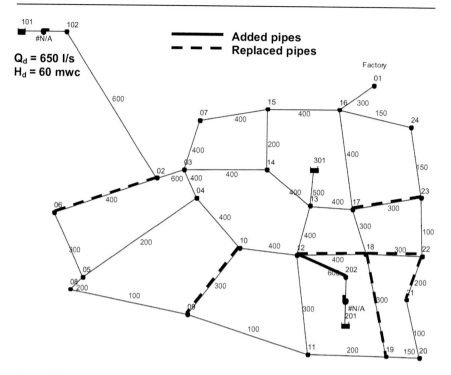

Figure 3.25 Alternative B: pipe diameters (mm) at the end of the design period

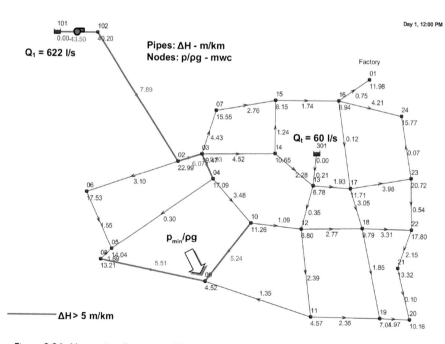

Figure 3.26 Alternative B: a ground-level tank, operation at 12:00 hours at the beginning of the design period

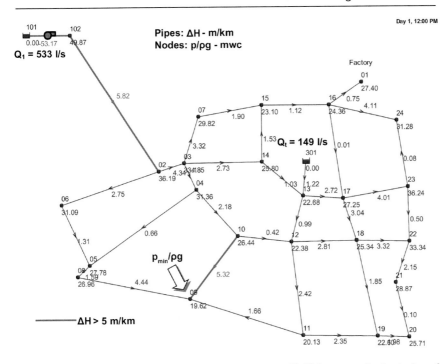

Figure 3.27 Alternative B: an elevated tank, operation at 12:00 hours at the beginning of the design period

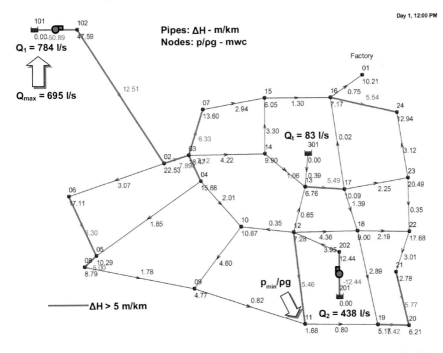

Figure 3.28 Alternative B: a ground-level tank, operation at 12:00 hours at the end of the design period

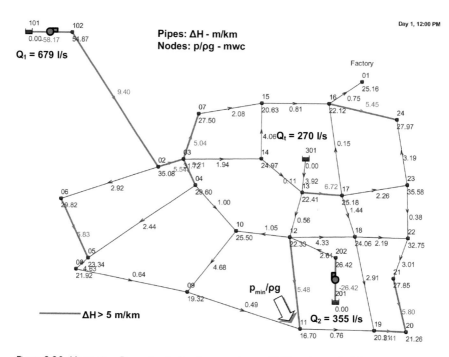

Figure 3.29 Alternative B: an elevated tank, operation at 12:00 at the end of the design period

Clarification:

- As the first approximation, the tank surface level has been assumed to be constant throughout the entire period of the simulation (the 'fixed head' node, which is modelled in EPANET as a reservoir). This gives an initial impression of the pressure distribution in the system. Two possibilities have been analysed:

  - a ground-level tank with a depth of 6 m (the fixed head = 28 + 6 = 34 msl), and
  - an elevated tank ± 20 m high incl. ± 2 m water depth (the fixed head = 28 + 22 = 50 msl).

- Based on the previous experience, the pump characteristics have been set as follows:

  - at the beginning of the design period: $H_{d,1}$ = 60 mwc, $Q_{d,1}$ = 460 l/s = Q_{maxd} at year 6, and
  - at the end of the design period: $H_{d,1}$ = 60 mwc, $Q_{d,1}$ = 650 l/s ($\approx Q_{max}$ of Source 1), and $H_{d,2}$ = 40 mwc, $Q_{d,2}$ = 250 l/s ($Q_{d,1} + Q_{d,2}$ = 900 l/s $\approx Q_{maxd}$ at year 30).

To provide a link with the main loop, the second source has been connected at Node 12.

Conclusions:

- Pipe investment for the network at the beginning of the design period is EUR 4,927,500. The additional cost for the pipes to be laid in approximately year 20 is EUR 1,455,000. Compared to Alternative A, the saving is mainly obtained by avoiding parallel pipes from the sources.
- There is a difference in the unit head loss distribution, depending on the level of the tank. This results from different quantities supplied from the tank and the pumping station(s). However, with a few exceptions, this does not have significant implications for the choice of pipe diameters.
- The ground-level tank draws too much water from the sources throughout the day. As a consequence, the required supply from Source 1 at the end of the design period exceeds its maximum capacity, which is 695 l/s. In addition, the nodal pressures in the system are too low.
- The elevated tank is showing much better hydraulic performance; the pumping is reduced due to increased water flow supplied from the tank. With the same pump characteristics, the pressures in the system are almost correct. However, erecting tanks with a large volume can be quite an expensive solution. The final decision on the tank height will be made after the balancing volume has been determined.

Comments:

- A balancing tank that is positioned too low receives more water than it can supply to the system. Such a tank will become full after some time. To prevent this, the pumping at the source can be reduced, which will slow down the filling of the tank. Another possibility would be to install an additional pumping station in the vicinity of the tank, which will empty its volume, thereby providing sufficient pressure.
- A balancing tank that is positioned too high will receive less water than it should supply to the system and therefore will soon become empty. To prevent this, the pumping head at the source should be increased.
- The amounts of water entering and leaving the tank can be calculated in advance, if the source is to supply the average flow throughout the day. A sophisticated way to quickly evaluate the tank position is to model it as an ordinary node with the balancing flows as its consumption. Such consumption would fluctuate between the maximum or minimum hour consumption reduced by the average flow supplied by the pump. It becomes negative when the tank is supplying the system and positive when its volume is being recovered. The pressure in the node, resulting from the simulations for two extreme consumptions, can give an impression about the range of water depths in the tank; it should therefore be reasonably high and within a range of a few metres.

3.4.4   Pumping heads and flows

The pumping heads and flows determined at the beginning and the end of the design period are summarised in tables 3.10 and 3.11 for both alternatives, respectively. As expected, more pumping will be involved in Alternative A than in the case of Alternative B.

Table 3.10 Alternative A: pumping stations

	Beginning of the design period			End of the design period		
	Q_d (l/s)	H_d (mwc)	$Q_{max,hour}$ (l/s)	Q_d (l/s)	H_d (mwc)	$Q_{max,hour}$ (l/s)
Source 1	280	60	338	560	60	645
Source 2	260	40	344	520	40	660

Table 3.11 Alternative B: pumping stations

	Beginning of the design period				End of the design period			
	Q_d (l/s)	H_d (mwc)	$Q_{max,hour}$ (l/s) ground	$Q_{max,hour}$ (l/s) elevated	Q_d (l/s)	H_d (mwc)	$Q_{max,hour}$ (l/s) ground	$Q_{max,hour}$ (l/s) elevated
Source 1	460	60	622	533	650	60	784	679
Source 2	–	–	–	–	250	40	438	355
Tank	–	–	60	149	–	–	83	271

Table 3.12 Preliminary pump selection

	Alternative A				Alternative B (elevated tank)			
	Q_p (l/s)	H_p (mwc)	No. of units begin	No. of units end	Q_p (l/s)	H_p (mwc)	No. of units begin	No. of units end
Source 1	100	60	3+1*	6+1	150	60	3+1	6+1
Source 2	90	40	3+1	6+1	130	40	–	2+1

* One unit is on stand-by.

In order to facilitate the demand variations over a period of 24 hours, the logical choice is to arrange the pumps in parallel. One possibility is shown in Table 3.12. The final choice of the pump type and number of units will take place after the network operation has been analysed.

Clarification:
- A hydraulically equivalent performance is reached between one pump of a particular Q_d and H_d, and n pumps of Q_d/n and H_d connected in parallel. For example, three units of 100 l/s against duty head of 60 mwc equal one unit of 300 l/s and the same duty head.

Application of booster stations

To illustrate the difficulties with pressure boosting in looped systems, file S6B2. NET has been created from S6B1.NET. The operation of the booster station has been tested in three different locations: from nodes 13 to 17, 13–12 and

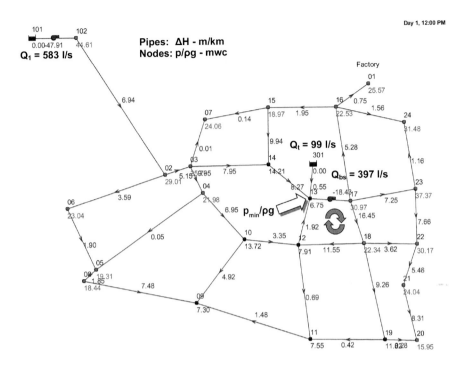

Figure 3.30 S6B2: a ground-level tank and booster 13–17, operation at 12:00 hours

14–301 (instead of Pipe 14–13). In all cases the same duty head and flow have been used: $H_{d,bs} = 30$ mwc, $Q_{d,bs} = 270$ l/s. The results of calculation are shown in figures 3.30–3.34.

Conclusions:

- Putting a booster station in the system causes a pressure drop in nodes at the suction side of the pump.
- The booster station tends to re-circulate the water along the loop to which it belongs (loop 12-13-17-18 in Figure 3.30 and the same loop together with 14-13-12-10-04-03 in Figure 3.32). This becomes obvious if one pipe is removed from the loop(s); the pressure downstream of the booster station will increase (figures 3.31 and 3.33).
- The pipe resistance increases in the vicinity of the booster station. Contrary to expectations, enlarging the diameter of these pipes makes no sense as it creates negative effects on the pressure in the system. The fact is that, with larger pipes, even more water will be pushed towards the booster station, which reduces the pumping head.

To analyse the relation between the booster pumping capacity and the pipe resistance, the pipes indicated in Figure 3.34 have been replaced with the next larger diameter from Table 3.4 (file S6B3.NET). As a result (shown in

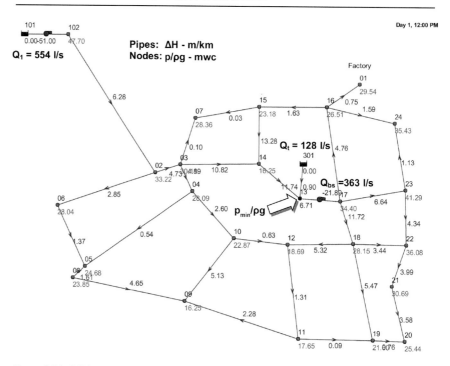

Figure 3.31 S6B2: booster 13–17, pipe 12–13 removed, at 12:00 hours

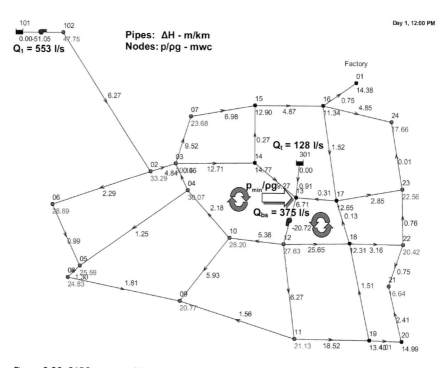

Figure 3.32 S6B2: a ground-level tank and booster 13–12, at 12:00 hours

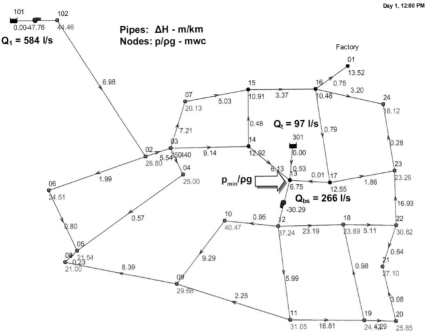

Figure 3.33 S6B2: a booster 13–12, pipes 04–10 and 17–18 removed

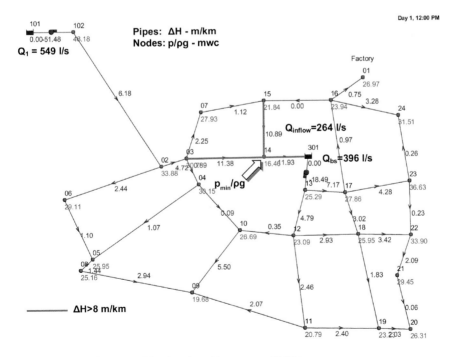

Figure 3.34 S6B2: a ground-level tank and booster, at 12:00 hours

Figure 3.35), the booster station increases the flow on account of the head (pressure). Installing much stronger pumps will reinstate the pressure, but will also initiate the old problem of water re-circulation and large pressure drops in some of the pipes. Operation of the system with doubled duty flow ($H_{d,bs} = 30$ mwc, $Q_{d,bs} = 540$ l/s) of the booster station is shown in Figure 3.36.

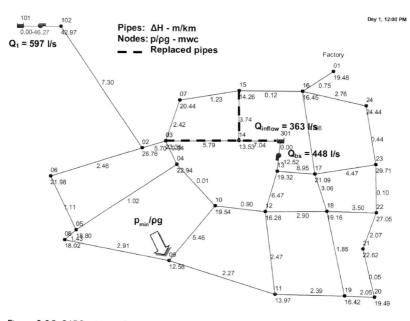

Figure 3.35 S6B3: pipe enlargement, operation at 12:00 hours

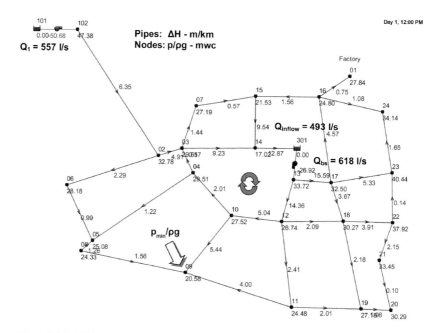

Figure 3.36 S6B3: booster pump enlargement, operation at 12:00 hours

Comments:

- Booster stations are useful if pressure has to be provided for higher zones of a distribution system, in which case they will be installed on a single line feeding a particular zone. They are of little hydraulic use in applications within looped networks, except to prevent water stagnation, which is not common in engineering practice.
- Three general observations regarding the modelling of booster stations in EPANET are:

    - Compared to the modelling of a source pumping station, the suction node of a booster station can be any node within the network. The flow direction in the pump will be from the first to the second node selected on the map after clicking on the pump button. The reversed flow is prevented by the programme (which shows $Q_{bs} = 0$).
    - A pipe replaced by a booster station has to be deleted first; otherwise the programme assumes two pipes in parallel where in fact one is with the booster pump.
    - In addition, the friction loss of the replaced pipe is also removed from the calculations (pumps as links do not generate any losses of energy in EPANET). If this loss is not negligible, the pump should be modelled by introducing a dummy node on either side (suction or pressure).

Due to all the listed deficiencies, Alternative B with a ground-level tank and booster station is therefore discounted with regard to the Safi network.

3.4.5 Storage volume

Based on the diurnal domestic consumption pattern from Figure 3.4 and the leakage levels as assumed in Table 3.8, the balancing volume at the beginning of the design period is calculated according to the principles explained in the section 'Storage design' in Chapter 4. The results of a spreadsheet calculation are shown in Table 3.13.

Table 3.13 Balancing storage volume

At year 6			Leakage (%)	10		V_{bal} (m³)	4373.0
Hour	p_f	Consumed (m³/h)	Leakage (m³/h)	Factory (m³/h)	Demand (m³/h)	ΔV (m³/h)	ΣV (m³/h)
1	0.68	989.6	110.0		1099.5	571.6	571.6
2	0.65	945.9	105.1		1051.0	620.1	1191.7
3	0.62	902.2	100.2		1002.5	668.6	1860.3
4	0.50	727.6	80.8		808.5	862.6	2722.9
5	0.58	844.0	93.8		937.8	733.3	3456.2
6	0.80	1164.2	129.4		1293.5	377.6	**3833.7**
7	1.08	1571.7	174.6	108.3	1854.6	−183.5	3650.2
8	1.10	1600.8	177.9	108.3	1887.0	−215.9	3434.4

(Continued)

(Continued)

At year 6			Leakage (%)	10		V_{bal} (m³)	4373.0
Hour	p_f	Consumed (m³/h)	Leakage (m³/h)	Factory (m³/h)	Demand (m³/h)	ΔV (m³/h)	ΣV (m³/h)
9	1.22	1775.4	197.3	108.3	2081.0	−409.9	3024.5
10	1.35	1964.6	218.3	108.3	2291.2	−620.1	2404.4
11	1.42	2066.4	229.6	108.3	2404.4	−733.3	1671.1
12	1.45	2110.1	234.5	108.3	2452.9	−781.8	889.3
13	1.10	1600.8	177.9	108.3	1887.0	−215.9	673.5
14	0.90	1309.7	145.5	108.3	1563.6	107.5	781.0
15	0.75	1091.4	121.3	108.3	1321.0	350.1	1131.0
16	0.65	945.9	105.1	108.3	1159.3	511.8	1642.8
17	0.73	1062.3	118.0	108.3	1288.7	382.4	2025.2
18	1.00	1455.2	161.7	108.3	1725.3	−54.2	1971.0
19	1.20	1746.3	194.0		1940.3	−269.2	1701.8
20	1.47	2139.2	237.7		2376.9	−705.8	996.0
21	1.45	2110.1	234.5		2344.6	−673.5	322.6
22	1.40	2037.3	226.4		2263.7	−592.6	−270.0
23	1.20	1746.3	194.0		1940.3	−269.2	**−539.2**
24	0.70	1018.7	113.2		1131.9	539.2	0.0
Average	**1.00**	**1455.2**	**161.7**		**1671.1**		

Clarification:
- The spreadsheet calculation has been conducted for the maximum consumption day conditions. The total balancing volume equals 3833.7 + 539.2 = 4372.9 ≈ 4400 m³.
- The same calculation for the demand at the end of the design period yields the balancing volume of 8196 ≈ 8200 m³.

Conclusions:
- In Alternative A, the reservoirs are located at the suction side of the source pumping stations that were designed to operate at similar capacity. Hence, the balancing volume can be shared evenly between the two sources. The total volume per tank, required at the beginning of the design period, is determined as follows:
  - balancing volume: 4400/2 = 2200 m³,
  - emergency volume for one (maximum consumption) hour in each tank: 2452.9 ≈ 2500 m³,
  - other provisions: approximately 0.6 m of the tank height.
    Assuming a circular cross-section of the tank, D=35 m, 0.6 m of water column is equivalent to 577.3 ≈ 600 m³. The total tank volume equals 2200 + 2500 + 600 = 5300 m³. For example, for a height of 5.5 m, V_{tot} ≈ 5290 m³. Two equal tanks yield the total storage volume in the system of approximately 10,600 m³.

- In the case of Alternative B, the balancing volume is located within the network. If the source pumping station works at constant average capacity, the reservoir volume in Source 1 will consist mainly of the emergency provision. Thus, at the beginning of the design period:

  - at Source 1, $V_{tot} = 2500 + 600 = 3100$ m^3 (for D = 30 m and H = 4.3 m, $V_{tot} = 3040$ m^3),
  - for the tank in the system, $V_{tot} = 4400 + 2500 + 600 = 7500$ m^3 (D = 40 m, H = 6.0 m, $V_{tot} = 7540$ m^3). The total storage volume in the system remains the same as in the case of Alternative A.

- During the design period, the emergency volume will be gradually converted into the balancing volume. If no action is taken, the deficit of the balancing volume 8200−4400 = 3800 m^3 would reduce the initial emergency volume of 5000 m^3 to only 1200 m^3 at the end of the design period. Thus, it is advisable to start extending the volume in good time; otherwise the emergency reserve will become effectively exhausted. An option is to construct the entire volume at the beginning of the design period, but care is to be taken about water stagnation due to initially large emergency volumes (post-chlorination may be required to prevent water quality deterioration).
- Based on the demand growth, it is estimated that some additional 10,000 m^3 of storage volume would be required at the end of the design period. An even share of this volume is planned between the two sources in Alternative A. A preliminary scenario for Alternative B is:

  - in Source 1, the tank volume increases from ± 3100 to ± 4600 m^3,
  - in Source 2, a new clear water tank of ± 4600 m^3 is to be built,
  - the balancing tank volume increases from ± 7500 to ± 11,400 m^3.

  Such a large volume of the elevated balancing tank changes the earlier conclusion about its better hydraulic performance compared to the ground-level tank; to construct a volume of ± 10,000 m^3 at a height of ± 20 m might be a relatively expensive solution, not to mention the aesthetic aspects if the structure is to be located in an urban area.

Preliminary dimensions of all tanks are summarised in tables 3.14 and 3.15. The dimensions at the end of the design period indicate the total required volumes. Depending on the selected form of the tanks (not necessarily circular cross-section) these dimensions should match the already existing design as much as possible in order to make the volume extension easier (i.e. by adding of the second compartment).

Table 3.14 Alternative A – storage tanks

	Beginning of the design period			End of the design period		
	D (m)	H (m)	V_{tot} (m^3)	D (m)	H (m)	V_{tot} (m^3)
Source 1	35	5.5	5290	50	5.5	10,800
Source 2	35	5.5	5290	50	5.5	10,800

Table 3.15 Alternative B – storage tanks

	Beginning of the design period			End of the design period		
	D (m)	H (m)	V_{tot} (m³)	D (m)	H (m)	V_{tot} (m³)
Source 1	30	4.3	3040	30	6.5	4600
Source 2	–	–	–	30	6.5	4600
Tank	40	6.0	7540	50	6.0	11,780

Comments:
• The balancing volume is exclusively dependent from the demand and its pattern of variation. It is in fact a property of a certain distribution area and not of the chosen supply scheme.
• In the direct pumping supply schemes, the balancing volume is located at the source(s) i.e. at the suction side of the source pumping stations. There, it evens out the constant production and variable pumping (Alternative A).
• In the combined supply schemes, the balancing volume is located within the distribution area and evens out the constant pumping and variable demand (Alternative B).

Reading:
• Volume 1, Section 4.2.3: 'Storage design'.

3.4.6   Summary of the preliminary hydraulic design

<u>Alternative A – direct pumping</u>

This alternative comprises:

• network layouts as shown in Figures 3.20 and 3.21 (EPANET input files S6A8 and S30A2),
• selection of the pump units as displayed in Table 3.12, and
• storage volume in the system as displayed in Table 3.14.

The following development strategy has been adopted:

• The network is left with larger diameters for reasons of reliability. Part of the reserve capacity will be exhausted before the system is extended.
• Extension of the pumping stations is tentatively planned around year 15. Laying of new pipes in the system will take place at the same time.
• For practical reasons, the clear water tanks will be immediately constructed with the volumes required at the end of the design period, which also adds to the reliability of supply.

<u>Alternative B – pumping and balancing storage</u>

The main characteristics of the system in this alternative are:

• network layout as shown in figures 3.24 and 3.25 (EPANET input files S6B1 and S30B1),

- selection of the pump units as displayed in Table 3.12, and
- storage volume in the system as displayed in Table 3.15.

The following development strategy has been adopted:

- The start of the supply from Source 2 is planned in year 21 at the latest, after the first source reaches its maximum capacity. The additional pipes should be in place at the same time.
- As an additional safety precaution, it is proposed to construct the clear water tank at Source 1 at the beginning with the volume that will be required at the end of the design period.
- The option with the ground-level balancing tank has been abandoned due to the pressure problems that cannot be solved by installing a booster station. It is not yet clear either the option with the elevated balancing tank should be accepted due to its excessively large volume and consequently expensive construction.

It appears at this stage that the present topography does not offer an effective way of using the balancing tank in the system, unless its (large) volume is elevated. Initially, this makes Alternative A appear simpler and more straightforward from the hydraulic point of view. Nevertheless, this conclusion is yet to be verified by further analyses of the system operation and a cost calculation.

The third possibility could be to keep the balancing volume at the sources, combined by a water tower with a volume of 500–1000 m³. The main task of the water tower would be to stabilise the operation of the pumps that in this case no longer operate at a constant capacity. Such an alternative is in essence a combination of alternatives A and B.

3.5   System operation

3.5.1   Regular operation

Alternative A – direct pumping

File S30A3.NET has been created based on the summary in Section 3.3.6. The test simulations of this layout show that sufficient pressure in the network can also be maintained with a pump duty head of 50 instead of 60 mwc at Source 1. Thus:

- Source 1: 6 units + 1 stand-by, $Q_d = 100$ l/s, $H_d = 50$ mwc, and
- Source 2: 6 units + 1 stand-by, $Q_d = 90$ l/s, $H_d = 40$ mwc.

The simulation starts with three units, 1001, 1002 and 1003, in operation at the first source and two units, 2001 and 2002, at the second source; for all other units, the **Initial Status** in the pump property editor is set as **CLOSED**. By checking the pressures in the system, hour by hour, each pump is further switched on or off when the pressure in any node drops outside the range of 20–60 mwc; this operation can be done manually or automatically. It is necessary to keep the pressure as close to 20 mwc for as long as possible.

Both the manual and automatic mode of pump operation are modelled in EPANET in the browser option **Data>>Controls** (double-click on **Simple**). For the above-selected pump units, the optimal schedule of manual operation is as shown below:

Link	ID	Setting	Condition
LINK	1003	CLOSED	AT TIME 2
LINK	1003	OPEN	AT TIME 6
LINK	2003	OPEN	AT TIME 6
LINK	1004	OPEN	AT TIME 7
LINK	2004	OPEN	AT TIME 7
LINK	1005	OPEN	AT TIME 9
LINK	2005	OPEN	AT TIME 9
LINK	1006	OPEN	AT TIME 11
LINK	2006	OPEN	AT TIME 11
LINK	1006	CLOSED	AT TIME 13
LINK	2006	CLOSED	AT TIME 13
LINK	1005	CLOSED	AT TIME 13
LINK	2005	CLOSED	AT TIME 13
LINK	1004	CLOSED	AT TIME 14
LINK	2004	CLOSED	AT TIME 14
LINK	1003	CLOSED	AT TIME 15
LINK	1003	OPEN	AT TIME 18
LINK	1004	OPEN	AT TIME 18
LINK	2004	OPEN	AT TIME 18
LINK	1005	OPEN	AT TIME 19
LINK	2005	OPEN	AT TIME 20
LINK	1006	OPEN	AT TIME 20
LINK	2006	OPEN	AT TIME 20
LINK	1006	CLOSED	AT TIME 23
LINK	2006	CLOSED	AT TIME 23
LINK	1005	CLOSED	AT TIME 24
LINK	2005	CLOSED	AT TIME 24
LINK	1004	CLOSED	AT TIME 24
LINK	2004	CLOSED	AT TIME 24
LINK	2003	CLOSED	AT TIME 24

Summarised per pump unit, the same schedule is shown in Table 3.16 (1 is On, 0 is Off). Pumps 1007 and 2007 have been left out of the operation, as stand-by units.

For such an operation, the ranges of pressures and velocities in the system are shown in figures 3.37 and 3.38. The supply ratio between the sources is given in Figure 3.39 (shown in EPANET as negative demand = source supply).

Table 3.16 Alternative A: pumping schedule for manual operation (S30A3)

Source 1	Schedule from 0–24	Source 2	Schedule from 0–24
1001	IIIIIIIIIIIIIIIIIIIIIIIIIII	2001	IIIIIIIIIIIIIIIIIIIIIIIIII
1002	IIIIIIIIIIIIIIIIIIIIIIIIII	2002	IIIIIIIIIIIIIIIIIIIIIIIIII
1003	II000011111111110001111111	2003	000000IIIIIIIIIIIIIIIIIIIII0
1004	0000000IIIIII10000IIIIII0	2004	0000000IIIIII10000IIIIII0
1005	00000000IIII0000000IIIII0	2005	00000000IIII000000IIIIII0
1006	000000000011000000011100	2006	000000000011000000011100
1007*	000000000000000000000000	2007*	000000000000000000000000

* Stand-by units.

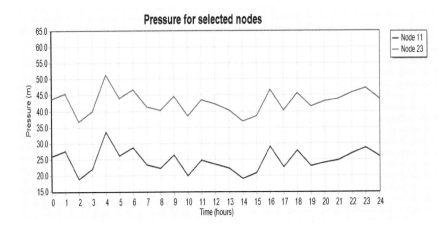

Figure 3.37 S30A3: pressure range in the network; manual pump operation (schedule in Table 3.16).

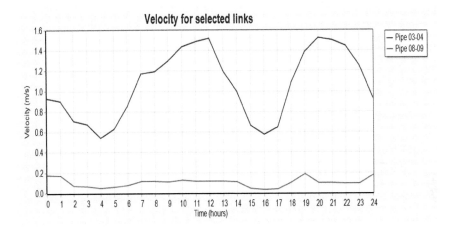

Figure 3.38 S30A3: velocity range in the network; manual pump operation (schedule in Table 3.16)

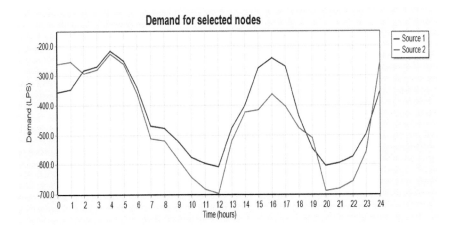

Figure 3.39 S30A3: supply from the sources; manual pump operation (schedule in Table 3.16)

Clarification

- Nodes 11 and 23 indicate the range of pressures in the system; the pressures in all the other nodes lay within this range.
- Pipes 08–09 and 03–04 show extreme velocities in the system; the velocities in all the other pipes lay within this range.
- Supply/source reservoirs are in EPANET always shown as nodes with negative demand.

Conclusions:

- A fairly balanced supply with stable pressures is provided by the proposed pumping regime.
- In very few cases, pressure slightly below 20 mwc has been tolerated. The minimum pressure of 18.89 mwc occurs in Node 11 at 02:00 hours. This is an hour of low demand and such pressure does not really affect the consumers. Switching on an extra unit at 02:00 hours in Source 1 would unnecessarily boost the pressure in Node 11 up to 30.01 mwc. Alternatively, an extra pump in Source 2 would give an even higher pressure of 31.59 mwc in Node 11.
- Constantly low velocities in a number of pipes suggest that these should be reduced in diameter unless additional capacity is required for the network operation in irregular situations.

Comments:

- Scheduling of pumps results in a minimum pressure that does not necessarily occur during the maximum consumption hour.
- The minimum pressure criterion should be maintained throughout the entire day. Much higher pressures during the low demand hours are not justified; they cause increased waste of pumping energy and water.
- Choosing the same model of pump for all the installed units has the advantage that the various units can implement the same pumping schedule on different days, which loads them more evenly. The stand-by unit can then also be employed. One possible adaptation of the regime is shown in Table 3.17.

Table 3.17 Alternative A: pumping schedule for adapted manual operation (S30A3)

Source 1	Schedule from 0–24	Source 2	Schedule from 0–24
1001	000000011111111111111111	2001	000000011111111111111111
1002	111111100000011111000000	2002	111111100000011111000000
1003	110000111111110001111111	2003	000000111111111111111110
1004	000000011111110000111110	2004	000000011111110000111110
1005	000000000111000000011110	2005	000000000111000000011110
1006	111111100001000000011100	2006	111111100001000000011100
1007	000000011111000000111111	2007	000000011111000000111111

In the automatic mode of operation, the pumps are controlled based on the reference pressure or the water level somewhere in the system. The usual monitoring point is at the discharge of the pumping station. In this exercise, Node 11, as the most critical pressure-wise, has been chosen. After a number of trials, the suggested pump control looks as follows (file S30A4.NET):

Link	ID	Setting	Condition
LINK	1003	OPEN	IF NODE 11 BELOW 37
LINK	1003	CLOSED	IF NODE 11 ABOVE 47
LINK	1004	OPEN	IF NODE 11 BELOW 32
LINK	1004	CLOSED	IF NODE 11 ABOVE 42
LINK	1005	OPEN	IF NODE 11 BELOW 27
LINK	1005	CLOSED	IF NODE 11 ABOVE 37
LINK	1006	OPEN	IF NODE 11 BELOW 22
LINK	1006	CLOSED	IF NODE 11 ABOVE 32
LINK	2003	OPEN	IF NODE 11 BELOW 37
LINK	2003	CLOSED	IF NODE 11 ABOVE 47
LINK	2004	OPEN	IF NODE 11 BELOW 32
LINK	2004	CLOSED	IF NODE 11 ABOVE 42
LINK	2005	OPEN	IF NODE 11 BELOW 27
LINK	2005	CLOSED	IF NODE 11 ABOVE 37
LINK	2006	OPEN	IF NODE 11 BELOW 22
LINK	2006	CLOSED	IF NODE 11 ABOVE 32

Pumps 1001, 1002, 2001 and 2002 are assumed to be constantly in operation. The simulation starts with three pumps switched at each source. This control regime will result in the pump operation as shown in Table 3.18.

For this schedule, the pressure and velocity range in the system are shown in figures 3.40 and 3.41, and the supply ratio in Figure 3.42.

Table 3.18 Alternative A: pumping schedule for automatic operation (S30A4)

Source 1	Schedule from 0–24	Source 2	Schedule from 0–24
1001	IIIIIIIIIIIIIIIIIIIIIIIIII	2001	IIIIIIIIIIIIIIIIIIIIIIIIII
1002	IIIIIIIIIIIIIIIIIIIIIIIIII	2002	IIIIIIIIIIIIIIIIIIIIIIIIII
1003	IIIIIIIIIIIIIIIIIIIIIIIIII	2003	IIIIIIIIIIIIIIIIIIIIIIIIII
1004	0000001IIIIIIIIII0IIIIIII0	2004	0000001IIIIIIIIIIIII0IIIIIII0
1005	00000001IIIIII00000IIIII0	2005	00000001IIIIII00000IIIII0
1006	00000000001II000000II1100	2006	00000000001II000000II1100
1007	0000000000000000000000000	2007	0000000000000000000000000

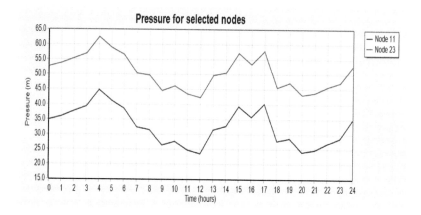

Figure 3.40 S30A4: pressure range in the network; automatic pump control (schedule in Table 3.18)

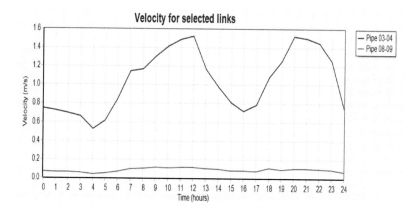

Figure 3.41 S30A4: velocity range in the network; automatic pump control (schedule in Table 3.18)

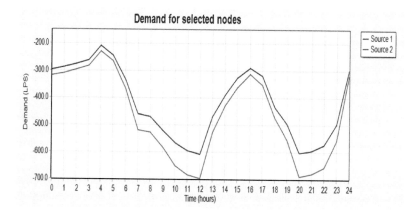

Figure 3.42 S30A4: supply from the sources; automatic pumps control (schedule in Table 3.18)

Clarification:
- The same control node (11) has been selected for both pumping stations for reasons of simplification, which yields the same schedule for both of them.

Conclusions:
- For specified controls, the model reacts by maintaining units 1001–1003 and 2001–2003 permanently switched on, while the other pumps are used during the peak demand hours only. To load all pumps evenly, groups 1004–1006 and 2004–2006 could take over the 24-hour operation every second day.
- A larger pressure variation is registered in the nodes over 24 hours than in the case of the manual operation.
- No significant change of the velocities is registered compared to the manual mode of operation.
- The supply ratio is nearly 50–50, which is preferable.

Comments:
- Modelling of the automatic pump operation yields the schedule as a result (as in Table 3.18), while the modelling of the manual pump operation requires such a schedule as an input (tables 3.16 and 3.17).

The selection of proper control pressures/levels may pose a problem. The following example shows how the system becomes sensitive to even slight modification of the control pressures:

Link	ID	Setting	Condition
LINK	1003	OPEN	IF NODE 11 BELOW 37
LINK	1003	CLOSED	IF NODE 11 ABOVE 42
LINK	1004	OPEN	IF NODE 11 BELOW 32
LINK	1004	CLOSED	IF NODE 11 ABOVE 37
LINK	1005	OPEN	IF NODE 11 BELOW 27
LINK	1005	CLOSED	IF NODE 11 ABOVE 32
LINK	1006	OPEN	IF NODE 11 BELOW 22
LINK	1006	CLOSED	IF NODE 11 ABOVE 27
LINK	2003	OPEN	IF NODE 11 BELOW 37
LINK	2003	CLOSED	IF NODE 11 ABOVE 42
LINK	2004	OPEN	IF NODE 11 BELOW 32
LINK	2004	CLOSED	IF NODE 11 ABOVE 37
LINK	2005	OPEN	IF NODE 11 BELOW 27
LINK	2005	CLOSED	IF NODE 11 ABOVE 32
LINK	2006	OPEN	IF NODE 11 BELOW 22
LINK	2006	CLOSED	IF NODE 11 ABOVE 27

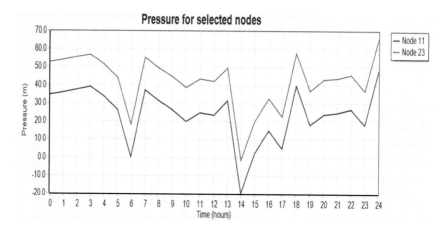

Figure 3.43 S30A5: pressure range in the network; modified settings of automatic pump operation

Figure 3.43 shows the pressures in the system (S30A5.NET).

Clarification:

- The diagram shows different pressure at 0 and 24 hours although the demand level is the same. The reason is the operation at 0 hours that is defined through the initial status of the pumps, while the operation at 24 hours depends on the specified control of pumps.

Conclusions:

- Unlike in S30A4, there is no overlap of the On and Off settings in the [CONTROLS] section of S30A5. This is the likely cause of the unstable operation and the low pressure in the system.

Simulations of the automatic pump operation may also show different results depending on the selected time step of the calculation. With the same pump controls as in S30A5, the hydraulic time step and the reporting time step have been reduced from one hour to 30 minutes (the browser option **Data>>Options>>Times** (click on **Hydraulic/Reporting Time Step**). Consequently, EPANET has applied linear interpolation of the hourly peak factors creating 48 values, each for every 30 minutes, which has influenced the automatic operation of the pumps. A refined pressure in the system is indicated in Figure 3.44 (file S30A6.NET).

If the **Pattern Time Step** is also reduced to 30 minutes, EPANET will then attribute the existing hourly peak factor values to the hydraulic time steps of 30 minutes consecutively. A manual input of 48 peak-factor values is needed in this case; otherwise, the available set of 24 hourly values will be utilised within the first 12 hours of the simulation run and therefore repeated twice causing wrong results (EPANET does not send any warning message in this situation!). Hence, a consistent choice of the time steps and the corresponding set of peak factors are needed in general. Depending on the mode of network operation, more accurate results can be occasionally obtained by choosing shorter time steps, which obviously requires longer simulation runs.

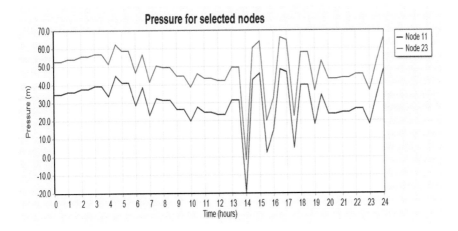

Figure 3.44 S30A6: pressure range in the network; automatic pump operation with time steps of 30 minutes

Alternative B – pumping and balancing storage

The simulation starts with the same system components as listed in Section 3.3.6 (input file S30B2.NET). As in the case of Alternative A, the pump duty head of 50 mwc has been adopted in Source 1. After a number of trial simulations, the tank D = 50 m with a depth of 6.5 m, slightly above the original estimate, has been elevated to 22 m (50 msl). Furthermore:

- The pump suction water levels in 101 and 201 have been set at ground level (modelled as reservoirs with a **Total Head** of 10.7 msl and 25.0 msl, respectively).
- The minimum depth of the elevated tank in 301 is 2.5 m and at the beginning of simulation the depth is estimated to be 3 m.
- The pumping stations are designed with the following units:

[PUMPS]

ID	Suction Node	Pressure Node	Duty Head (mwc)	Duty Flow (l/s)		
1001	101	102	50	150	; PST1	- unit 1
1002	101	102	50	150	;	- unit 2
1003	101	102	50	150	;	- unit 3
1004	101	102	50	150	;	- unit 4
1005	101	102	50	150	;	- unit 5
1006	101	102	50	150	;	- unit 6
1007	101	102	50	150	; PST1	- stand-by
2001	201	202	40	130	; PST2	- unit 1
2002	201	202	40	130	;	- unit 2
2003	201	202	40	130	; PST2	- stand-by

All the pumps in both sources, except the stand-by pumps, are switched on and operate for 24 hours at a more or less constant regime i.e. average flow; the demand variation is balanced from the tank. Consequently, no control commands are required for the pumps in this mode of operation. The pressure range in the system is shown in Figure 3.45 and the tank water depth variation in Figure 3.46.

Conclusions:

• The system operation is stable. The minimum pressure of 20.80 mwc appears in Node 11 at 12:00 hours. Switching a few pumps off during the periods of high pressure (low demand) can further reduce the pressure. For example, operating four instead of six pumps in Source 1 until 06:00 hours will have

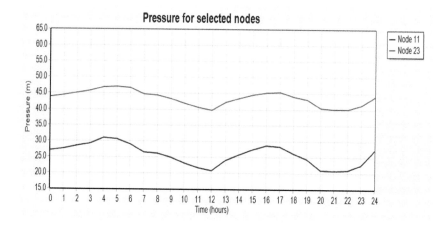

Figure 3.45 S30B2: pressure range in the network; 22 m-high balancing tank, D = 50 m, H = 6.5 m

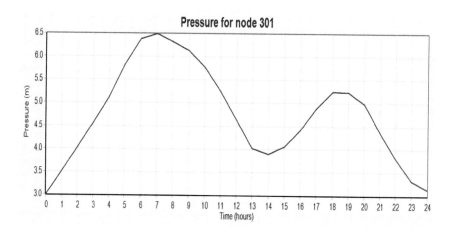

Figure 3.46 S30B2: variation of the tank depth; 22 m-high balancing tank, D = 50 m, H = 6.5 m

no significant implication on the patterns in the above two figures except that the pressure will slightly drop (in Node 11 at 12:00 hours, $p/\rho g = 20.47$ mwc).

A further test is done with smaller pumps and an elevated tank of 20 m (S30B3.NET):

[PUMPS]

ID	Suction Node	Pressure Node	Duty Head (mwc)	Duty Flow (l/s)
1001	101	102	50	130 ; 150->130
1002	101	102	50	130
1003	101	102	50	130
1004	101	102	50	130
1005	101	102	50	130
1006	101	102	50	130
1007	101	102	50	130
2001	201	202	40	120
2002	201	202	40	120
2003	201	202	40	120

The results are shown in figures 3.47 and 3.48.

Conclusions:
- The system is stable although the pressure drops slightly below 20 mwc. The most critical value is 18.53 mwc in Node 11 at 12:00 hours.

Comments:
- For the given network layout, efficient operation of the balancing tanks is reached as a result of matching the pump operation with the tank elevation and its size. While doing this, the following three cases are possible:

CASE I – THE TANK RECEIVES TOO MUCH WATER

This situation is simulated in the file S30B3–1.NET by setting the balancing volume 10 m lower, i.e. from 20 to 10 m height. The water depth variation of the tank is shown in Figure 3.49. It is clear from the figure that the tank will overflow by the next day already if the same operation is maintained. The following design options are suggested provided sufficient pressure can be maintained in the system:

- the tank bottom should be raised to a higher elevation, and/or
- smaller pumps should be used at the sources, and/or
- some pumps should be switched off in periods of low demand.

CASE 2 – THE TANK LOSES TOO MUCH WATER.

This is simulated in S30B3–2.NET by setting the balancing volume at 30 m high, instead of 20 m. The water depth variation in the tank resulting from this action is shown in Figure 3.50. This case contrasts with the above.

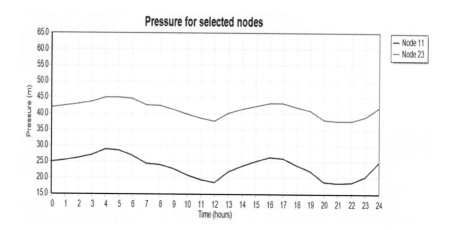

Figure 3.47 S30B3: smaller pumps and 20 m-high tank, D = 50 m, H = 6.5 m

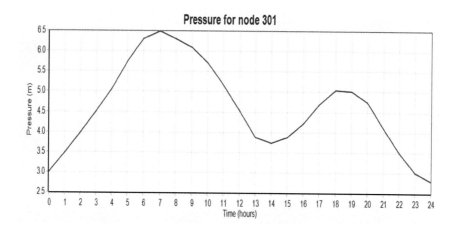

Figure 3.48 S30B3: smaller pumps and 20 m-high tank, D = 50 m, H = 6.5 m

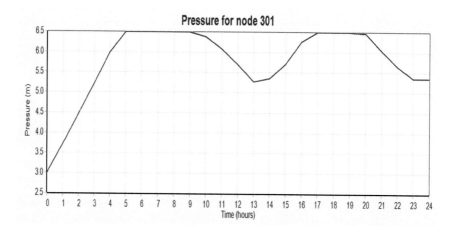

Figure 3.49 S30B3: 10 m-high balancing tank, D = 50 m, H = 6.5 m (Case 1)

CASE 3 – THE TANK IS BALANCING TOO QUICKLY.

To demonstrate this, the volume of the tank has been reduced from D = 50 m to D = 25 m, keeping the elevation at H = 20 m. The results are shown in Figure 3.51 (S30B3–3.NET). The remedies are:

- to increase the tank volume i.e. the cross-section area (diameter) or/and the available depth, and/or
- to adjust the pumping schedule; if the balancing volume has been reduced, the pumps can no longer operate at constant (average) flow.

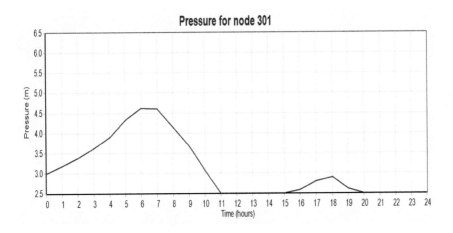

Figure 3.50 S30B3: 30 m-high balancing tank, D = 50 m, H = 6.5 m (Case 2)

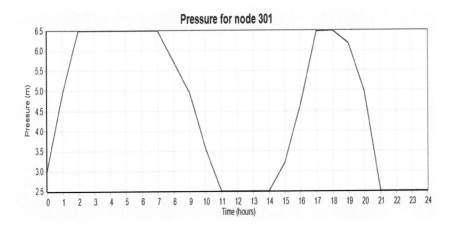

Figure 3.51 S30B3: 20 m-high balancing tank, D = 25 m, H = 6.5m (Case 3)

When the elevated tank is reduced in volume, it loses the demand balancing function and becomes in fact a water tower.

Water towers provide stable pressures in the system and at the same time prevent too frequent switching of the pumps. The pumping schedule in this case becomes similar to those for the direct supply conditions. To satisfy the pumping flow that exceeds Q_{avg} in some periods of the day, stronger pumps are needed than those operated in combination with balancing tanks.

Usually, the pump operation will be controlled automatically depending on the registered water level in the tank. To simulate these conditions, a 20 m-high water tower of diameter 12 m, minimum and maximum depth of 0.2 m and 6.5 m, respectively (making the volume approximately 730 m³), has been simulated in file S30B4.NET. The initial water depth at the beginning of the simulation was set at 1.0 m and the following pump control is suggested by taking the same pumps from file S30B2.NET:

[CONTROLS]

Link	ID	Setting	Condition
LINK	1003	OPEN	IF NODE 301 BELOW 4.0
LINK	1003	CLOSED	IF NODE 301 ABOVE 6.0
LINK	1004	OPEN	IF NODE 301 BELOW 3.0
LINK	1004	CLOSED	IF NODE 301 ABOVE 5.0
LINK	1005	OPEN	IF NODE 301 BELOW 2.0
LINK	1005	CLOSED	IF NODE 301 ABOVE 4.0
LINK	1006	OPEN	IF NODE 301 BELOW 1.0
LINK	1006	CLOSED	IF NODE 301 ABOVE 3.0
LINK	2002	OPEN	IF NODE 301 BELOW 3.0
LINK	2002	CLOSED	IF NODE 301 ABOVE 6.0
LINK	2003	OPEN	IF NODE 301 BELOW 2.0
LINK	2003	CLOSED	IF NODE 301 ABOVE 5.0
LINK	2004	OPEN	IF NODE 301 BELOW 1.0
LINK	2004	CLOSED	IF NODE 301 ABOVE 4.0

Two pumps at Source 1 and one pump at Source 2 are switched on at the beginning of the simulation. Based on the above controls, the pumps will operate further according to the schedule shown in Table 3.19. As a consequence of the control

Table 3.19 Alternative B: pumping schedule for automatic operation controlled from the water tower (S30B4)

Source 1	Schedule from 0–24	Source 2	Schedule from 0–24
1001	IIIIIIIIIIIIIIIIIIIIIIIII	2001	IIIIIIIIIIIIIIIIIIIIIIIII
1002	IIIIIIIIIIIIIIIIIIIIIIIII	2002	I0000001IIIIIIII00IIIIIII
1003	I0III000IIIIIIIII0IIIIIII	2003	I0000000IIIIIII000000IIIII
1004	I0000001IIIIIII0001IIIII0	2004	I00000000001I00000001II0
1005	I00000001011I00000011II0	2005	000000000000000000000000
1006	I000000000011I00000001100		
1007	000000000000000000000000		

regime, two additional units at Source 2 will be needed. The pressure range in the system is shown in Figure 3.52, while the volume variation in the water tower is shown in Figure 3.53.

Conclusions:
- More sudden changes of the water tower depth are caused by the pump operation. This has no serious implications on the pressure in the system.
- The level in the tank at the end of the day does not match the level at the beginning of the day. This is not a problem because the tank does not have a demand-balancing role. In an automatic mode of operation switching the pumps on or off can adjust any extremely low or high level.

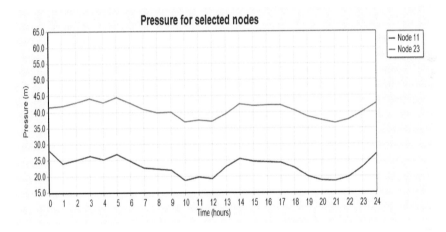

Figure 3.52 S30B4: pressure range in the network; 20 m-high water tower, V = 730 m³

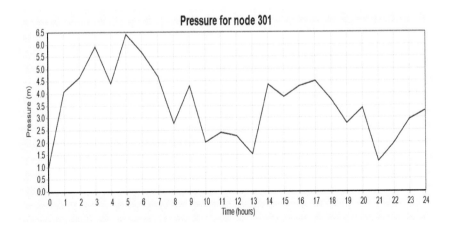

Figure 3.53 S30B4: variation of the tank depth; 20 m-high water tower, V = 730 m³

3.5.2 Factory supply under irregular conditions

One of the design requirements is to provide a reliable supply for the factory. Under normal supply conditions the minimum pressure at that point is ± 27 mwc, in both alternatives. Obviously, the most critical situation will be caused by the pipe burst 01–16, which is rather easy to solve because another pipe in parallel could be added to it. More difficult to predict are the consequences of the failure of other pipes connecting Node 16. Furthermore, the supply requirement during firefighting in the factory has been examined in this section.

In all cases it is assumed that the disaster takes place at the worst moment i.e. during the maximum consumption hour. As a remedy, both pumping stations are set at full capacity, including the stand-by units.

Alternative A – direct pumping

The regular operation of the S30A3 layout is shown in Figure 3.54. The failure of pipes 15–16, 16–24 and 16–17 respectively are shown in figures 3.55–3.57.

Clarification:
*   While simulating a pipe failure, it is assumed that the line is closed i.e. being repaired. This is an acceptable approach bearing in mind that the real quantities of water lost from the system are difficult to predict. Hence, the pipe should be simply disconnected from the rest of the system by changing the **Initial Status** in its property editor from **OPEN** to **CLOSED**, or writing the corresponding command in the Control Editor.

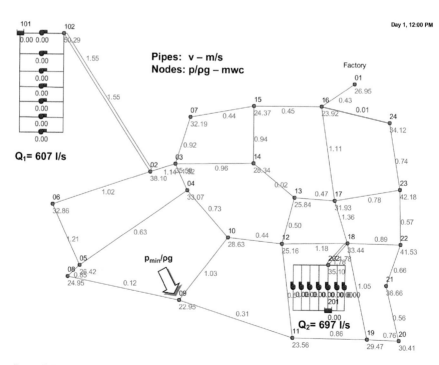

Figure 3.54 S30A3: manual operation at 12:00 hours

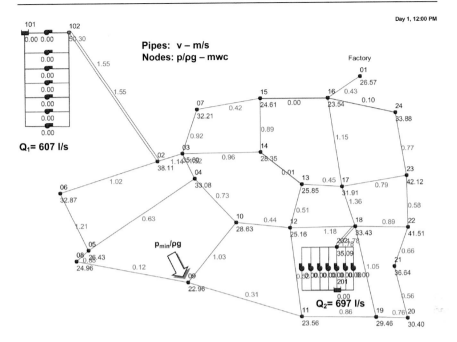

Figure 3.55 S30A3–1: burst 15–16, operation at 12:00 hours

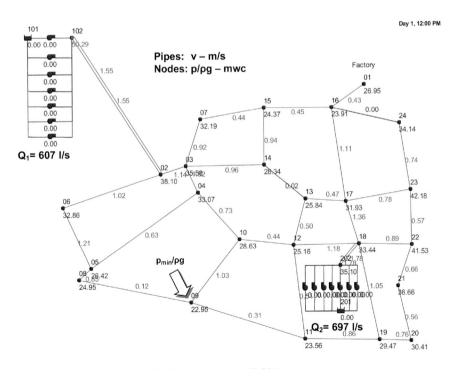

Figure 3.56 S30A3–2: burst 16–24, operation at 12:00 hours

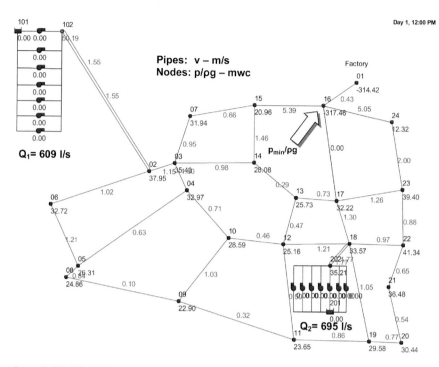

Figure 3.57 S30A3–3: burst 16–17, operation at 12:00 hours

Conclusions:

- There is hardly any effect on pressure in the system as a result of the burst of pipes 15–16 and 16–24. These are small diameters pipes (D = 100 mm) with low flows during regular operation and can be disconnected without affecting the rest of the system.
- Failure of pipe 16–17 is a much more serious problem. This pipe is on the main path from Source 2 and if closed, the water will move to the alternative routes towards the factory. In this case, pipes 15–16–24 become much too small and a severe pressure drop (velocities above 4 m/s!) will be caused. Increasing the pumping head at the sources does not help and enlarging at least one of the pipes will be necessary.

Figure 3.58 shows the solution by replacing the diameter of 15–16 with D = 300 mm. In both pumping stations, the stand-by units are switched on.

Conclusions:

- The pressure in critical points has been restored at a somewhat lower level, however it is still above the required minimum of 20 mwc.

To simulate fire demand at the factory node, an additional quantity of 180 m³/h = 50 l/s has been assigned to Node 01 (total 80.09 l/s) in the S30A3 layout. The results of the calculation are shown in Figure 3.59.

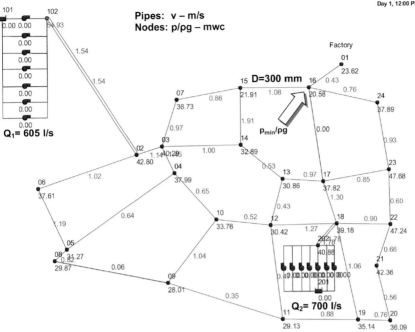

Figure 3.58 S30A3–4: remedy, operation at 12:00 hours

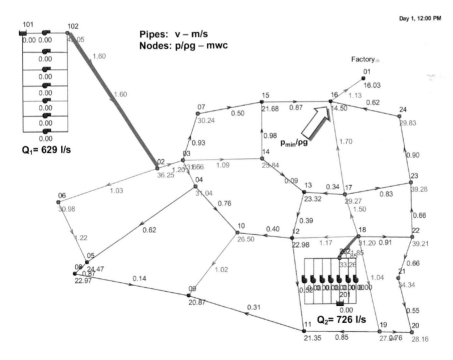

Figure 3.59 S30A3: fire demand at the factory, operation at 12:00 hours

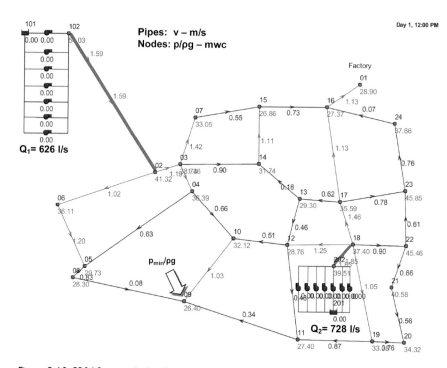

Figure 3.60 S30A3: remedy for fire demand, operation at 12:00 hours

Conclusions:

* As a result of the increased demand in Node 01, the pressure in this node drops to 16 mwc, while a minimum 30 mwc has been stipulated. Except for nodes 01 and 16, the rest of the system is not affected; the bottleneck is again somewhere in the area nearby the factory.

Replacing pipes 07–15–16 with diameter D = 300 mm can solve the problem. An operation with this modification of the system layout is shown in Figure 3.60. The stand-by pumps in both pumping stations are turned on.

Conclusions:

* The pressure in the factory is slightly below the required 30 mwc (28.90 mwc). Further computer runs would show that reaching the minimum of 30 mwc requires an additional pumping unit in both pumping stations. This is considered to be rather an expensive option in relation to the gain in the pressure.

Comments:

* It can be assumed that the fire is limited to a certain number of hours. The fire demand specified in this exercise is slightly exaggerated for educational

purpose i.e. to realise the impact on the rest of the network. Pressure problems during firefighting can also be solved by local measures, such as the operation of district valves, booster installation at the factory, etc. It is normal to assume that consumers in the vicinity of the object will be temporarily affected during firefighting. Nonetheless, the inconvenience caused during a few hours is usually acceptable compared to a large investment in pipes and pumps that would rarely function to the full capacity.

Alternative B – pumping and balancing storage

The simulation of disasters in distribution systems with balancing tanks takes into consideration the tank level at the moment of disaster. Obviously, the balancing pattern of the tank is going to be disrupted, which may affect the pressure in the system.

The pipe burst near the factory has been simulated keeping the 20 m-high balancing tank and manually operated pumps (file S30B3.NET). The regular operation is shown in Figure 3.61.

The same burst scenarios have been analysed as in the case of Alternative A. To consider the level in the tank, the bursts have been assumed to occur during the maximum consumption hour, with the stand-by pumps switched on at the same

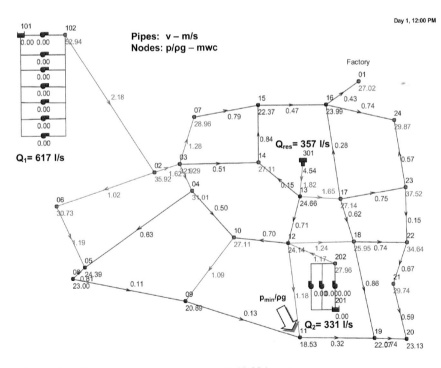

Figure 3.61 S30B3: elevated tank, operation at 12:00 hours

point in time. The sample input format for the control lines to simulate such an operation is as follows:

[CONTROLS]

Link	ID	Setting	Condition
LINK	31	CLOSED	AT TIME 12
LINK	2003	OPEN	AT TIME 12

Clarification:
- Figures 3.62, 3.64 and 3.66 show the effect of the burst that happened at 12:00 hours, while figures 3.63, 3.65 and 3.67 show the situation after switching on the stand-by pump at the second source, also at 12:00 hours.

Conclusions:
- In none of the cases did the burst cause severe pressure problems. Pipes 15–16–17 already belong to the secondary mains loop and further enlargement is therefore not required. Due to sufficient reserve capacity in the system, the stand-by pump at Source 2 is able to easily restore the minor drop of pressure.

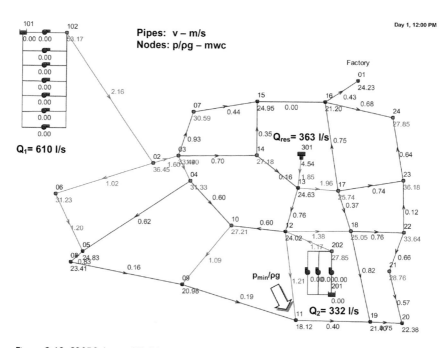

Figure 3.62 S30B3: burst 15–16, operation at 12:00 hours

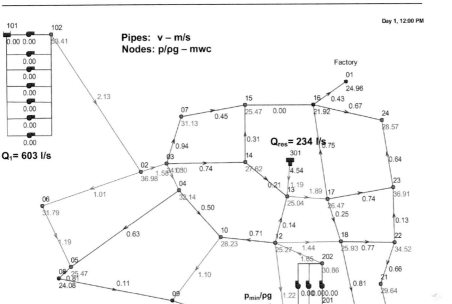

Figure 3.63 S30B3: remedy, operation at 12:00 hours

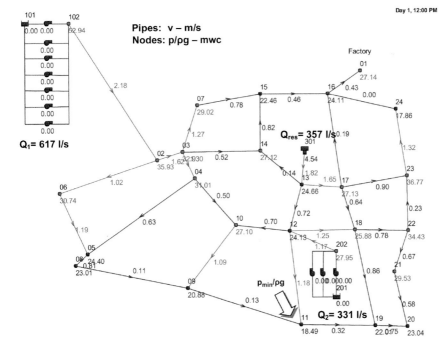

Figure 3.64 S30B3: burst 16–24, operation at 12:00 hours

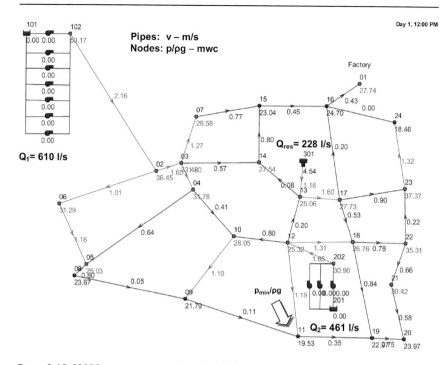

Figure 3.65 S30B3: remedy, operation at 12:00 hours

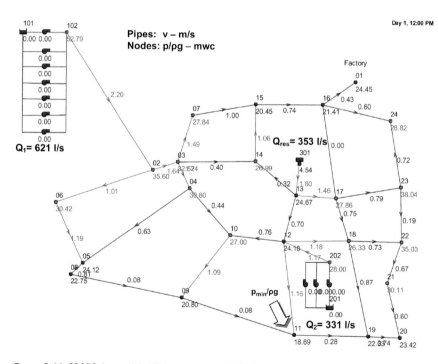

Figure 3.66 S30B3: burst 16–17, operation at 12:00 hours

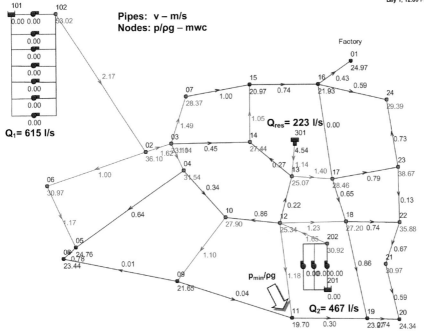

Figure 3.67 S30B3: remedy, operation at 12:00 hours

Fire demand at the factory is analysed in a similar way as the pipe burst, assuming that the fire breaks out exactly during the maximum consumption hour. To achieve this condition in the model, the demand pattern of the factory has been adjusted to simulate the disaster at 12:00 hours (the peak factor has been changed from 1.0 to 2.66, which increases the regular factory demand from approximately 30 to 80 l/s). The simulation results are shown in Figure 3.68.

Conclusions:
- No serious loss of pressure will be observed in the system. However, the pressure in the factory itself drops significantly below the required 30 mwc to 23.98 mwc; thus some remedying is necessary. In the first place the pipes 01–16 and 13–17 should be enlarged from 300 to 400 and 400 to 500 mm, respectively. In particular, the second pipe was generating a large velocity (head loss) and this measure was meant to reduce its resistance. In addition, two extra pumps should be added to the pumping station in Source 2 (total 4 + 1). This remedy, shown in Figure 3.69, brings the pressure in the factory close to 30 mwc, but is altogether complex and rather expensive.

A better solution would have been reached if, instead of two extra units in Source 2, a parallel pipe of 600 mm had been laid next to 102–02, and the stand-by pump in Source 1 switched on. The pressure would have grown substantially but the maximum supply capacity of Source 1 would have been exceeded in this situation. This solution, shown in Figure 3.69, is therefore unacceptable.

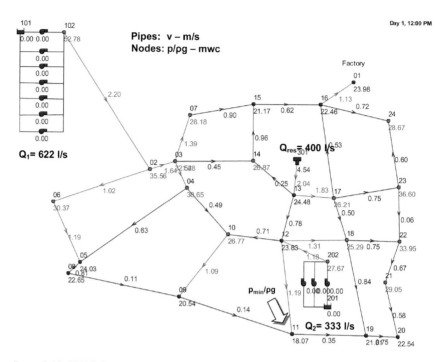

Figure 3.68 S30B3: fire demand, operation at 12:00 hours

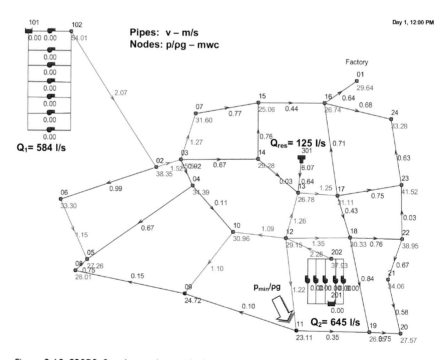

Figure 3.69 S30B3: fire demand remedy 1, operation at 12:00 hours

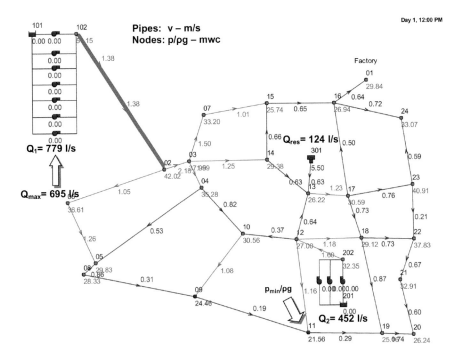

Figure 3.70 S30B3: remedy 2, operation at 12:00 hours

3.5.3   Reliability assessment

Reliability of the two alternative networks has been tested by assuming a single pipe failure and running the simulations at 75% of the demand on the maximum consumption day except for the factory where the demand of 30.09 l/s has remained constant.

Effects of the failure have been analysed for each pipe in the system. The nodes where the pressure drops below 20 mwc during the maximum consumption hour, as a result of the failure, have been identified.

The direct pumping layout from the S30A3 file has been tested, while in the case of Alternative B this was the layout from the S30B3 file. In both cases, the new input files have been created, by dropping the demand (through the demand multiplier) to 75% and switching the stand-by pumps on. The new file names are S30A7 and S30B5, respectively. The calculation results are shown in tables 3.20 and 3.21.

Clarification:
* The tables indicate the nodes with pressure below 20 mwc, the minimum pressure value, node number, and pressure in the factory. Furthermore the supply flows are shown for Source 1, Source 2 and the reservoir (Alternative B).
* The negative value for the reservoir indicates the inflow (refilling).

Table 3.20 Alternative A: reliability assessment (S30A7)

Pipe burst	Nodes with $p/\rho g < 20$ mwc	$p_{min}/\rho g$ (mwc) in node (x)	$p_{fact}/\rho g$ (mwc)	Q_1 (l/s)	Q_2 (l/s)
02–03*	None	39.76 (11)	43.27	440	546
02–06	05, 06, 08	−10.05 (08)	43.43	439	547
03–04	None	34.78 (09)	42.63	415	571
04–05	None	40.22 (11)	43.78	450	536
05–06	05, 08	16.58 (08)	43.55	443	543
05–08	None	27.06 (08)	43.70	447	539
08–09	None	40.20 (11)	43.76	449	537
09–10	09	−37.98 (09)	43.89	452	534
04–10	None	38.06 (09)	43.22	431	555
10–12	None	38.77 (09)	44.10	459	527
09–11	None	40.32 (11)	43.77	449	537
11–12	None	39.19 (11)	43.76	448	538
12–18	None	36.73 (12)	44.05	474	512
11–19	None	22.73 (11)	43.83	454	532
18–19	11, 19, 20	−9.96 (11)	43.86	459	527
19–20	20	19.96 (20)	43.75	449	537
20–21	None	40.13 (11)	43.77	449	537
21–22	21	16.95 (21)	43.79	449	537
18–22	None	34.53 (21)	43.17	451	535
22–23	None	40.22 (11)	43.65	449	537
17–18	None	28.78 (16)	31.82	515	471
12–13	None	40.15 (11)	43.84	449	537
03–14	None	37.33 (15)	41.61	419	567
03–07	07	−393.98 (07)	42.84	443	543
07–15	None	40.20 (11)	43.74	449	537
14–15	15	−83.19 (15)	40.62	450	536
13–14	None	40.28 (11)	43.87	451	535
13–17	None	39.48 (15)	44.55	462	524
17–23	None	26.16 (24)	42.70	448	538
15–16	None	40.20 (11)	43.46	449	537
16–17	01, 16	−200.07 (16)	−197.73	451	535
16–24	None	40.21 (11)	43.65	449	537
23–24	24	−13.79 (24)	41.68	449	537
102–02*	None	35.33 (09)	39.46	374	612
202–18*	None	35.89 (11)	39.59	503	483
Pipe burst	Nodes with $p/\rho g < 20$ mwc	$p_{min}/\rho g$ (mwc) in node (x)	$p_{fact}/\rho g$ (mwc)	Q_1 (l/s)	Q_2 (l/s)

* Assumes the burst of one of the pipes in parallel.

Table 3.21 Alternative B: reliability assessment (S30B5)

Pipe burst	Nodes with $p/\rho g < 20$ mwc	$p_{min}/\rho g$ (mwc) in node (x)	$p_{fact}/\rho g$ (mwc)	Q_1 (l/s)	Q_2 (l/s)	Q_r (l/s)
02–03	None	23.02 (11)	28.31	175	456	356
02–06	05, 06, 08	−54.44 (08)	31.80	573	441	−27
03–04	None	24.29 (11)	32.00	551	448	−12
04–05	None	25.31 (11)	31.81	589	437	−40
05–06	05, 08	−11.29 (08)	31.80	578	439	−31
05–08	08	−82.22 (08)	31.81	585	438	−37
08–09	None	25.28 (11)	31.80	588	437	−39
09–10	09	−261.73 (09)	31.72	592	435	−41
04–10	None	24.67 (11)	31.87	567	444	−25
10–12	None	25.45 (11)	31.74	601	431	−46
09–11	None	25.21 (11)	31.81	588	437	−39
11–12	11	7.21 (11)	31.43	588	433	−35
12–18	None	22.72 (19)	29.96	585	422	−20
11–19	None	26.24 (11)	31.75	588	436	−38
18–19	19, 20	14.68 (19)	32.04	589	439	−42
19–20	20	3.61 (20)	31.79	588	437	−39
20–21	None	25.20 (11)	31.81	588	437	−39
21–22	21	0.08 (21)	31.84	588	437	−39
18–22	20, 21, 22	−24.24 (21)	31.82	588	437	−39
22–23	None	25.30 (11)	31.81	588	437	−39
17–18	None	24.99 (11)	31.96	588	439	−41
12–13	None	27.04 (13)	32.56	571	408	8
03–14	None	25.94 (11)	32.21	546	430	11
03–07	None	21.31 (15)	26.85	549	435	2
07–15	None	23.85 (15)	29.03	562	436	−12
14–15	None	25.23 (11)	31.61	589	437	−40
13–14	None	25.73 (11)	32.25	566	433	−13
13–17	None	23.36 (11)	27.30	603	446	−63
17–23	23, 24	8.79 (23)	31.68	589	437	−40
15–16	None	24.95 (11)	29.73	577	438	−28
16–17	None	25.33 (11)	31.04	590	437	−41
16–24	None	25.29 (11)	31.84	588	437	−39
23–24	None	25.31 (11)	31.77	588	437	−39
102–02	05, 08, 09, 15	18.77 (08)	24.64	0	521	465
202–12	None	20.72 (11)	29.83	618	0	368
Pipe burst	Nodes with $p/\rho g < 20$ mwc	$p_{min}/\rho g$ (mwc) in node (x)	$p_{fact}/\rho g$ (mwc)	Q_1 (l/s)	Q_2 (l/s)	Q_r (l/s)

Conclusions:

- Both layouts show a fairly good reliability level. The effects of the pipe bursts are minimal for the majority of the nodes. In a few cases the pressure will be affected in the vicinity of the burst location. The most critical pipes appear to be 02–06, 18–19 and 16–17 in the case of the Alternative A layout and 02–06, 18–19, 17–23 and 102–02 in the case of the Alternative B layout.
- The direct pumping alternative shows higher pressures in general, which gives the impression that this network has more reserve capacity i.e. more room for reduction of the pipe diameters than the other one. This conclusion will be taken into account while deciding the final layout of this alternative.

As a comparison, the initially developed layout with the branched structure of the secondary mains, shown in Figure 3.16 (S6A7), has been tested for reliability as well. The results are shown in Table 3.22.

Table 3.22 Alternative A: reliability assessment for branched secondary mains (S6A7–1)

Pipe burst	Nodes with $p/\rho g$ < 20 mwc	$p_{min}/\rho g$ (mwc) in node (x)	$p_{fact}/\rho g$ (mwc)	Q_1 (l/s)	Q_2 (l/s)
02–03	01, 03, 04, 07, 09–20	2.95 (15)	15.81	106	413
02–06	None	22.59 (08)	41.61	274	245
03–04	None	26.57 (11)	33.87	211	308
04–05	None	35.88 (11)	41.93	277	242
05–06	None	32.09 (08)	41.72	275	244
05–08	08	−32.16 (08)	41.81	276	243
08–09	None	35.82 (11)	41.88	277	242
09–10	09	−154.27 (09)	41.94	277	242
04–10	None	28.00 (11)	35.08	220	299
10–12	None	34.12 (11)	40.28	262	257
09–11	None	35.66 (11)	41.87	277	242
11–12	None	25.99 (11)	41.83	276	243
12–18	None	31.92 (13)	43.78	299	220
11–19	None	33.59 (11)	41.90	277	242
18–19	19	19.11 (19)	41.95	278	241
19–20	20	19.22 (20)	41.82	277	242
20–21	None	35.87 (11)	41.87	277	242
21–22	None	35.74 (11)	41.88	277	242
18–22	None	35.66 (11)	41.63	277	242
22–23	None	35.82 (11)	41.82	277	242
17–18	01, 16, 17, 23, 24	−172.42 (16)	−172.39	284	235
12–13	None	24.08 (08)	41.67	277	242
03–14	None	29.21 (15)	41.29	274	245
03–07	None	26.72 (15)	41.31	275	244
07–15	None	34.88 (15)	41.60	276	243

Pipe burst	Nodes with $p/\rho g < 20$ mwc	$p_{min}/\rho g$ (mwc) in node (x)	$p_{fact}/\rho g$ (mwc)	Q_1 (l/s)	Q_2 (l/s)
14–15	None	34.85 (15)	41.62	276	243
13–14	None	35.79 (11)	41.86	276	243
13–17	None	35.82 (11)	41.83	277	242
17–23	None	35.80 (11)	41.81	277	242
15–16	None	35.80 (11)	41.77	276	243
16–17	01, 16, 24	−176.49 (16)	−173.46	280	239
16–24	None	26.42 (24)	42.08	277	242
23–24	None	35.81 (11)	41.86	277	242
102–02	01–24	−32.25 (15)	−6.89	0	519
202–18	01–24	−49.86 (11)	−45.29	519	0
Pipe burst	Nodes with $p/\rho g < 20$ mwc	$p_{min}/\rho g$ (mwc) in node (x)	$p_{fact}/\rho g$ (mwc)	Q_1 (l/s)	Q_2 (l/s)

Conclusions:

- Large parts of the network will lose pressure if pipes 02–03, 17–18, 102–02 and 202–18 burst. A critical pipe is also 16–17. If the branched layout of the secondary pipes has to be maintained, laying of the parallels is inevitable in the case of these listed pipes.

3.6   Final layouts

3.6.1   Alternative A – direct pumping

Beginning of the design period

The following adjustments of the layout proposed in Section 3.3.6 have been introduced based on the conclusions regarding the system operation:

- The diameters of a few, mainly peripheral, pipes have been reduced.
- To improve the supply reliability of the factory:
  - pipe 15–16 has been enlarged to D = 300 mm,
  - parallel pipes, D = 300 mm, have been laid along route 01–16.
- In order to maintain the system reliability in general, parallel pipes, D = 500 mm, have been laid along route 202–18.
- Slightly smaller pump units have been proposed for the pumping station in Source 1.

As in the conclusions of Section 3.3.6:

- the second source has been immediately connected to the system, and
- the storage volume has been immediately constructed for the demand at the end of the design period.

Table 3.23 Alternative A: beginning of the design period

	Pumping stations			Storage volume		
	Q_p (l/s)	H_p (mwc)	No. of Units	D (m)	H (m)	V_{tot} (m³)
Source 1	100	50	3+1	50	5.5	10,800
Source 2	90	40	3+1	50	5.5	10,800

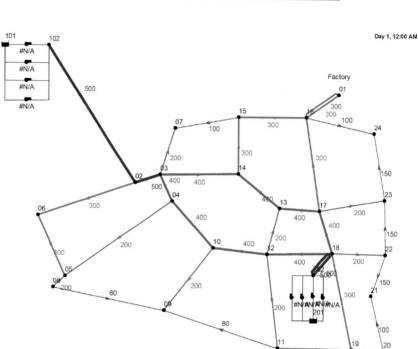

Figure 3.71 Alternative A: pipe diameters (mm) at the beginning of the design period

Table 3.24 Alternative A: pumping schedule for manual operation (S6A9)

Source 1	Schedule from 0–24	Source 2	Schedule from 0–24
1001	IIIIIIIIIIIIIIIIIIIIIIII	2001	IIIIIIIIIIIIIIIIIIIIIIII
1002	00000001IIIIIIIIIIIIIII0	2002	III0001IIIII0001IIIIIIII
1003	00000000001III00000011I0	2003	000000001II00000II100000
1004*	000000000000000000000000	2004*	000000000000000000000000

* Stand-by units.

The main characteristics of the pumping stations and the storage volume are given in Table 3.23.

The final layout of the system at the beginning of the design period is shown in Figure 3.71 (file S6A9.NET). For the manual operation as shown in Table 3.24,

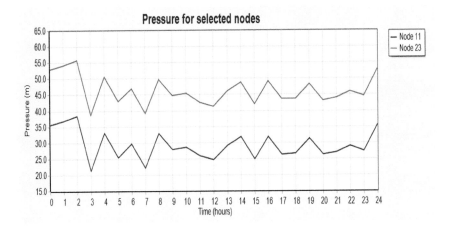

Figure 3.72 S6A9: pressure range in the network; manual pump control (schedule in Table 3.24)

the range of pressures in the system on the maximum consumption day is shown in Figure 3.72.

The registered pressure in the factory during the maximum consumption hour is 30.38 mwc. The following pressures are observed in the case of the pipe burst (the stand-by pumps are 'Off'):

Burst pipe	$p_{fact}/\rho g$ (mwc)
15–16	28.79
16–17	24.73
16–24	30.53
01–16 (single)	30.18

The pressure during the firefighting in the factory is 34.12 mwc (the stand-by pumps = 'On'). The reliability assessment for the 75% demand conditions shows insufficient pressure during the following pipe bursts (the stand-by pumps are 'On'):

Burst pipe	Nodes with $p/\rho g$ < 20 mwc	$p_{min}/\rho g$ (mwc)	$p_{fact}/\rho g$ (mwc)
05–08	08	−31.44	48.51
09–10	09	−144.73	48.55
102–02	09, 13 and 15	17.24	23.63

Except for Node 13, all those indicated are peripheral nodes. The table shows that the effects of the listed pipe bursts are localised to very small parts of the system and the situation can therefore be considered as satisfactory. The pressure in the factory will also not be affected.

End of the design period

The following system extension is planned to satisfy the demand at the end of the design period (file S30A8.NET):

Pipe	D_{old} (mm)	D_{new} (mm)
102–02	500	2×500
02–03	500	2×500
02–06	300	400
09–10	200	300
11–19	200	300
18–19	300	400

Three additional pump units of the same size as the existing ones are planned in each pumping station. For the manual operation as shown in Table 3.25, the range of pressures in the system during the maximum consumption day is shown in Figure 3.73.

Table 3.25 Pumping schedule – Alternative A, manual operation (S30A8)

Source 1	Schedule from 0–24	Source 2	Schedule from 0–24
1001	IIIIIIIIIIIIIIIIIIIIIIII	2001	IIIIIIIIIIIIIIIIIIIIIIII
1002	IIIIIIIIIIIIIIIIIIIIIIII	2002	IIIIIIIIIIIIIIIIIIIIIIII
1003	II00001IIIIIIII000IIIIIII	2003	000000IIIIIIIIIIIIIIIIIII0
1004	00000001IIIIII0000IIIIII0	2004	0000000IIIIII0000IIIII0
1005	000000000IIII000000III1I0	2005	000000000IIII000000III1I0
1006	00000000000II00000000II100	2006	00000000000II00000000II100
1007	000000000000000000000000	2007	000000000000000000000000

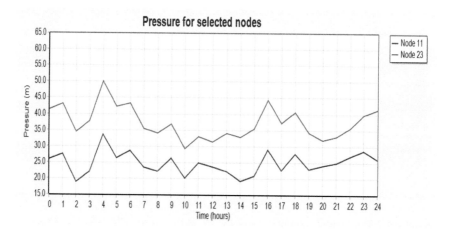

Figure 3.73 S30A8: pressure range in the network; manual pump control (schedule in Table 3.25)

Pressure in the factory during the maximum consumption hour is 26.15 mwc. The following pressures are observed in the case of the pipe burst:

Burst pipe	Stand-by pumps	$p_{fact}/\rho g$ (mwc)
15–16	Off	23.80
16–17	On	19.32
16–24	Off	26.61
01–16 (single)	Off	25.95

The pressure during the firefighting in the factory reaches $p_{fact}/\rho g = 25.23$ mwc, with the stand-by pumps turned on. This pressure is below the required minimum of 30 mwc and local pressure boosting is suggested.

The reliability assessment for the 75% demand conditions shows insufficient pressure during the following pipe bursts (the stand-by pumps are 'On') file S30A8–1.NET:

Burst pipe	Nodes with $p/\rho g$ < 20 mwc	$p_{min}/\rho g$ (mwc)	$p_{fact}/\rho g$ (mwc)
02–06	05, 06 and 08	−54.97	43.25
05–06	05 and 08	−5.56	43.36
05–08	08	−317.41	43.53
09–10	09	−893.27	43.68
18–19	11, 19 and 20	−18.03	43.01
19–20	20	3.20	43.48
21–22	21	17.26	43.62
18–22	21	19.53	42.93
03–07	07	−395.09	41.12
17–23	24	15.50	43.54
23–24	24	−12.71	42.96

Here as well, the pressure drop is localised to a few peripheral nodes with the factory pressure not being affected by the disaster. Hence, the system reliability can be considered as acceptable.

3.6.2 Alternative B – Pumping and balancing storage

Beginning of the design period

The following adjustments of the layout presented in Section 3.3.6 have been introduced: a few, mainly peripheral pipes, have been reduced in diameter.

- To improve the supply reliability of the factory, parallel pipes of D = 300 mm have been laid along route 01–16.
- The balancing tank, D = 40 m, H = 6.5 m, has been adopted. The height of the tank has been set to 20 m.
- Smaller pumping units are proposed in Source 1. One extra unit, out of three planned for the extension, will be installed immediately for more flexible supply in irregular situations.

As in the conclusions of Section 3.3.6, the storage volume at Source 1 has been immediately constructed for the demand at the end of the design period.

The information on the pumps and the storage volume is given in Table 3.26.

The final layout of the system at the beginning of the design period is shown in Figure 3.74 (file S6B4.NET).

For constant average operation of the pumping station, the range of pressures in the system on the maximum consumption day is shown in Figure 3.75 and the volume variation of the balancing tank in Figure 3.76.

Pressure in the factory during the maximum consumption hour is 29.69 mwc. The following pressures are observed in the case of the pipe burst (the stand-by pumps are turned off):

Burst pipe	$p_{fact}/\rho g$ (mwc)
15–16	27.17
16–17	29.34
16–24	29.72
01–16 (single)	29.50

Table 3.26 Alternative B: beginning of the design period

	Pumping stations			Storage volume		
	Q_p (l/s)	H_p (mwc)	No. of Units	D (m)	H (m)	V_{tot} (m³)
Source 1	130	50	4+1	30	6.5	4600
20 m-high tank	–	–	–	40	6.5	8170

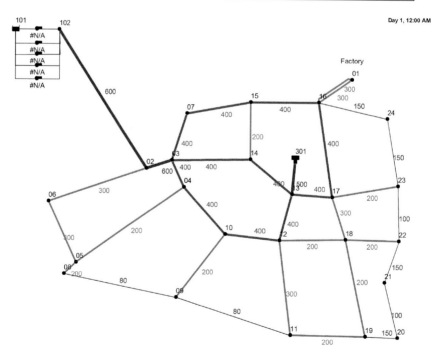

Figure 3.74 S6B4: pipe diameters (mm) at the beginning of the design period

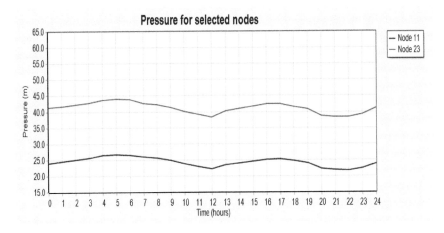

Figure 3.75 S6B4: pressure range in the network; 20 m-high balancing tank, D = 40 m, H = 6.5 m

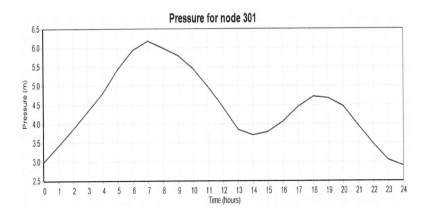

Figure 3.76 S6B4: variation of the tank depth; 20 m-high balancing tank, D = 40 m, H = 6.5 m.

The pressure during the firefighting in the factory is $p_{fact}/\rho g$ = 28.59 mwc (the stand-by pumps are 'On').

The reliability assessment for the 75% demand conditions shows insufficient pressure during the following pipe bursts (the stand-by pumps are turned on):

Burst pipe	Nodes with $p/\rho g$ < 20 mwc	$p_{min}/\rho g$ (mwc)	$p_{fact}/\rho g$ (mwc)
02–06	05, 06 and 08	13.19	32.80
05–08	08	−47.94	32.78
09–10	09	−159.89	32.73
11–12	11 and 19	10.34	32.46

End of the design period

Compared to the summary in Section 3.3.6, the extent of renovation towards the end of the design period has been reduced. The following reconstruction is planned (file S30B6.NET):

Pipe	D_{old} (mm)	D_{new} (mm)
02–06	300	400
09–10	200	300
11–12	300	400
12–18	200	400
18–19	200	300

- two additional pump units of the same type as the existing ones are going to be installed in the pumping station of Source 1,
- the pumping station in Source 2 is going to be built with three units of the following duty flow and head: $Q_d = 120$ l/s, $H_d = 40$ mwc,
- the clear water reservoir, $V = 4600$ m³, is going to be constructed in Source 2,
- the diameter of the balancing tank is to be increased from $D = 40$ to 50 m, which effectively means the construction of an additional compartment, $D = 30$ m, besides the existing one. The total additional volume is accidentally equal to those of the clear water reservoirs in the two sources ($V = 4600$ m³).

For the average flow supply from the pumping stations, the system operation on the maximum consumption day is shown in figures 3.77 and 3.78.

Pressure in the factory during the maximum consumption hour is 27.28 mwc. The following pressures are observed in the case of the pipe burst (the stand-by pumps are 'Off'):

Burst pipe	$p_{fact}/\rho g$ (mwc)
15–16	24.43
16–17	24.01
16–24	27.49
01–16 (single)	27.09

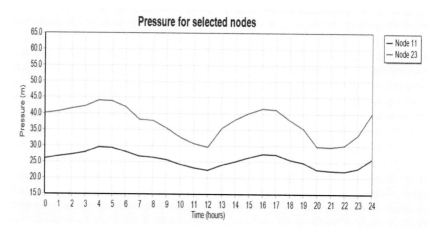

Figure 3.77 S30B6: pressure range in the network; 20 m-high balancing tank, D = 50 m, H = 6.5 m

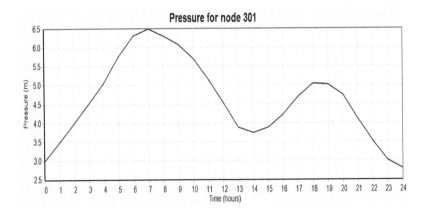

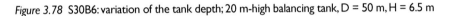

Figure 3.78 S30B6: variation of the tank depth; 20 m-high balancing tank, D = 50 m, H = 6.5 m

During the firefighting in the factory, the pressure on the maximum consumption hour is $p_{fact}/\rho g = 26.28$ mwc (stand-by pumps are 'On'). This pressure is below the required minimum of 30 mwc and local pressure boosting is advised.

The reliability assessment for the 75% demand conditions (stand-by pumps are 'On') shows insufficient pressure during the following pipe bursts:

Burst pipe	Nodes with $p/\rho g < 20$ mwc	$p_{min}/\rho g$ (mwc)	$p_{fact}/\rho g$ (mwc)
06	05, 06 and 08	−66.86	32.06
05–06	05 and 08	−17.45	32.06
05–08	08	−329.39	32.06
09–10	09	−904.88	32.03
11–12	11	4.89	31.59
18–19	19	18.72	32.23
19–20	20 and 21	−11.42	32.02
21–22	21	1.04	32.07
18–22	20, 21 and 22	−28.40	32.02
17–23	23 and 24	6.00	31.93
102–02	05, 08, 09 and 15	18.57	24.77

3.6.3   Phased development

Running the simulation of the layouts proposed at the beginning of the design period for the demand conditions at the end of the design period suggests the point in time when the system extension should take place.

Alternative A:
*   With the stand-by pumps switched on, the S6A9 layout will show a considerable drop of pressure for the demands above ± 850 l/s (3060 m³/h). This demand corresponds to the maximum consumption hour demand around year 15–16 (see Table 3.8). Hence, the proposed year of the system extension

is year 15. As an additional safeguard, five (instead of four) pumps will be installed in both pumping stations at the beginning of the design period, to facilitate the pressure until the extension will take place. The two remaining units are left for the second phase.

Alternative B:

• In the case of layout S6B4, the pressure problems start for the demands around ± 950 l/s (3420 m³/h), which is the maximum consumption hour demand around year 19–20. The reason for this is the reservoir itself; it facilitates the pressure as long as it is not empty. To prevent this, the final decision has been made to build the pumping station with the full number of seven units already at the beginning of the design period. This should help to maintain the balancing role of the reservoir in the years before the extension will take place. The proposed year of the system extension is year 19.

Tables 3.27 and 3.28 show the summary of the phased development of the system for the two alternatives.

3.6.4   Cost analyses

Based on the prices listed in Tables 3.3 and 3.4, the following cost calculations of the proposed layouts have been conducted.

Table 3.27 Alternative A: phased development

Pipe diameter	Beginning of the design period	Construction at year 15	End of the design period
	Total length (m)		
D = 80 mm	2700	–	2700
D = 100 mm	2200	–	2200
D = 150 mm	2550	–	2550
D = 200 mm	6850	–	6850
D = 300 mm	5950	1650	7600
D = 400 mm	4800	2100	6900
D = 500 mm	3500	2000	5500
D = 600 mm	–	–	–
Total pipes (m)	28,550	5750	34,300
Pumping stations	Number of units/Q_p (l/s)/H_p (mwc)		
Source 1	5 / 100 / 50	2 / 100 / 50	7 / 100 / 50
Source 2	5 / 90 / 40	2 / 90 / 40	7 / 90 / 40
Storage	Volume (m³)		
Source 1	10,800	–	10,800
Source 2	10,800	–	10,800
Total volume (m³)	21,600	–	21,600

Table 3.28 Alternative B: phased development

Pipe diameter	Beginning of the design period	Construction at year 19	End of the design period
	Total length (m)		
D = 80 mm	2700	–	2700
D = 100 mm	1200	–	1200
D = 150 mm	2750	–	2750
D = 200 mm	7250	–	7250
D = 300 mm	4050	2050	6100
D = 400 mm	7100	2950	10,050
D = 500 mm	100	–	100
D = 600 mm	2000	–	2000
Total pipes (m)	27,150	5000	32,150
Pumping stations	Number of units/Q_p (l/s)/H_p (mwc)		
Source 1	7 / 130 / 50	–	7 / 130 / 50
Source 2	–	3 / 120 / 40	3 / 120 / 40
Storage	Volume (m³)		
Source 1	4600	–	4600
Source 2	–	4600	4600
Water tower	8170	4600	12,770
Total volume (m³)	12,770	9200	21,970

Investment costs

The investment costs of the final layouts are summarised in Table 3.29 for Alternative A and Table 3.30 for Alternative B.

Clarification:
*   The installed capacity of the pumping stations has been assumed as $Q_{max} = 1.5 \times Q_d$, with all units (including stand-by) in operation. Note that the price is calculated for the capacities converted from l/s into m³/h.
*   Investment price of the water tower includes the cost of the supporting structure.

Assuming that 60 % of the investment needed at the beginning of the design period will be borrowed immediately, 40 % in year 4 and the entire sum for the second phase in the year when the extension will take place, the total values of the investments brought to year 6 are (the loan interest is 8%):

Alternative A:

$0.6 \times 13,485,481/1.08^{-5} + 0.4 \times 13,485,481/1.08^{-2} + 4,026,430/1.08^9 = $ EUR 20,194,761

The repayment of the loan starts at year 6. The annual instalments to be paid are:

$20,194,761 \times (0.08 \times 1.08^{25})/(1.08^{25}-1) = $ EUR 1,891,821

Table 3.29 Alternative A: investment costs, rounded in Euros (EUR)

Pipe diameter	Beginning of the design period		Investment at year 15	
	Total length (m)	Total price (EUR)	Total length (m)	Total price (EUR)
D = 80 mm	2700	162,000	–	–
D = 100 mm	2200	154,000	–	–
D = 150 mm	2550	229,500	–	–
D = 200 mm	6850	890,500	–	–
D = 300 mm	5950	1,071,000	1650	297,000
D = 400 mm	4800	1,248,000	2100	546,000
D = 500 mm	3500	1,085,000	2000	620,000
D = 600 mm	–	–	–	–
Total pipes	28,550	4,840,000	5750	1,463,000
Pumping stations	Installed capacity (m³/h)	Total price (EUR)	Installed capacity (m³/h)	Total price (EUR)
Source 1	5 × 540 = 2700	2,780,104	2 × 540 = 1080	1,335,700
Source 2	5 × 486 = 2430	2,555,377	2 × 486 = 972	1,227,730
Total PST	5130	5,335,481	2052	2,563,430
Reservoirs	Total volume (m³)	Total price (EUR)	Total volume (m³)	Total price (EUR)
Source 1	10,800	1,655,000	–	–
Source 2	10,800	1,655,000	–	–
Total reservoirs	21,600	3,310,000	–	–
Total A	–	13,485,481	–	4,026,430

Table 3.30 Alternative B: investment costs, rounded in Euros (EUR)

Pipe diameter	Beginning of the design period		Investment at year 19	
	Total length (m)	Total price (EUR)	Total length (m)	Total price (EUR)
D = 80 mm	2700	162,000	–	–
D = 100 mm	1200	84,000	–	–
D = 150 mm	2750	247,500	–	–
D = 200 mm	7250	942,500	–	–
D = 300 mm	4050	729,000	2050	369,000
D = 400 mm	7100	1,846,000	2950	767,000
D = 500 mm	100	31,000	–	–
D = 600 mm	2000	720,000	–	–
Total pipes	27,150	4,762,000	5000	1,136,000
Pumping stations	Installed capacity (m³/h)	Total price (EUR)	Installed capacity (m³/h)	Total price (EUR)
Source 1	7 × 702 = 4914	4,488,675	–	–
Source 2	–	–	3 × 648 = 1944	2,137,603
Total PST	4914	4,488,675	1944	2,137,603
Reservoirs	Total volume (m³)	Total price (EUR)	Total volume (m³)	Total price (EUR)
Source 1	4600	725,000	–	–
Source 2	–	–	4600	725,000
Water tower	8170	1,750,700	4600	1,001,000
Total reservoir	12,770	2,475,700	9200	1,726,000
Total B	–	11,726,375	–	4,999,603

Alternative B:

$0.6 \times 1\,1,726,375/1.08^{-5} + 0.4 \times 11,726,375/1.08^{-2} + 4,999,603/1.08^{13} = $ EUR 17,647,336

The annual instalments to be paid are:

$17,647,336 \times (0.08 \times 1.08^{25})/(1.08^{25} - 1) = $ EUR 1,653,381

Conclusions:
* Alternative B costs less than A. The reason is the lower initial investment and later start of the second phase.
* The cost of the elevated balancing tank appears not to be the predominant factor. Alternative A has lower investment costs for the storage but more expensive pumping stations.

Reading:
* See notes, Volume 1, Section 4.1.2: 'Economic aspects'.

Operation and maintenance costs

The operation and maintenance costs are calculated as a percentage of the raw investment costs (Table 3.4). The average value is taken although a trend of increase might be expected towards the end of the design period.

Alternative A:
  Distribution system: $0.005 \times (4,840,000 + 1,463,000) = $ EUR 29,735
  Pumping stations: $0.020 \times (5,335,481 + 2,563,430) = $ EUR 157,978
  Reservoirs: $0.008 \times 3,310,000 = $ EUR 26,480
  TOTAL: EUR 214,193 per year.

Alternative B:
  Distribution system: $0.005 \times (4,762,000 + 1,136,000) = $ EUR 29,490
  Pumping stations: $0.020 \times (4,488,675 + 2,137,603) = $ EUR 132,526
  Reservoirs: $0.008 \times (2,475,700 + 1,726,000) = $ EUR 33,614
  TOTAL: EUR 195,630 per year.

Water price increase

From Table 3.8, the average water demand/production at year 6 is 31,977 m³/d = 11,671,605 m³/y.

At the end of the design period, the annual production equals $61,344 \times 365 = $ 22,390,560 m³/y.

The total quantity of water supplied during the design period equals 406,381,510 m³ or on average 16,255,260 m³ per year.

Due to the loan repayment and the O&M costs, the average water price increase is going to be:

Alternative A:

$(1,891,821 + 214,193)/16,255,260 =$ EUR 0.13 per m³

Alternative B:

$(1,653,381 + 195,630)/16,255,260 =$ EUR 0.11 per m³

3.6.5   Summary and conclusions

<u>Design and construction</u>

The design layout B is less expensive than in the case of Alternative A; it requires lower initial investments and a later extension of the system. The distribution network is purposely left with surplus capacity at the beginning of the design period in order to provide a good supply in the final years before the reconstruction of the system takes place.

Construction of the large elevated tank may prove to be a structural or even an aesthetic problem, bearing in mind the location that is close to the centre of the town. Otherwise, the price does not seem to be a limitation.

<u>Operation and maintenance</u>

Operation of the system B is stable, with mild variations in pressures throughout the day. Moreover it is reasonable to expect that the pumping costs in this scheme of supply will be lower than in Alternative A. The maintenance costs are also lower compared to Alternative A although the price difference is small.

<u>Supply to the factory</u>

Here as well, Alternative B shows better performance than A. The layout provides sufficient pressures in the factory regardless of calamities in the system. The small deficit of pressure during firefighting should be overcome by local measures.

<u>Reliability</u>

Both A and B layouts are fairly reliable in case of failure in the system.

<u>Overall conclusion</u>

Given all the facts, the final choice in this exercise is Alternative B.

Network rehabilitation exercise

4.1 Learning objectives and set-up

This assignment concerns the rehabilitation of an existing water distribution net-work. The main objective of this exercise is to convert a single-source gravity system into a combined system, where the second source and pumping station are to be added to provide a stable hydraulic operation over a period of 20 years. The pressures in the network are also to be controlled by the appropriate location and setting of a pressure-reducing valve (PRV). Compared to the design exercise in Appendix 3, in this exercise more attention is given to the modelling of bal-ancing storage, and to the design of pumping stations as described in Chapter 4 (Volume 1). In addition, water loss is introduced by modelling it as a separate 'consumption' category, separating the pressure-driven pattern from the diurnal patterns of other consumption categories. A single rehabilitation alternative is to be analysed for the present situation and the future demand scenario with the phased development of the network halfway through the design period i.e. in year 10.

When lectured, the exercise contact time consists of four sessions of three hours (excluding the software introduction!), with the following set-up:

Session 1 – Current demand scenario
- Case introduction
- Demand calculation
- Network rehabilitation: sizing of pipe diameters, calculation of tank volume
- Network operation: pressure range, demand balancing, PRV settings.

Session 2 – Future demand scenario
- Calculation of the demand growth, connection of new source
- System redesign by adding pumps and storage; possible relocation of the PRV
- Pump scheduling, pressure range, demand balancing, final PRV settings.

Session 3 – Pumping station design
- Selection of pumps and motors
- Layout of pumping station
- Calculation of hydraulic losses and NPSH.

Session 4 – Cost calculations
- Investment costs for network rehabilitation and upgrade
- Operation and maintenance costs
- Loan repayment through water tariff increase.

The future network operation is feasible because of the additional water source, which must be connected at the point in time when the capacity of the current source is exceeded. To provide a variety of solutions, each student will work with an individual data set combining one of the six different locations of the second source (A to F), with an individual number (IN), which is assigned with values between 1.3 and 1.8, and which is used in the calculation of particular demand.

Here as well, the elaborated rehabilitation alternatives are analysed with the help of EPANET software (Version 2), used throughout the tutorial. For this purpose, a model of the case network has been prepared, which consists of a group of EPANET input files answering particular questions in the exercise. All the file names follow uniform coding: 'NTxy.NET', where:

- NT stands for the name of the town (the raw EPANET file given to the students at the beginning of the work is named NAMETOWN.NET; they are requested to rename it, leading to a unique set of files submitted for evaluation, and eliminating the possibility of misplacing or duplicating any of them),
- x is the number of the question but without the dots; e.g. for Question 3.1.5, x = 315,
- y is the index of specific alternatives considered in the process of answering the question.

The full version of the EPANET software and the case network files mentioned in the text are available in the electronic materals attached to the book. General information relevant for this exercise, which includes the EPANET programme installation and abbreviated instructions for use, is given in Appendix 8. In addition to the network modelling software, the electronic materials include spreadsheet applications and documents prepared to facilitate faster handling of the input data, which also includes a catalogue of fixed speed ETANORM pumps from the German manufacturer KSB; these are available for the final pump selection and design of the pumping station.

As is the case in Appendix 3, the results of calculations are presented in the tables and figures. These may be accompanied by:

- *Clarifications* that explain some (calculation) details from the table/figure,
- *Conclusions* that suggest further steps based on the calculation results,
- *Comments* that elaborate points in a wider context of the problem,
- *Reading* that refers to the chapters of the main text in Volume 1 that are related to the problem.

The output of the work is to be described in a design exercise report that is submitted for evaluation, together with the prescribed set of EPANET input files prepared by the student. Although not compulsory, an oral exam can be organized to further enlighten and defend the selected design alternatives. Based on the average student performance, the estimated study load for this exercise is 56 hours, or two credits in the European Credit Transfer System (ECTS), which includes the contact time.

The solved example presented in this appendix contains partly modified data where IN = 2.0 and the number of consumers connected to the network in the district SUBURBS is 3700. Both of these inputs have higher values than in the students' assignment, which is done in order to amplify the problems and make solutions more obvious, as well as to prevent these being copied by the students. The second source, to mitigate the future demand increase, is located in the district CAMPUS (area D).

Upon completion of the exercise, the student should be able to:

- perform computer-aided hydraulic calculations using EPANET software and
- predict the consequences of demand growth on the hydraulic performance of a particular water transport and distribution system;
- analyse the implications of various operational modes of pumping stations and
- evaluate the investment and operational costs for the selected network layout and supply scheme;
- propose a preliminary design for a water distribution network, including the pumping station and the balancing storage, which can satisfy the prescribed service level in a cost-effective manner.

4.2   Case introduction – Nametown

NAMETOWN is a small university town of approximately *35,000* inhabitants, which is located in the hilly area of NAME valley. A rapid expansion of the area around the town is expected over the coming years because of the growth in the number of students at the university. In addition, a hotel is going to be built in the new residential district of the town (called NEW TOWN).

The existing gravity water distribution network, shown in Figure 4.1, presently covers approximately *32,000* consumers, or 91% of the total population. An additional *3000* inhabitants of the SUBURBS district, who currently use private wells, are also to be connected to the network to avoid the risk of pollution caused by improper sanitation.

The network combines loops and branches and is composed of mixed pipe materials and mixed age: *60*-year-old cast iron (CI), galvanized steel (GS) of between *25* and *50* years of age, and polyethylene (PE) and ductile iron (DI) of between *15* and *25* years. The network rehabilitation and expansion is planned to meet the demand forecast. An international loan has been approved for this purpose.

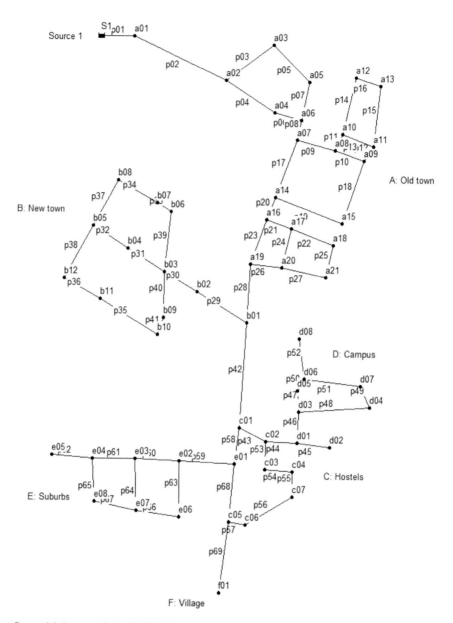

Figure 4.1 Layout of the NAMETOWN gravity network

4.2.1 Topography

The network stretches throughout the valley, in an area with ground elevations ranging between *9.66* and *82.48* meters above sea level (*msl*), which is shown in Figure 4.2. The five districts, A to E, have been developed at different elevations.

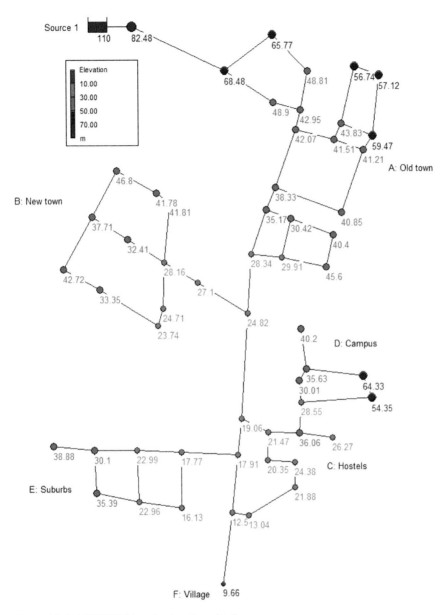

Figure 4.2 NAMETOWN node elevations (msl)

The elevation differences within the district range between *11.88* m, in the case of the HOSTELS district, and up to *54.14* m for OLD TOWN.

4.2.2   Supply source and distribution system

The source of supply, located at *Source 1*, is groundwater stored after simple treatment in a reservoir from where it is supplied by gravity. The fixed water surface level in the reservoir is assumed at *110* msl, as indicated in Figure 4.2. To maintain the groundwater levels, the maximum production capacity of the source is estimated at *90* l/s. To increase the network reliability, the second source can be activated at the location *New Source* (based on the individual data!).

The distribution system is composed of loops and branches with pipe diameters ranging between *25* and *300* mm, as shown in Figure 4.3. The pipe lengths vary between *85* and *727* m, while the absolute roughness coefficient (*k*) has values

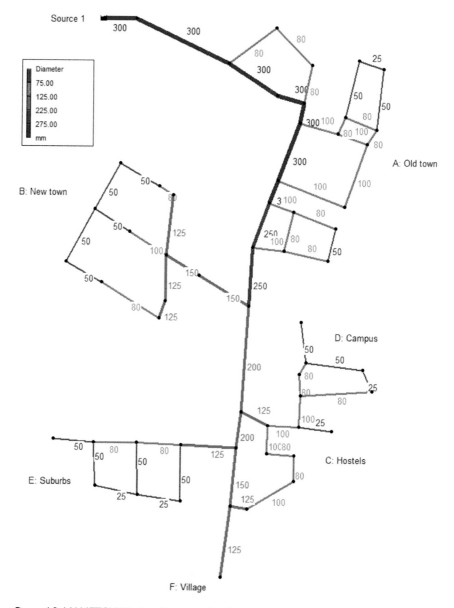

Figure 4.3 NAMETOWN pipe diameters (mm)

between *0.5* and *2.0* mm, mostly based on the age of the pipes. The overall information about the pipes is given in Table 4.1. The connection between the new hotel and the system is planned to be in node *b12* located in the NEW TOWN district.

Regarding the local situation (soil conditions, local manufacturing), polyethylene (PE) is chosen as the pipe material for the network rehabilitation. Because of this selection, it is assumed that the roughness of the new pipes will remain low throughout the design period. The accepted higher *k* value of *0.25* mm also compensates for the impact of local losses in the network.

Table 4.1 Pipe data

NAMETOWN – current situation

Pipe name	Material age	Node from	Node to	Diameter (mm)	Length (m)	k value (mm)
p01	CI-60	SI	a01	300	220	2.0
p02	CI-60	a01	a02	300	657	2.0
p03	PE-15	a02	a03	80	380	0.5
p04	CI-60	a02	a04	300	371	2.0
p05	PE-15	a03	a05	80	325	0.5
p06	CI-60	a04	a06	300	180	2.0
p07	PE-15	a06	a05	80	249	0.5
p08	CI-60	a06	a07	300	125	2.0
p09	GS-50	a07	a08	100	256	2.0
p10	GS-50	a08	a09	100	200	2.0
p11	PE-15	a08	a10	80	111	0.5
p12	PE-15	a09	a11	80	106	0.5
p13	PE-15	a10	a11	80	215	0.5
p14	PE-15	a10	a12	50	369	0.5
p15	PE-15	a11	a13	50	386	0.5
p16	PE-15	a13	a12	25	168	0.5
p17	CI-60	a07	a14	300	390	2.0
p18	GS-50	a09	a15	100	420	2.0
p19	GS-50	a14	a15	100	459	2.0
p20	DI-25	a14	a16	300	153	1.0
p21	PE-25	a16	a17	100	168	1.0
p22	PE-15	a17	a18	80	291	0.5
p23	DI-25	a16	a19	250	295	1.0
p24	PE-15	a17	a20	80	255	0.5
p25	PE-15	a18	a21	50	206	0.5
p26	PE-25	a19	a20	100	202	1.0
p27	PE-15	a20	a21	80	289	0.5
p28	DI-25	a19	b01	250	372	1.0
p29	GS-25	b01	b02	150	376	1.0
p30	GS-25	b02	b03	150	248	1.0
p31	PE-15	b03	b04	100	278	0.5
p32	PE-15	b04	b05	50	270	0.5
p33	PE-15	b06	b07	80	106	0.5
p34	PE-15	b07	b08	50	290	0.5
p35	PE-15	b10	b11	80	439	0.5

(Continued)

Table 4.1 (Continued)

NAMETOWN – *current situation*

Pipe name	Material age	Node from	Node to	Diameter (mm)	Length (m)	k value (mm)
p36	PE-15	b11	b12	50	269	0.5
p37	PE-15	b08	b05	50	333	0.5
p38	PE-15	b05	b12	25	380	0.5
p39	PE-25	b03	b06	125	380	1.0
p40	PE-25	b03	b09	125	289	1.0
p41	PE-25	b09	b10	125	112	1.0
p42	DI-15	b01	c01	200	727	0.5
p43	PE-15	c01	c02	125	184	0.5
p44	PE-15	c02	d01	100	201	0.5
p45	PE-15	d01	d02	25	214	0.5
p46	PE-15	d01	d03	100	195	0.5
p47	PE-15	d03	d05	80	136	0.5
p48	PE-15	d03	d04	80	461	0.5
p49	PE-15	d04	d07	25	148	0.5
p50	PE-15	d05	d06	80	85	0.5
p51	PE-15	d06	d07	50	368	0.5
p52	PE-15	d06	d08	50	256	0.5
p53	PE-15	c02	c03	100	177	0.5
p54	PE-15	c03	c04	80	177	0.5
p55	PE-15	c04	c07	80	159	0.5
p56	PE-15	c06	c07	100	349	0.5
p57	PE-15	c05	c06	125	108	0.5
p58	DI-15	c01	e01	200	168	0.5
p59	PE-15	e01	e02	125	361	0.5
p60	PE-15	e02	e03	80	289	0.5
p61	PE-15	e03	e04	80	277	0.5
p62	PE-15	e04	e05	50	266	0.5
p63	PE-15	e02	e06	50	354	0.5
p64	PE-15	e03	e07	50	325	0.5
p65	PE-15	e04	e08	50	272	0.5
p66	PE-15	e06	e07	25	280	0.5
p67	PE-15	e07	e08	25	282	0.5
p68	DI-15	e01	c05	150	366	0.5
p69	PE-15	c05	f01	125	452	0.5

4.2.3 *Water demand and leakage*

The water consumption in the area consists of three categories: mostly domestic but different at the university campus and at the student accommodation (hostels). In addition, variable leakage percentages apply: from *12* to *25* %, which is mostly attributable to the age of the network. The structure of the average (i.e. base) node demand/leakage and allocation of demand patterns is given in Table 4.2, which also shows the number of consumers supplied from each node at the average specific demand of *100* l/c/d.

The total average demand is currently *45.727* l/s. This demand has to be increased by a multiplier of *Individual number* (*IN*), based on the individual data, which includes the effects of seasonal variations i.e. reflects the demand on the maximum consumption day. Furthermore, it is expected that the population in the districts will grow exponentially each year of the design period, as shown in Table 4.3. In addition, the uniform specific

Table 4.2 Structure of the base node demand and leakage

Node name	Elevation (msl)	Demand pattern	District	NAMETOWN – current situation					
				Population	q (l/c/d)	Qc (l/s)	Leaks (%)	Ql (l/s)	Qd (l/s)
a01	82.48	Domestic	Old town	0	0	0.000	0	0.000	0.000
a02	68.48	Domestic	Old town	868	100	1.005	25	0.335	1.340
a03	65.77	Domestic	Old town	768	100	0.889	25	0.296	1.185
a04	48.90	Domestic	Old town	960	100	1.111	25	0.370	1.481
a05	48.81	Domestic	Old town	656	100	0.759	25	0.253	1.012
a06	42.95	Domestic	Old town	320	100	0.370	25	0.123	0.494
a07	42.07	Domestic	Old town	704	100	0.815	25	0.272	1.086
a08	41.51	Domestic	Old town	368	100	0.426	25	0.142	0.568
a09	41.21	Domestic	Old town	544	100	0.630	25	0.210	0.840
a10	43.83	Domestic	Old town	208	100	0.241	25	0.080	0.321
a11	59.47	Domestic	Old town	304	100	0.352	25	0.117	0.469
a12	56.74	Domestic	Old town	416	100	0.481	25	0.160	0.642
a13	57.12	Domestic	Old town	288	100	0.333	25	0.111	0.444
a14	38.33	Domestic	Old town	976	100	1.130	25	0.377	1.506
a15	40.85	Domestic	Old town	1216	100	1.407	25	0.469	1.877
a16	35.17	Domestic	Old town	704	100	0.815	25	0.272	1.086
a17	30.42	Domestic	Old town	496	100	0.574	25	0.191	0.765
a18	40.40	Domestic	Old town	592	100	0.685	25	0.228	0.914
a19	28.34	Domestic	Old town	1024	100	1.185	25	0.395	1.580
a20	29.91	Domestic	Old town	304	100	0.352	25	0.117	0.469
a21	45.60	Domestic	Old town	528	100	0.611	25	0.204	0.815
b01	24.82	Domestic	New town	960	100	1.111	12	0.152	1.263
b02	27.10	Domestic	New town	544	100	0.630	12	0.086	0.715
b03	28.16	Domestic	New town	1200	100	1.389	12	0.189	1.578
b04	32.41	Domestic	New town	752	100	0.870	12	0.119	0.989
b05	37.71	Domestic	New town	480	100	0.556	12	0.076	0.631
b06	41.81	Domestic	New town	912	100	1.056	12	0.144	1.199
b07	41.78	Domestic	New town	672	100	0.778	12	0.106	0.884
b08	46.80	Domestic	New town	592	100	0.685	12	0.093	0.779
b09	24.71	Domestic	New town	96	100	0.111	12	0.015	0.126
b10	23.74	Domestic	New town	96	100	0.111	12	0.015	0.126
b11	33.35	Domestic	New town	832	100	0.963	12	0.131	1.094
b12	42.72	Domestic	New town	1424	100	1.648	12	0.225	1.873
c01	19.06	Hostels	Hostels	448	100	0.519	15	0.092	0.610
c02	21.47	Hostels	Hostels	112	100	0.130	15	0.023	0.153
c03	20.35	Hostels	Hostels	112	100	0.130	15	0.023	0.153
c04	24.38	Hostels	Hostels	176	100	0.204	15	0.036	0.240
c05	12.50	Hostels	Hostels	272	100	0.315	15	0.056	0.370
c06	13.04	Hostels	Hostels	640	100	0.741	15	0.131	0.871
c07	21.88	Hostels	Hostels	336	100	0.389	15	0.069	0.458
d01	36.06	Campus	Campus	176	100	0.204	15	0.036	0.240
d02	26.27	Campus	Campus	160	100	0.185	15	0.033	0.218

(Continued)

Table 4.1 (Continued)

Node name	Elevation (msl)	Demand pattern	District	NAMETOWN – current situation					
				Population	q (l/c/d)	Qc (l/s)	Leaks (%)	Ql (l/s)	Qd (l/s)
d03	28.55	Campus	Campus	288	100	0.333	15	0.059	0.392
d04	54.35	Campus	Campus	560	100	0.648	15	0.114	0.763
d05	30.01	Campus	Campus	336	100	0.389	15	0.069	0.458
d06	35.63	Campus	Campus	336	100	0.389	15	0.069	0.458
d07	64.33	Campus	Campus	240	100	0.278	15	0.049	0.327
d08	40.20	Campus	Campus	768	100	0.889	15	0.157	1.046
e01	17.91	Domestic	Suburbs	256	100	0.296	20	0.074	0.370
e02	17.77	Domestic	Suburbs	160	100	0.185	20	0.046	0.231
e03	22.99	Domestic	Suburbs	192	100	0.222	20	0.056	0.278
e04	30.10	Domestic	Suburbs	336	100	0.389	20	0.097	0.486
e05	38.88	Domestic	Suburbs	528	100	0.611	20	0.153	0.764
e06	16.13	Domestic	Suburbs	224	100	0.259	20	0.065	0.324
e07	22.96	Domestic	Suburbs	256	100	0.296	20	0.074	0.370
e08	35.39	Domestic	Suburbs	144	100	0.167	20	0.042	0.208
f01	9.66	Domestic	Village	4000	100	4.630	20	1.157	5.787
			TOTAL:	31,860		36.875		8.852	45.727

Table 4.3 Annual population growth in the districts of NAMETOWN

District		A Old town	B New town	C Hostels	D Campus	E Suburbs	F Village
Annual population growth (%)		1.3	2.0	1.6	1.8	1.5	3.0

consumption of currently *100* l/c/d will grow for another *46–20IN*[1] l/c/d. As a result of the network rehabilitation, the current leakage percentages are expected to have stabilized at a uniform value of *17* % in each node, by the end of the design period.

The diurnal patterns for the consumption categories are shown in figures 4.4 to 4.8 and in Table 4.4, which includes the new hotel. The leakage variation described by the leakage peak factors simulates its relation to the pressure (normally higher overnight than during the day).

Finally, the new hotel is going to be built for the capacity of *400–100IN*[2] beds and specific consumption of *300* litres per bed per day. No seasonal variations apply in this case.

---

1 The text in italics refers to a formula to be used to diversify the data sets within the group. E.g. for IN = 1.8, the specific consumption by the end of the design period will be 100 + (46−36) = 110 l/c/d. Equally, for IN = 1.3, the specific consumption becomes 120 l/c/d.

2 In the same way as for the calculation of the specific demand growth, following the above equation, for IN = 1.8, the total number of hotel beds will be 400−180 = 220. Equally for IN = 1.3, the number of beds is 270. As a consequence, the node demand in b12 will differ.

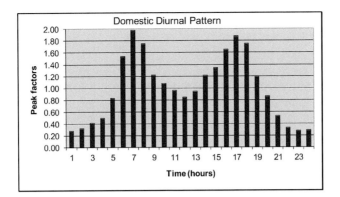

Figure 4.4 Domestic consumption pattern

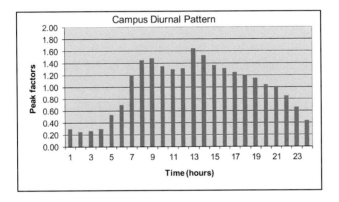

Figure 4.5 Consumption pattern at the university campus

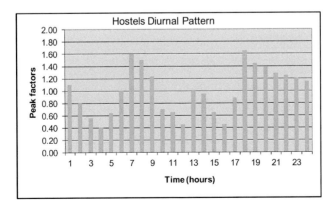

Figure 4.6 Consumption pattern in the student hostels

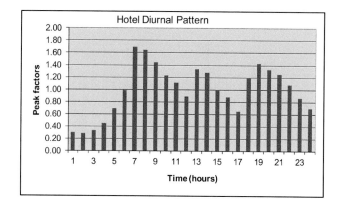

Figure 4.7　Consumption pattern at the new hotel

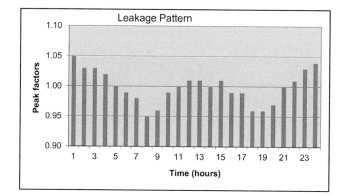

Figure 4.8　Leakage pattern

Table 4.4　Diurnal consumption patterns, including leakage variation

NAMETOWN – diurnal patterns

Hours	Domestic	Campus	Hostels	Hotel	Leakage
1	0.28	0.30	1.10	0.30	1.05
2	0.32	0.25	0.80	0.28	1.03
3	0.41	0.27	0.56	0.33	1.03
4	0.50	0.30	0.40	0.45	1.02
5	0.83	0.53	0.64	0.69	1.00
6	1.54	0.70	1.00	0.99	0.99
7	1.98	1.20	1.60	1.70	0.98
8	1.75	1.45	1.50	1.65	0.95
9	1.22	1.48	1.23	1.45	0.96
10	1.08	1.35	0.70	1.23	0.99

NAMETOWN – *diurnal patterns*

Hours	Domestic	Campus	Hostels	Hotel	Leakage
11	0.96	1.30	0.65	1.12	1.00
12	0.85	1.32	0.45	0.89	1.01
13	0.95	1.65	1.01	1.34	1.01
14	1.21	1.53	0.95	1.28	1.00
15	1.35	1.37	0.65	1.00	1.01
16	1.65	1.32	0.45	0.88	0.99
17	1.88	1.25	0.88	0.65	0.99
18	1.75	1.20	1.65	1.20	0.96
19	1.20	1.15	1.44	1.43	0.96
20	0.86	1.04	1.38	1.33	0.97
21	0.53	1.01	1.28	1.26	1.00
22	0.33	0.85	1.25	1.08	1.01
23	0.28	0.66	1.20	0.87	1.03
24	0.29	0.43	1.15	0.70	1.04

Table 4.5 Investment costs for the equipment, in Euros (EUR)

No.	Component	EUR	Per
1.1	Laying PE pipe D = 50 mm	45.00	m
1.2	Laying PE pipe D = 80 mm	72.00	m
1.3	Laying PE pipe D = 100 mm	90.00	m
1.4	Laying PE pipe D = 150 mm	135.00	m
1.5	Laying PE pipe D = 200 mm	180.00	m
1.6	Laying PE pipe D = 250 mm	225.00	m
1.7	Laying PE pipe D = 300 mm	270.00	m
1.8	Laying PE pipe D = 400 mm	360.00	m
1.9	Laying PE pipe D = 500 mm	450.00	m
1.10	Laying PE pipe D = 600 mm	540.00	m
2.	Pumping station	$5000 \times Q^{0.8}$	$Q = Q_{max}$ m^3/h
3.	Reservoir	$300{,}000 + 150 \times V$	$V = V_{tot}$ m^3
4.	Support structure for H m elevated tank	$3 \times H \times V$	$V = V_{tot}$ m^3
5.	Pressure-reducing valve (PRV)	$1000 + 30 \times D$	$D = D_{PRV}$ mm

4.2.4 Financial considerations

An international loan has been requested and approved with a payback period of *20* years and interest rate of *8* %. The loan is intended only for construction of the distribution part of the network, which does not include the tertiary network or service connections.

The equipment will be procured according to the prices listed in Table 4.5. For laying pipes in parallel, a *double* price as indicated under 1.1–1.10 should be considered.

Table 4.6 Annual operation and maintenance costs

No.	Component	% of investment
I.	Distribution pipes	0.5
2.	Pumping station	2.0
3.	Storage	0.8

The preliminary cost of the pumping station takes the maximum installed capacity into consideration, while the cost of the tanks will be based on the total volume. Elevated tanks will have an approximated additional cost of the supporting structure proportional to the height and the volume. Finally, the preliminary cost of the PRV is based on the selected diameter.

Annual operation and maintenance costs (O&M) are determined as a percentage of the investment costs, as shown in Table 4.6. For the current layout, before any renovation takes place, the annual O&M costs can be assumed at EUR *100,000*, which will increase by another EUR *30,000* by the end of the design period.

As all cost calculations are to be done in Euros, the effects of local inflation may be neglected.

4.3   Design steps and corresponding questions

4.3.1   Current demand scenario

3.1.1   Connect 3000 new consumers in the district SUBURBS to the network, assuming an average specific consumption of *100* l/c/d. Assume a homogeneous population distribution of the new consumers and the same diurnal demand pattern within the district. Adapt the base demand and average leakage (flow) of nodes e01 to e08 accordingly. What will the percentage of increase in the average demand in the SUBURBS district be after these actions?

3.1.2   Based on the assigned IN value, calculate the total number of beds in the new hotel and attach it to node b12. Adapt the calculated base demand of the hotel to negate the effect of seasonal variations during the simulations. Also, neglect the increase of leakage in node b12 as a result of the demand increase caused by the hotel. What will the percentage of increase in the average demand in NEW TOWN be after these actions?

3.1.3   Use your own IN value and adjust the general demand multiplier; run the 24-hour extended period simulation (EPS) for the scenario of the maximum consumption day, with a time interval of one hour. *Plot the 24-hour diagram* of the demand variation for the entire system. What will be the demand in the maximum consumption hour and at what time will it appear? Equally, what will be the demand in the minimum consumption hour and at what time will it appear?

3.1.4   Based on the results of the simulation under 3.1.3, *plot the network layout* showing the values of the pressures and hydraulic gradients ('Unit Headloss' in EPANET), in the maximum consumption hour; *use two separate figures*, respectively. Analyse these values and identify the list of critical nodes and pipes in the network. <u>What will be the most affected district of the NAMETOWN and what will the degree of service interruption be (low and/or negative pressures)</u>?

3.1.5   Based on the conclusions drawn under 3.1.4, make a *pipe* rehabilitation plan and adapt the network model so that the bottlenecks can be removed and the pressures maintained above a threshold of *25* mwc at all times. Make a list of the new pipe diameters, run the EPS and plot the layout of the modified network showing the values of the pressures and hydraulic gradients in the maximum consumption hour. Furthermore, *plot the 24-hour pressure diagram* of the two nodes with the lowest and the highest pressure in the network, respectively. Finally, *plot the 24-hour velocity diagram* of the two pipes with the lowest and the highest velocity in the network, respectively. <u>What will the range of pressures and velocities in the network be during the maximum consumption day</u>?

3.1.6   Using the diagram produced under 3.1.3, calculate the balancing volume of the reservoir at source S1. Assume constant production/inflow over 24 hours. <u>What will the total volume be if the emergency provision of three hours of average flow on the maximum consumption day has been planned in addition to the balancing volume (neglecting the other provisions)</u>? Size the reservoir cross-section area and height, and model the source S1 as a set-up of a supply node connected to the balancing tank with a dummy pipe (instead of the existing reservoir). Run the hydraulic simulation and *plot the 24-hour diagram* showing the tank depth variation. <u>What is the depth of the water (the volume percentage) at the beginning/end of the day (at midnight)</u>?

3.1.7   Analyse the (higher) pressures in the lower parts of the system, and provide one pressure-reducing valve (PRV) that will minimise the downstream pressures, however not below the threshold. Run the EPS and *plot the network layout* showing the valve operation. <u>What will be the most suitable current location and setting of the PRV</u>?

NOTE: The EPANET files used in questions 3.1.3, 3.1.5 and 3.1.7 have to be saved and submitted (electronically) with the final report. Use a uniform filename pattern by replacing NAMETOWN.NET with your own name or a name of your choice, and the question number. E.g. TRITOWN313.NET, etc.

4.3.2   *Future demand scenario*

3.2.1   Using your own IN value, calculate the demand of NAMETOWN at the end of the design period, by taking into consideration the annual population increase per district (indicated in Table 4.3), the growth of unit consumption per capita and the leakage percentages, as specified above. <u>What will the percentage increase in average demand in the NAMETOWN be compared to the current situation</u>?

3.2.2   Adapt the network model for the demand scenario at the end of the design period and run EPS. Based on the simulation results, *plot the network layout* showing the values of the pressures and hydraulic gradients in the maximum consumption hour. Analyse these values and identify the list of critical nodes and pipes in the network. How severe is the degree of service interruption (low and/or negative pressures)?

3.2.3   Connect the second source reservoir (S2) to the network. The bottom elevation of the reservoir should be estimated based on the elevations of the neighbouring nodes. Set the diameter of the new PE pipe connection at *150* mm and estimate its length proportional to the length of the other pipes. Run another EPS. Based on the simulation results, *plot the network layout* showing the values of the pressures and hydraulic gradients in the maximum consumption hour. How much has the situation improved after connecting the second source?

3.2.4   Based on the conclusions from 3.2.2 and 3.2.3, propose the most suitable location for the pumping station in the network. Select the duty flow(s) and head(s), the number of pump units and their scheduling so that the available choice of KSB ETANORM pumps shown in Figure 4.9 can be fitted later. Due to their age, *it is also planned to replace all the remaining CI pipes*, regardless of their hydraulic performance. In case of severe problems, combine the best possible choice of the pumps with an extended plan of renovation. This can include: (1) additional pipe replacement, (2) changes to the PRV settings and/or position, (3) a balancing tank at a suitable location in the network, or (4) a combination of these measures. Adapt the network model accordingly, so that the bottlenecks can be removed and the pressures maintained above the threshold of *20* mwc throughout the maximum consumption day. Following the balance of supply between the two sources, calculate the volume and dimensions of the reservoir in S2. Then model the second source which is also a tank (such as in the case of S1) and run the EPS. *Plot the layout of the modified network* showing the values of the pressures and hydraulic gradients in the maximum consumption hour. Furthermore, *plot the 24-hour diagram* showing the depth variation in both source tanks. What will be the range of minimum and maximum depth, and the balance of supply between the two sources? What is the range of pressures and velocities in the network during the maximum consumption day?

NOTE: The EPANET files used in questions 3.2.2, 3.2.3 and 3.2.4 have to be saved and submitted (electronically) with the final report.

4.3.3   Pumping station design

3.3.1   Based on the results under 3.2.4, select the required pump type, impeller size, and the number of units that can supply the daily demand variation in the network. The specifications for the selected KSB ETANORM pump units are available in Annex 1. Adapt the model by replacing the duty flow(s) and head(s) with the exact pair of points from the selected

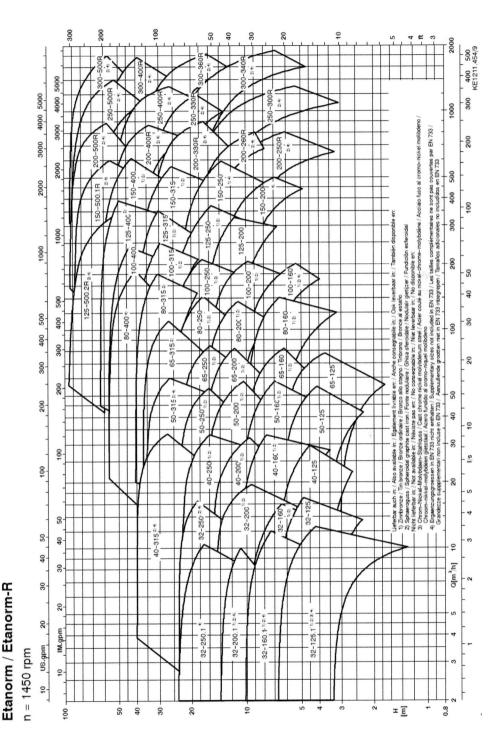

Figure 4.9 The KSB ETANORM pump range (n = 1450 rpm)

diagrams and verify the choice by running another EPS. *Plot a 24-hour pressure diagram* of the two nodes with the lowest and the highest pressures in the network, respectively. Furthermore, *plot a 24-hour diagram* of the two pipes with the lowest and the highest velocity in the network, respectively. Finally, *plot a 24-hour diagram* of the flows for all the pumps in the pumping station. What will be the total energy consumption for the pumping station on the maximum consumption day?

3.3.2   Calculate the required power for the pumps and driving devices. Based on this result, select the motor size (available from the pump specification details in the original KSB catalogue) that is to be used with each pump unit. Using the pump specification details, *determine the pump inlet and outlet diameters* along with the required dimensions of the foundation plate. *Estimate the minimum surface area* that is needed to accommodate the pump unit and the motor, allowing sufficient space around the pumps to enable installation and routine maintenance.

3.3.3   It is assumed that electricity will be used for the pump operation. *Size a transformer* for this purpose. In addition, *decide on the type and number of stand-by units*, and *size the diesel-driven generator* that is to be used in irregular situations.

3.3.4   Based on the type and number of the selected pumps, *draw a schematic layout of the pump arrangement*, indicating the pipe lengths, the position of the pump units, and the major valves and fittings. An example for a station with three pumps is shown in Figure 4.10. *Determine tentative*

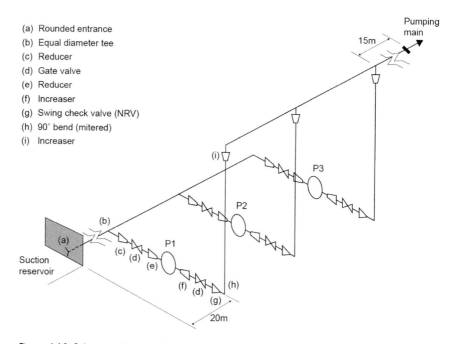

Figure 4.10 Schematic layout of the pumping station

diameters of the main pipes shown in the figure and *calculate the total friction loss* for the worst-case scenario. Assume the length of the feeder main is *20* m, and the length of the discharge header between this pump and the exit point of the pumping station is *15* m. The pipe lengths between the pumps on the suction and the pressure side should be estimated based on the space determined under 3.3.2. Assume the absolute roughness, k, of all pipes to be *0.25* mm.

3.3.5 *Observe the locations where the major local losses occur and calculate these* (tank inlet, major bends, branches, valves, reducers and enlargers) under the worst-case scenario. Add *20%* provision to account for the local losses omitted in this calculation (other bends, measuring equipment, pipe joints, etc.). <u>What will the total head loss be in the pumping station?</u>

3.3.6 *Enlarge the pipe diameters where necessary* in order to ensure that the total head loss is within an acceptable range (for this exercise, assume maximum *1.0* mwc). *Check the required NPSH* for the selected pumps and determine the maximum allowed elevation of the pump axis that satisfies this requirement.

NOTE: Each pipe in the layout should be listed with its length, diameter, and the maximum velocity. All points where the local losses have been calculated should be shown with the head loss value. Furthermore, the EPANET file used in Question 3.3.1 must be saved and submitted (electronically) with the final report.

4.3.4 Cost calculations

4.3.1 Summarising all the interventions in the network proposed under 3.1.7 and 3.3.1, *calculate the total investment cost. Calculate the present worth of the investment and capital recovery factor* (annuity). For this, assume that the investments to improve the current situation are made *immediately*, while the investments that should provide a satisfactory level of service at the end of the design period are made after *10* years.

4.3.2 *Calculate the costs of operation and maintenance* at the beginning (after the initial renovation has taken place), and at the end of the design period. Assume the pumping energy costs by applying a uniform electricity tariff of *0.15* EUR/kWh (at the end of the design period).

4.3.3 With production costs of EUR *0.45* per m³ at the beginning, and EUR *0.60* per m³ at the end of the design period (tax included in both cases), *calculate the total cost of water* at the beginning and the end of the design period, including operation, maintenance and the loan repayment of the distribution network.

4.4 The layout of the design report

The text of the final report should be concise, yet include *discussions and justification* for the selected approach in the network design. The case introduction should be limited only to the summary of the input data based on the individual data set.

Make sure that you answer all the questions and include all the diagrams and electronic versions of the EPANET files requested in the previous section, *not more, not less*. The checklist of the figures to be enclosed is:

Current demand scenario – the beginning of the design period (Session 1)

1. a 24-hour diagram of the demand variation for the entire system on the maximum consumption day. (3.1.3)
2. the layout of the network showing the values of the pressures in the maximum consumption hour. (3.1.4)
3. the layout of the network showing the values of the hydraulic gradients ('Unit Headloss' in EPANET) in the maximum consumption hour. (3.1.4)
4. a 24-hour pressure diagram of the two nodes with the lowest and the highest pressures in the network on the maximum consumption day, respectively. (3.1.5)
5. a 24-hour velocity diagram of the two pipes with the lowest and the highest velocities in the network, respectively. (3.1.5)
6. a 24-hour diagram showing the volume/depth variation of the source reservoir in S1. (3.1.6)
7. the layout of the network pressures showing the PRV operation on the maximum consumption day. (3.1.7)

Future demand scenario – the end of the design period (Session 2)

8. the layout of the network showing the values of the pressures in the maximum consumption hour. (3.2.2)
9. the layout of the network showing the values of the hydraulic gradients in the maximum consumption hour. (3.2.2)
10. the layout of the network showing the values of the pressures in the maximum consumption hour. (3.2.3)
11. the layout of the network showing the values of the hydraulic gradients in the maximum consumption hour. (3.2.3)
12. the layout of the network showing the values of the pressures in the maximum consumption hour. (3.2.4)
13. the layout of the network showing the values of the hydraulic gradients in the maximum consumption hour. (3.2.4)
14. a 24-hour diagram showing the depth variation in both source tanks. (3.2.4)

Pumping station design – the end of the design period (Session 3)

15. a 24-hour pressure diagram of the two nodes with the lowest and the highest pressures in the network, respectively. (3.3.1)
16. a 24-hour velocity diagram of the two pipes with the lowest and the highest velocities in the network, respectively. (3.3.1)
17. a 24-hour diagram of flows for all the pumps in the pumping station. (3.3.1)
18. a schematic layout of the pump arrangement, indicating the pipe lengths, the position of the pump units, and the major valves and fittings; each pipe in the layout should be indicated with its length, diameter, and the maximum velocity; all the points where the local losses have been calculated should be shown with the head loss value. (3.3.4)

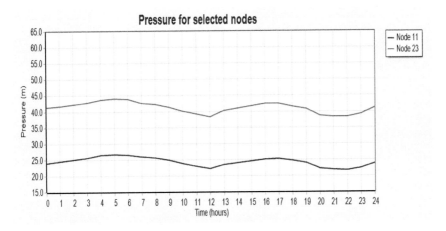

Figure 4.11 Example of the EPANET results for a time series

NOTE: All the figures except no. 18 can be produced as direct output from EPA-NET by the '**Edit>>Copy to . . .** ' menu option. If the diagrams are saved as files, the 'Metafile' format will consume less memory than the 'Bitmap'. One example of the direct transfer of results from EPANET to MS Word is shown in Figure 4.11. 24-hour diagrams can also be produced by exporting the results in tabular format to MS Excel, and making the diagrams there (after pasting the data from Clipboard). Furthermore, figure no. 18 can be pasted from Spreadsheet Lesson 6–6 together with the corresponding table showing the results of head-loss calculations.

Finally, the electronic versions of *seven* EPANET files (NET format) related to the following questions are to be submitted with the final report: 3.1.3, 3.1.5, 3.1.7, 3.2.2, 3.2.3, 3.2.4, and 3.3.1. Make sure that each file is given a unique name that clearly indicates the question and the author of the solution, as suggested above.

4.5 The tutorial

The example solved in this section is based on IN = 2.0 while the number of consumers connected to the network in the SUBURBS district is 3700. The second source, to mitigate the future demand increase, is located in the CAMPUS district (area D). All other data are used as specified in the case introduction in Section 4.2.

4.5.1 Base demand modelling (Question 3.1.1)

The SUBURBS district currently has 2096 consumers supplied from the distribution system, which results in a total base demand of 3.032 l/s including the leakage, as can be seen in Table 4.7.

Table 4.7 Structure of the current base demand in the SUBURBS district

Node name	Elevation (msl)	Demand pattern	District	NAMETOWN – *current situation*					
				Population	q (l/c/d)	Q_c (l/s)	Leaks (%)	Q_l (l/s)	Q_d (l/s)
e01	17.91	Domestic	Suburbs	256	100	0.296	20	0.074	0.370
e02	17.77	Domestic	Suburbs	160	100	0.185	20	0.046	0.231
e03	22.99	Domestic	Suburbs	192	100	0.222	20	0.056	0.278
e04	30.10	Domestic	Suburbs	336	100	0.389	20	0.097	0.486
e05	38.88	Domestic	Suburbs	528	100	0.611	20	0.153	0.764
e06	16.13	Domestic	Suburbs	224	100	0.259	20	0.065	0.324
e07	22.96	Domestic	Suburbs	256	100	0.296	20	0.074	0.370
e08	35.39	Domestic	Suburbs	144	100	0.167	20	0.042	0.208
			TOTAL:	2096		2.426		0.606	3.032

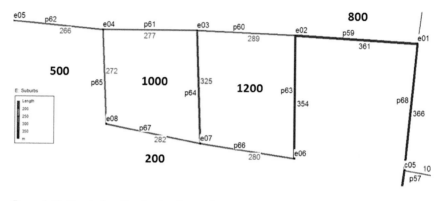

Figure 4.12 Population distribution (example calculation)

To connect additional consumers in the SUBURBS district, the simplified process of the node demand modelling is described in Volume 1, Section 4.3.4. In combination of loops and branches, the assumption of uniform distribution of house connections alongside the pipes can be expected, as illustrated in Figure 4.12.

The arbitrary distribution of 3700 inhabitants as shown in the figure will lead to the calculation of the number of consumers supplied in nodes e01 to e08, based on the following assumptions:

- 500 consumers supplied from pipes p62 and p65;
- 1000 consumers supplied from pipes p65, p61, p64 and p67;
- 200 consumers supplied from pipe p67 (located outside the loop);
- 1000 consumers supplied from pipes p64, p60, p63 and p66;
- 800 consumers supplied from pipe p59.

The number/length of pipes involved that share the total number of consumers outside the loops will depend on the knowledge about the situation in the field.

The distribution in the above figure, proportional to the related pipe lengths, would look as follows:

- 247 consumers out of 500 are supplied by p62 and 253 by p65; consequently, 123.5 consumers will be allocated to node e05, 123.5 + 126.5 to node e04 and 126.5 to node e08;
- 235 consumers out of 1000 are supplied by p65, 240 by p61, 281 by p64 and 244 by p67; consequently, 117.5 consumers will each be allocated to nodes e08 and e04, 120 to nodes e04 and e03, 140.5 to nodes e03 and e07 and 122 to e07 and e08;
- an additional 100 consumers will each be allocated to nodes e07 and e08;
- 313 consumers out of 1200 are supplied by p64, 278 by p60, 340 by p63 and 269 by p66; consequently, 156.5 consumers will each be allocated to nodes e07 and e03, 139 to nodes e03 and e02, 170 to nodes e02 and e06 and 134.5 to e06 and e07;
- an additional 400 consumers will be allocated to node e02 and 400 to node e01.

Finally, the allocation of 3700 consumers to nodes e01 to e08 looks as follows: e01–400, e02–709, e03–556, e04–487, e05–124, e06–305, e07–653, and e08–466.

Assuming the same specific demand for the additional consumers, the total demand in the district after their connection to the centralised supply will be 8.385 l/s, as shown in Table 4.8. This results in a demand increase of (8.385–3.032)/3.032/0.01 = 176.6% i.e. more than twice.

Clarifications:
- The above demands have been calculated using the spreadsheets available with this exercise. The spreadsheet named *'SUBURBS-population growth'* allocates the population numbers randomly spread over the area, as demonstrated in Figure 4.12, to the nodes e01 to e08. Based on the allocated population in each node, the second spreadsheet named *'NAMETOWN-nodes and pipes'* calculates the nodal consumption, leakage and demand, for selected

Table 4.8 Structure of the base demand in the SUBURBS district after connecting 3700 consumers

Node name	Elevation (msl)	Demand pattern	District	NAMETOWN – current situation					
				Population	q (l/c/d)	Q_c (l/s)	Leaks (%)	Q_l (l/s)	Q_d (l/s)
e01	17.91	Domestic	Suburbs	656	100	0.759	20	0.190	0.949
e02	17.77	Domestic	Suburbs	869	100	1.006	20	0.251	1.257
e03	22.99	Domestic	Suburbs	748	100	0.866	20	0.216	1.082
e04	30.10	Domestic	Suburbs	823	100	0.953	20	0.238	1.191
e05	38.88	Domestic	Suburbs	652	100	0.755	20	0.189	0.943
e06	16.13	Domestic	Suburbs	529	100	0.612	20	0.153	0.765
e07	22.96	Domestic	Suburbs	909	100	1.052	20	0.263	1.315
e08	35.39	Domestic	Suburbs	610	100	0.706	20	0.177	0.883
			TOTAL:	5796		6.708		1.677	8.385

specific consumption and leakage percentages. This spreadsheet is also used to determine the values shown in tables 4.1 and 4.2.

- Minor deviations in the total values shown in tables 4.7 and 4.8 originate from the rounding of individual values.

Conclusions:

- Because of the uniform values of specific demand and leakage percentage, the demand growth percentage is equivalent to the population growth percentage in this case, which usually is not the situation when varied demand categories are supplied from the same network.
- A sharp increase in the demand in the SUBURBS district, after connecting 3700 consumers, is also very likely to have an impact on a few other districts, demanding more significant network renovation to mitigate the pressure losses in the network.

Comment:

- The presented demand modelling approach offers a good starting point in the model building process. More detailed water demand analyses are needed at a later stage because the simplifications introduced in this step may still result in an insufficiently accurate demand distribution in the model.

Reading:
Volume 1, Section 4.3.4: 'Nodal demands'.

4.5.2   Modelling of hotel demand (Question 3.1.2)

The hotel is planned in node b12 in the NEW TOWN district. This district currently has 8560 consumers supplied from the distribution system, which results in a total base demand of 11.258 l/s including the leakage, as can be seen in Table 4.9.

Table 4.9 Structure of the current base demand in the NEW TOWN district

Node name	Elevation (msl)	Demand pattern	District	Population	q (l/c/d)	Q_c (l/s)	Leaks (%)	Q_l (l/s)	Q_d (l/s)
b01	24.82	Domestic	New town	960	100	1.111	12	0.152	1.263
b02	27.10	Domestic	New town	544	100	0.630	12	0.086	0.715
b03	28.16	Domestic	New town	1200	100	1.389	12	0.189	1.578
b04	32.41	Domestic	New town	752	100	0.870	12	0.119	0.989
b05	37.71	Domestic	New town	480	100	0.556	12	0.076	0.631
b06	41.81	Domestic	New town	912	100	1.056	12	0.144	1.199
b07	41.78	Domestic	New town	672	100	0.778	12	0.106	0.884
b08	46.80	Domestic	New town	592	100	0.685	12	0.093	0.779
b09	24.71	Domestic	New town	96	100	0.111	12	0.015	0.126
b10	23.74	Domestic	New town	96	100	0.111	12	0.015	0.126
b11	33.35	Domestic	New town	832	100	0.963	12	0.131	1.094
b12	42.72	Domestic	New town	1424	100	1.648	12	0.225	1.873
			TOTAL:	8560		9.907		1.351	11.258

The hotel capacity for IN = 2.0 will be 400–100×2.0 = 200 beds, which at an average consumption of 300 l/d per bed increases the base demand for 200×300/24/3600 = 0.694 l/s. Because of the multiplication of all the nodes by the general demand multiplier, that multiplication can be annulled by dividing the base demand of the hotel by the same multiplier; the seasonal variations do not apply in this case, as mentioned in the introduction to the assignment. Thus, the base demand of the hotel is then modelled in EPANET as 0.694/2.0 = 0.347 l/s. Yet, on the average day and with a fully occupied hotel, the base demand in the NEW TOWN increases from 11.258 l/s to 11.952 l/s, which is a growth of 6.2 %. The base demand of node b12 will then grow from 1.873 l/s to 2.567 l/s, or for 37 %.

Clarification:
- Because the seasonal demand variations do not apply in the case of the hotel, the growth percentages would be reduced if the total demand on the maximum consumption day is analysed. The hotel baseline demand of 0.347 l/s that is corrected by the general demand multiplier is used in this case, resulting in the growth percentages of ((11.258 + 0.347)/11.258–1) ×100 = 3.1 % and 18.5 %, respectively.

Conclusion:
- Hotel construction in the NEW TOWN district moderately affects the demand in this area and the rest of the network. Localised pressure drops are therefore expected, requiring the renovation mostly in the vicinity of node b12.

4.5.3   The range of network demand in the current scenario (Question 3.1.3)

In this exercise, the general demand multiplier that represents the seasonal demand variations i.e. the scenario of maximum consumption day is introduced in EPANET by the '**Project**>>**Analysis Options . . .** >>**Demand Multiplier**' menu option. This will display the corresponding property editor that looks as shown in Figure 4.10a (left). For the selected IN value of 2.0 and assuming no impact from the hotel on the leakage increase in the NEW TOWN district, the property editor of the demand in node b12 will look as in Figure 4.10b (right). It opens after double-clicking on the node, and right-clicking again in the cell **Demand Categories**.

The diurnal time pattern is to be created for the hotel from the consumption peak factors shown in Table 4.4, which is done by introducing these values in the EPANET Data Browser. Following this step, the model to run the first simulation of the current demand scenario has been finalised.

This simulation run (file NT313a) delivers a system flow as shown in Figure 4.14 (viewed in EPANET by the '**Report**>>**Graph . . .** >>**System Flow**' menu option).

The same result can be observed in tabular form, looking at the time series for the source node (S1) or the link (p01) that connects the source with the network (menu option '**Report**>>**Table . . .** >> **Time Series for Node/Link**'). The results

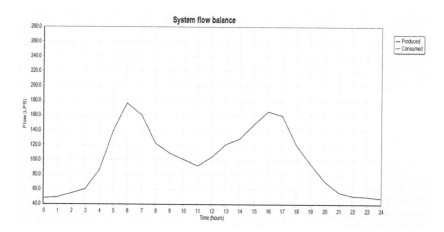

Hydraulics Options	
Property	Value
Flow Units	LPS
Headloss Formula	D-W
Specific Gravity	1
Relative Viscosity	1
Maximum Trials	40
Accuracy	0.001
If Unbalanced	Continue
Default Pattern	Domestic
Demand Multiplier	2.0
Emitter Exponent	0.5
Status Report	No
CHECKFREQ	2
MAXCHECK	10
DAMPLIMIT	0

Junction b12

Property	Value
*Junction ID	b12
X-Coordinate	492740.82
Y-Coordinate	1995777.38
Description	New town
Tag	
*Elevation	42.72
Base Demand	1.648
Demand Pattern	Domestic
Demand Categories	3

Demands for Junction b12

	Base Demand	Time Pattern	Category
1	1.648	Domestic	
2	0.225	Leakage	
3	0.347	Hotel	
4			
5			
6			

OK   Cancel   Help

Figure 4.13 The general demand multiplier (a) left, and the demand in node b12 (b) right, for IN = 2.0

Figure 4.14 NT313a: the demand variation of NAMETOWN on the maximum consumption day

for S1 will show the negative demand, which depicts the source supply. In either of these cases, the simulation shows the range of the demands between 48.21 l/s at 00:00/24:00 hours, and 176.61 l/s at 06:00 hours.

Clarifications:
• The shape of the curve in Figure 4.14 reflects the mixture of all the consumption categories and the leakage patterns. The green curve showing the total

demand overlaps the read curve showing the production because the patterns are identical.

• Nodal demands in EPANET are calculated as a product of the base demands, the general demand multiplier (without any exception!) and the hourly peak factors if the diurnal pattern has been defined and assigned to the node (in the junction property editor).

• If the effect of the general demand multiplier is to be cancelled in some nodes, their base demands should always be divided by the same multiplier before being put in the property editor (as suggested above, in the hotel case).

• The (diurnal) demand pattern is specified in the pattern editor, with peak factors assigned to the selected time period i.e. the hydraulic time step. These values represent the hourly values only if the full series of 24 has been defined, and if the EPS duration has been defined for 24 hours at the hydraulic time step of one hour, leading to 24 snapshot calculations.

• If the total number of peak factors is smaller than the number of snapshot calculations, the peak factor series will be restarted from the beginning as many times as needed to complete the full duration of the EPS.

• EPANET executes the most common EPS of 24 hours at the hydraulic time step of one hour by showing the results that include 25 values, where the first one (at 00:00 hours) and the last one (at 24:00 hours) are actually the same. So, EPANET assigns the first defined peak factor to the snapshot at 00:00 hours (not at 01:00 hours!). The actual consequence is that the maximum peak demands in EPANET results appear one hour earlier than specified in figures 4.4–4.8 and Table 4.4. This should be noted but no specific intervention is needed in this exercise (often neglected in general). Yet, the first or the last value need to be removed when calculating the average flows for the day.

• EPANET applies the default diurnal pattern ('Domestic' in this exercise) even if this has not been specified in the property editor of the newly introduced nodes.

Conclusion:
• A relatively wide range of diurnal flows is likely to result in a wide range of pressures during the day, also amplified by the wide range of elevations in the network.

4.5.4 The range of pressures and gradients in the current scenario (Question 3.1.4)

This question takes into consideration the same simulation as in the previous question. The snapshot results during the maximum consumption hour (at 06:00) on the maximum consumption day are shown in Figures 4.15 for the pressures, and in 4.16 for the hydraulic gradients.

Clarification:
• 'Unit Headloss' in EPANET has the same meaning as the hydraulic gradient being expressed in m/km. Hydraulic gradients are commonly expressed without dimensions. In this case the unit headloss of, say, 10 m/km is equivalent to S = 0.010.

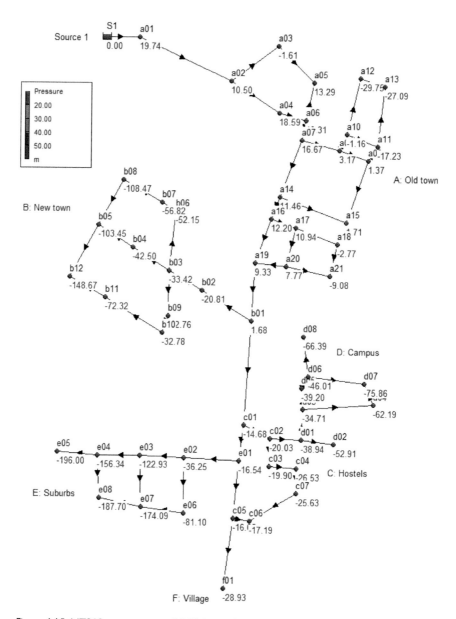

Figure 4.15 NT313a: pressures at 06:00 hours (mwc)

Conclusions:

- Figure 4.15 shows that the network is considerably under-designed for the given demand scenario, barely having nodes with positive pressure, let alone above the required minimum of 20 mwc. None of the nodes actually satisfies this pressure criterion during the maximum consumption hour.

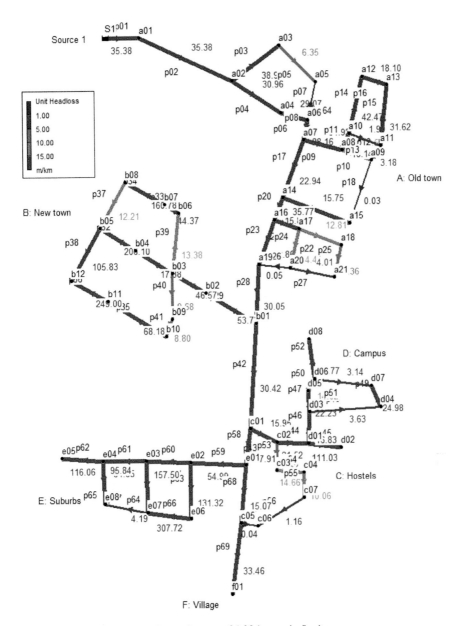

Figure 4.16 NT313a: hydraulic gradients at 06:00 hours (m/km)

- The most critical situation is in the SUBURBS district, with the minimum pressure of -196.00 mwc, which is a logical consequence of connecting 3700 consumers without increasing the conveying capacity of the network. The pressure in node b12, where the hotel is planned for construction, is -148.67 mwc.

- The low pressures in general originate from too small pipes generating high friction losses, which is reflected in very high values of hydraulic gradients, as shown in Figure 4.16. The figure shows just a few peripheral pipes having gradients below 10 m/km, which could be considered acceptable, given their small diameters and low flows they are carrying. The highest value of the hydraulic gradient of 307.72 m/km is registered in pipe p66 in the SUBURBS district. The hydraulic gradients in the pipes connecting the hotel in node b12 are also extremely high (p36–249.00 m/km, p38–105.84 m/km).

Comments:

- Low pressures and high hydraulic gradients complement each other and are two sides of the same problem, which is the low conveying capacity of the network caused by too small pipe diameters for the given demand scenario. Raising the heads by pumping will improve the pressures but with extreme hydraulic gradients the cost of that pumping will be too high. Enlarging the pipes in order to mitigate the hydraulic scenario shown in figures 4.15 and 4.16 is a far more viable option.
- The overview of the critical nodes and pipes can be made using the menu option 'View>>**Query** . . . , which opens the editor to filter the results based on the specified criteria. Figure 4.17 shows the nodes with negative pressure (left) and the pipes with a hydraulic gradient above 10 m/km (right), both shown in red.

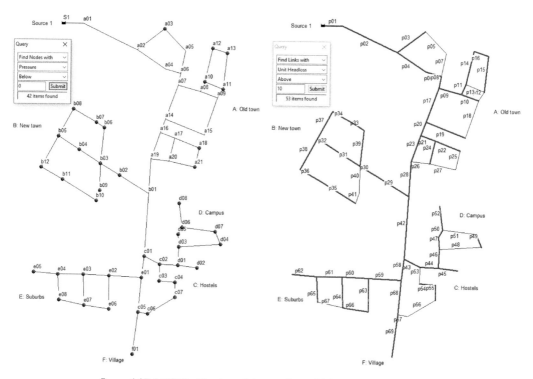

Figure 4.17 NT313a: filtering of the results at 06:00 hours

4.5.5 Renovation plan in the current demand scenario (Question 3.1.5)

High demand in the network demands a radical renovation plan already in place at the beginning of the design period. Just to check the effect of dealing with the most important route going through the middle of the network (p01-p02-p04-p06-p08-p17-p20-p23-p28-p42-p58-p68-p69), these 13 pipes have been enlarged with the first available larger diameter in Table 4.5. Consequently, the first seven pipes along the route were enlarged from 300 to 400 mm, the following two from 250 to 300 mm, the following two from 200 to 250 mm, and the final two from 150 to 200 mm and from 100 to 150 mm, respectively. The results of hydraulic simulation based on this modification (file NT315a) show significant improvement, as can be seen in Figure 4.18, but the critical parts of the network are still significantly underserved; actually, without water.

Conclusions:

- The intervention shown in Figure 4.18 has reduced the number of nodes with negative pressure from 42 in the original network to 9 (out of a total of 57 nodes). At the same time, the number of pipes with a hydraulic gradient above 10 m/km was reduced from 53 to 37 (out of a total of 69 pipes).
- Despite these improvements, the network is still largely under-designed with significant problems in the SUBURBS and NEW TOWN districts.

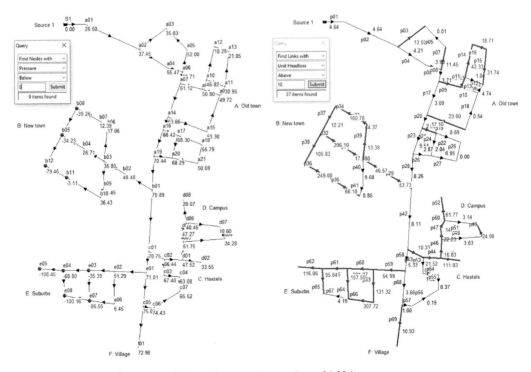

Figure 4.18 NT315a: enlargement of the main route: the results at 06:00 hours

- Any further plans for renovation should focus on the remaining pipes in these areas, starting from the upstream side. A diameter increase that lowers the hydraulic gradient to below 10 m/km would be considered a good starting point in arriving at the final list of pipes.

After running several trials, the final list of pipes that are enlarged in the network is shown in Table 4.10 and Figure 4.19 (illustrated in red). All the new pipes are modelled with a uniform k value of 0.25 mm, which is used to easily identify the enlarged pipes. The EPANET file with this scenario is named NT315b.

The results of the hydraulic simulations for this renovation plan are shown in the figures 4.20 to 4.23.

Table 4.10 Pipe renovation list (material PE, k = 0.25 mm)

NAMETOWN – *renovated pipes for the current demand scenario*

Pipe name	Node from	Node to	New D (mm)	Length (m)	Old D (mm)
p01	S1	a01	400	220	300
p02	a01	a02	400	657	300
p04	a02	a04	400	371	300
p06	a04	a06	400	180	300
p08	a06	a07	400	125	300
p14	a10	a12	80	369	50
p15	a11	a13	80	386	50
p17	a07	a14	400	390	300
p20	a14	a16	400	153	300
p23	a16	a19	300	295	250
p28	a19	b01	300	372	250
p29	b01	b02	200	376	150
p30	b02	b03	200	248	150
p31	b03	b04	150	278	100
p32	b04	b05	100	270	50
p38	b05	b12	80	380	50
p42	b01	c01	250	727	200
p43	c01	c02	150	184	125
p44	c02	d01	150	201	100
p45	d01	d02	50	214	25
p46	d01	d03	150	195	100
p47	d03	d05	100	136	80
p49	d04	d07	50	148	25
p50	d05	d06	100	85	80
p52	d06	d08	80	256	50
p58	c01	e01	250	168	200
p59	e01	e02	150	361	125
p60	e02	e03	150	289	80
p61	e03	e04	100	277	80
p62	e04	e05	80	266	50
p63	e02	e06	100	354	50
p64	e03	e07	100	325	50
p65	e04	e08	80	272	50
p66	e06	e07	50	280	25
p67	e07	e08	50	282	25

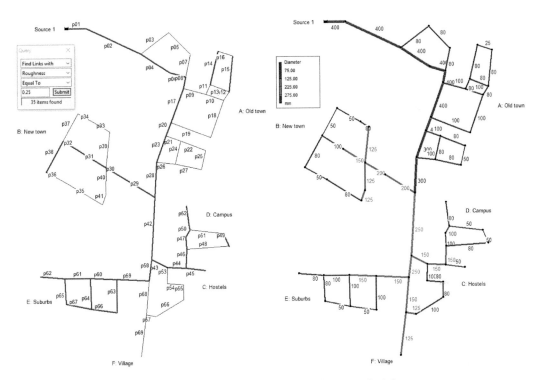

Figure 4.19 NT313a: plan of the renovation (left) with new pipe diameters (right)

Conclusions:

- In total 35 out of 69 pipes have been replaced with a larger diameter. The length of the new pipes is 10,090 m, which equals 52.5 % of the total network length of 19,224 m. This implicates a major renovation in the gravity supplied network in order to mitigate the huge pressure problems.
- Resulting from this renovation, the minimum pressure of 21.78 mwc appears in node d07. Strictly speaking, this value is below the required minimum of 25 mwc. In that sense, the critical node has an isolated problem, because it is surrounded by nodes with a much higher pressure; the origin of such pressures is in the lower elevations of these nodes compared to d07.
- Further improvement of the pressure in d07 would increase the pressures in general, which is considered unreasonable for the time being, especially knowing that the location of the second source will be in the CAMPUS district. Thus, the problem is temporary and of limited duration. The critical value shown in Figure 4.20 appears in the maximum consumption hour of the maximum consumption day and as such it is the absolute minimum pressure to be expected throughout the year.
- The highest pressure currently occurs in node e01, which is 71.09 mwc at 06:00 hours and reaches 90 mwc during the night time, as Figure 4.22 shows.

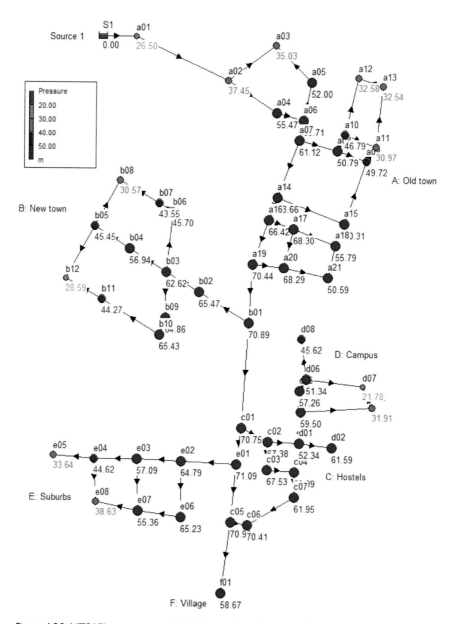

Figure 4.20 NT315b: pressures at 06:00 hours (mwc)

Hence, the network operates with a pressure range of nearly 50 mwc through-out the day, which originates from the fixed topography in the area and the gravity supply.

- Thus, the situation in Figure 4.20 shows generally high pressures. This is the main reason to postpone the replacement of some pipes with high hydraulic gradients until the demand grows more significantly. The implication of this decision is that the velocities in some pipes will be rather high. Figure 4.23 shows the maximum velocity reaching nearly 1.7 m/s (pipe p69).

Comments:
- Higher velocities in gravity systems, in the absence of pumping costs, may be seen as positive from the perspective of water quality (better conveyance,

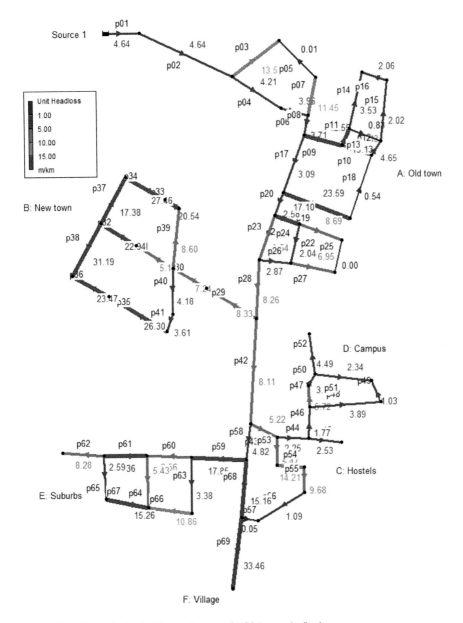

Figure 4.21 NT315b: hydraulic gradients at 06:00 hours (m/km)

reducing the sediment accumulation in the pipes). Care should be however taken when expanding these networks where pipes with higher velocities/ friction losses become serious bottlenecks.

• Starting from the recommended range of the design criteria discussed in Volume 1, Section 4.2.1, it should be obvious that it is impossible to satisfy these criteria unanimously in all the nodes and pipes. This is clearly to be seen in figures 4.22 and 4.23. These design criteria therefore serve more as guidelines than as rules; some compromise will constantly be needed. It is therefore most important to be consistent in the further development phases of the network.

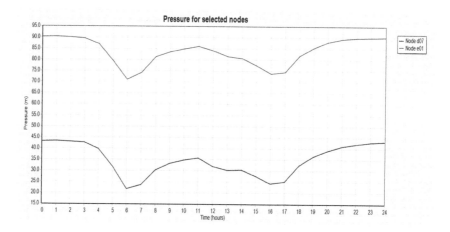

Figure 4.22 NT315b: pressure range on the maximum consumption day

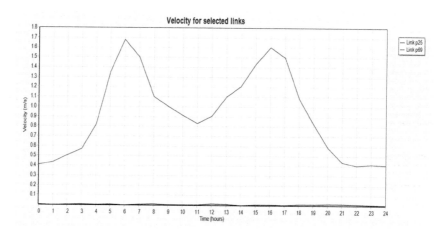

Figure 4.23 NT315b: velocity range on the maximum consumption day

4.5.6 The balancing volume of the source tank (Question 3.1.6)

S1 is currently the only source in the network, and it supplies the entire demand by gravity. The reservoir S1 is therefore a balancing tank fed by the average flow for 24 hours. The balancing volume of this tank serves to calculate the total volume and the tank dimensions. Based on the overall system flow shown in Figure 4.14, shown for the answer to Question 3.1.3, the calculation follows the theory discussed in Volume 1, Section 4.2.3. The corresponding Spreadsheet Lesson 8–10 (see Appendix 7) can be used as an aid, with the time series results for pipe p01 that connects the tank. Furthermore, other tank provisions apart from the balancing and emergency volumes can be neglected in this exercise. As required, the emergency volume is planned at three hours of average flow supply. The results of the spreadsheet calculation are shown in Figure 4.24.

Clarifications:

- The input data in the 'Tank Out' column of the spreadsheet are taken from the tabular results for the flow in pipe p01. The flow rate at 00:00 hours has been removed from the analysis because it is identical to the value at 24:00. The flow values originally presented in l/s in EPANET are converted into m³/h in the spreadsheet (the conversion factor is 3.6).
- The average flow calculated from the demand variation has been thereafter filled in the 'Tank In' column and was also used for the calculation of the emergency provision.

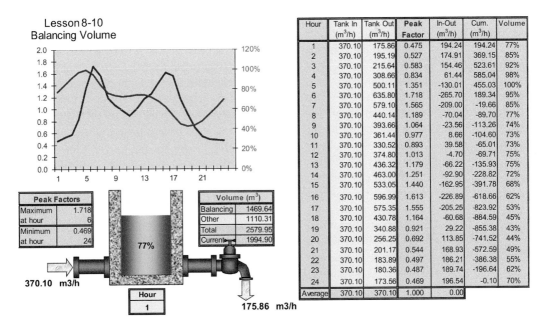

Hour	Tank In (m³/h)	Tank Out (m³/h)	Peak Factor	In-Out (m³/h)	Cum. (m³/h)	Volume
1	370.10	175.86	0.475	194.24	194.24	77%
2	370.10	195.19	0.527	174.91	369.15	85%
3	370.10	215.64	0.583	154.46	523.61	92%
4	370.10	308.66	0.834	61.44	585.04	98%
5	370.10	500.11	1.351	-130.01	455.03	100%
6	370.10	635.80	1.718	-265.70	189.34	95%
7	370.10	579.10	1.565	-209.00	-19.66	85%
8	370.10	440.14	1.189	-70.04	-89.70	77%
9	370.10	393.66	1.064	-23.56	-113.26	74%
10	370.10	361.44	0.977	8.66	-104.60	73%
11	370.10	330.52	0.893	39.58	-65.01	73%
12	370.10	374.80	1.013	-4.70	-69.71	75%
13	370.10	436.32	1.179	-66.22	-135.93	75%
14	370.10	463.00	1.251	-92.90	-228.82	72%
15	370.10	533.05	1.440	-162.95	-391.78	68%
16	370.10	596.99	1.613	-226.89	-618.66	62%
17	370.10	575.35	1.555	-205.25	-823.92	53%
18	370.10	430.78	1.164	-60.68	-884.59	45%
19	370.10	340.88	0.921	29.22	-855.38	43%
20	370.10	256.25	0.692	113.85	-741.52	44%
21	370.10	201.17	0.544	168.93	-572.59	49%
22	370.10	183.89	0.497	186.21	-386.38	55%
23	370.10	180.36	0.487	189.74	-196.64	62%
24	370.10	173.56	0.469	196.54	-0.10	70%
Average	370.10	370.10	1.000	0.00		

Lesson 8-10
Balancing Volume

Peak Factors	
Maximum	1.718
at hour	6
Minimum	0.469
at hour	24

Volume (m³)	
Balancing	1469.64
Other	1110.31
Total	2579.95
Current	1994.90

370.10 m3/h

77%

Hour
1

175.86 m3/h

Figure 4.24 A balancing volume calculation in S1 (Spreadsheet Lesson 8–10)

Conclusions:

- The spreadsheet results show an average demand of 370.10 m³/h = 102.81 l/s. This value is higher than the maximum of the source in S1, which is 90 l/s, the reason being the extraordinarily high value of IN = 2.0 (the students work with the maximum IN value of 1.8).
- Assuming that this average flow can be delivered, the balancing volume is calculated at V_{bal} = 1469.64 m³. With an emergency volume of 1110.31 m³, the total volume of the tank will be V_{S1} = 2579.95 ≈ 2580 m³.
- The initial volume of the tank at the beginning of the day is 1994.90 ≈ 1995 m³, which is approximately 77 % of the total volume.
- The emergency volume takes 43 % of the total volume of the tank. This volume is reached at 19:00 hours. The rest of the day, the volume will be recovered up to the same level as at the beginning of the day.

Comments:

- Calculating the balancing volume for the demand on the maximum consumption day will create additional emergency provision on any other day.
- In reality, the demand follows less precise patterns from one day to another, and also the initial levels at 00:00 and 24:00 hours will rarely be identical. The target is to make them similar by adapting operational parameters. Gravity systems in this respect offer less flexibility than the pumped supply schemes; the valves will normally be used to balance the demand and supply.

Reading:
Volume 1, Section 4.2.3: 'Storage design'.

With properly selected tank dimensions, the same volume variation pattern can be simulated in the computer model. The source reservoir S1 should be remodelled as EPANET 'Tanks', with the actual diameter (assumed to be a round bottom) and specified minimum, maximum and initial depths, based on the above calculations using the spreadsheet. The bottom level of the tank can be adjusted by deducting the maximum depth from the fixed reservoir level which is set at 110 msl. The tank should be further fed with a constant[3] flow from a node with negative base demand simulating the average supply. The friction loss of the connecting dummy pipe is irrelevant; this pipe only provides the feed to the tank, and should be therefore taken out of the cost calculations. Equally, the elevation and the pressure of the supply node make no difference for the overall hydraulic performance of the network that is only influenced by the water level in the tank. The only important aspect there is to ensure that the base negative demand combined with the demand pattern and the general multiplier reflects the actual average demand in the network. This technique has been applied and the results for the depth variation of the tank in S1 are shown in Figure 4.25 (EPANET file NT316a).

---

3 A separate demand pattern needs to be introduced, simulating constant flow; for instance, with the name 'Source'. To simulate constant supply, the 'Source' pattern can be defined by a single peak factor, which EPANET executes repetitively throughout the entire period of the simulation.

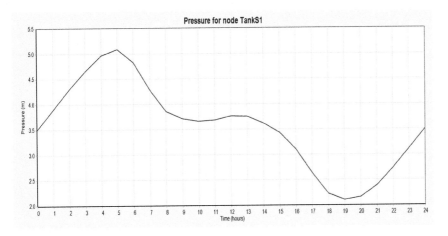

Figure 4.25 NT316a: source tank volume variation (S1)

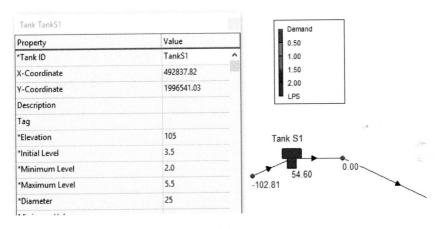

Figure 4.26 NT316a: source tank modelling (S1)

Clarifications:

*   For the bottom elevation of the tank specified as the elevation above the reference level (usually in msl), the pressure in the y axis of the tank diagram shows the actual tank depth.
*   Clicking the right mouse button on the graph opens the editor, which allows the fonts and the scale of the axes to be adapted.
*   The tank has been modelled with a diameter of 25 m and a maximum depth of 5.5 m, leading to the total volume of $V_{S1} = 2699.81 \approx 2670$ m³; thus, slightly bigger than the one calculated by the Spreadsheet Lesson 8–10.

- The bottom elevation of the tank was set at 105 msl, while the minimum level was set at 2.0 m.
- At the beginning of the simulation the initial depth was 3.5 m, showing 64 % of the volume (lower than in the spreadsheet, but this assumes the value at 00:00/24:00, and in a slightly bigger tank in the model).
- The modelling parameters indicated above are shown in Figure 4.26.

Conclusions:
- The tank reaches exactly the same initial level of 3.50 m, at the end of the simulation, at 24:00 hours.
- The maximum water depth of 5.09 m is reached at 05:00 hours, while the minimum depth of 2.10 m is reached at 19:00 hours. Both points in time coincide with the balancing pattern shown in the spreadsheet in Figure 4.24.

Comments:
- When the tank balancing volume is well designed it normally assumes that the tank is full before the morning peak hour, and is at the emergency volume after the afternoon peak demand hour. Both can be observed in figures 4.24 and 4.25.
- When modelling the source as a 'Tank' in EPANET, the pressure in the network will be influenced by the actual level of water in the tank at every hour. In many cases, this pressure variation will not be significant compared to the modelling of the source as a 'Reservoir'; the range is set by the decision on the tank minimum and maximum levels. The pressure range with S1 modelled as a 'Tank', shown in Figure 4.27, can be compared with the situation shown in Figure 4.22 where S1 was modelled as a 'Reservoir'.

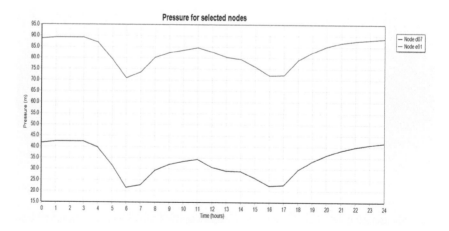

Figure 4.27 NT316a: pressure range on the maximum consumption day (S1 = 'Tank')

4.5.7 Preliminary location of the pressure-reducing valve (Question 3.1.7)

A pressure-reducing valve (PRV) is planned in the network in an attempt to reduce the pressures, especially in the lower part of the network where they are relatively high, also during the maximum consumption hour. Given the topography in the area and the network configuration, the highest impact should be reached by positioning the PRV somewhere along the main pipe route (p01-p02-p04-p06-p08-p17-p20-p23-p28-p42-p58-p68-p69) and not too close to the source.

PRVs are typically installed to control the downstream pressures of a network zone at a certain pre-set value, which is the downstream setting of the valve. When the upstream pressure is below this setting, the valve is fully opened; otherwise it reduces the downstream pressure to the setting. In the event of the downstream pressure building up and consequently reversing the flow direction, the PRV operates as a non-return valve (NRV) or check valve (CV) i.e. it will shut off. Taking all this into consideration makes little sense to place a PRV on a loop of pipes because the upstream and the downstream sides of the valve will not be hydraulically separated (as illustrated in details in Appendix 3).

It is thus very important to clearly define the upstream and the downstream nodes of the valve in the modelling process. Secondly, any valve in EPANET is a link with a virtual length; thus, it generates no friction losses. Therefore replacing an ordinary pipe between two nodes with any type of valve means neglecting the pipe friction losses over potentially considerable length, which is incorrect. To avoid this, a dummy node can be introduced on the downstream side of the valve, having the same elevation as the upstream node of the pipe where the valve is installed. This dummy node (named here PRVds) is then connected with the downstream node of the pipe, fully maintaining its physical properties.

Modelling of the PRV has been tested in four different locations: (1) at node b01 of pipe p42 (file NT317a), (2) at node c01 of pipe p58 (file NT317b), (3) at node a14 of pipe p20 (file NT317c), and (4) at node a19 of pipe p28 (file NT317d). In all these cases, a dummy node was introduced (PRVds) that was connected to the downstream node of these pipes (c01, e01, a16, and b01, respectively). The PRV diameters have been reduced to the first lower value in Table 4.5, compared to the diameter of the pipe in which they are installed. The initial PRV settings in all cases were 60 mwc. All four locations are indicated in Figure 4.28. The results of the hydraulic simulations are shown in figures 4.29–4.37.

Clarifications:
* Because the model is in the demand-driven mode, the reduction of the pressure by the PRV has no impact whatsoever on the demands in the network. Consequently, the demand as well as the balancing curve of the tank in the source will always be the same regardless of the location or the setting of the PRV. The same will be with the pressures upstream of the PRV that are higher than i.e. not influenced by the valve setting.

- The PRV operation in the 'Time Series' graph can be illustrated by showing the PRV flow, which replicates the total downstream demand, after the valve, or by 'Unit Headloss', which shows the pressure difference between the upstream and downstream nodes of the PRV. In that sense the unit m/km is wrong because the PRV has no length; the result represents the PRV local loss of energy, in mwc.
- Figures 4.29 and 4.30 show the summary of the PRV flows and head losses for the EPANET simulation runs conducted for the four PRV positions indicated in Figure 4.28.

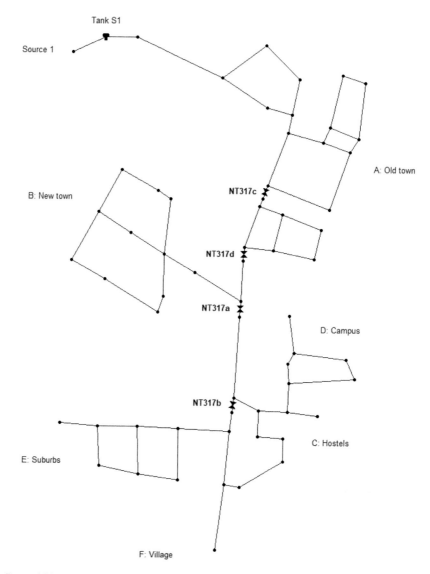

Figure 4.28 Four alternative PRV locations (files NT317a-d)

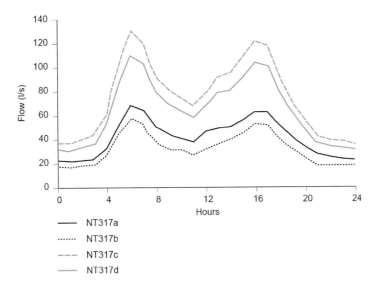

Figure 4.29 Four alternative PRV flows (files NT317a-d)

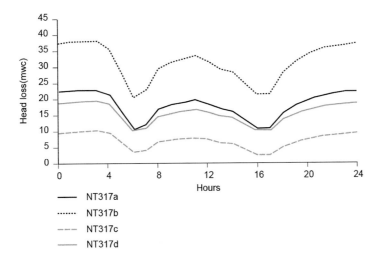

Figure 4.30 Four alternative PRV head losses (files NT317a-d)

Conclusion (figures 4.29 and 4.30):
• Not surprisingly, the most upstream PRV (file NT317c) controls the largest portion of the network demand but with the lowest PRV head loss, compared to the most downstream PRV (file NT317b) controlling the smallest area but with the largest PRV head loss. This is the direct consequence of the selected PRV settings that are limited by the topography and the location of the most critical pressure point in the CAMPUS district (at node d07). Further elaboration is in figures 4.31 to 4.37.

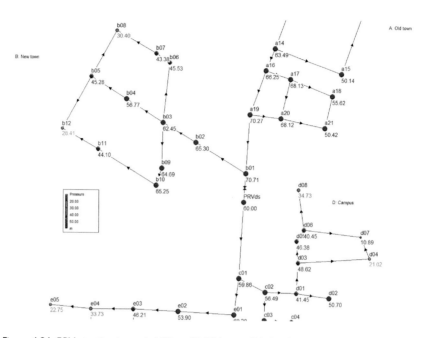

Figure 4.31 PRV starting in node b01, at 06:00 hours (file NT317a)

Conclusions (PRV location a):

- The PRV starting in node b01 controls the downstream areas C, D, E and F (file NT317a, Figure 4.31). Before the PRV with a setting of 60 mwc was placed, the pressure in node c01 was 70.75 mwc, while the pressure in the most critical node d07 was 21.78 mwc (from Figure 4.20); modelling the source as a balancing tank instead of the reservoir would yield the values 70.58 mwc and 21.60 mwc, respectively (obtained by running the hydraulic simulations of file NT316a).
- The PRV starting level of 60 mwc reduces the pressure in node c01 to 59.86 mwc, while the pressure in the most critical node d07 becomes 10.89 mwc; the latter is considered to be too low. Hence, with this location for the PRV the possibilities for pressure reduction are limited due to the location of the most critical point.

Conclusions (PRV location b):

- The PRV starting in node c01 controls the downstream areas C, E and F (file NT317b) i.e. it is located downstream of the CAMPUS district and the most critical pressure in node d07 should be therefore less dependent on the PRV setting. This is visible in the results shown in Figure 4.32 where the pressure in d07 equals 21.27 mwc. However, further increase of the pipe diameters in the SUBURBS district needs to be implemented because of the consequent pressure drop in this area.
- Nevertheless, this PRV is located on a loop, and the effect of the setting is therefore not fully utilised in the downstream part of the network.

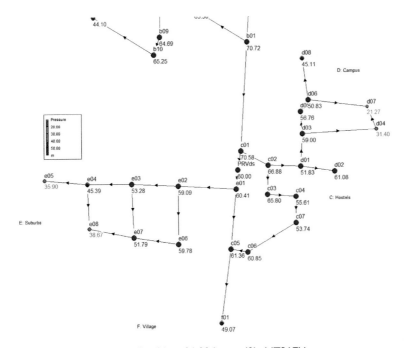

Figure 4.32 PRV starting in node c01, at 06:00 hours (file NT317b)

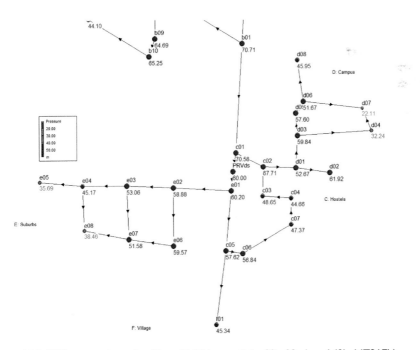

Figure 4.33 PRV starting in node c01, at 06:00 hours, link c02-c03 closed (file NT317b)

Conclusion (PRV location b, closed link):

- An improvement can be reached by closing the link between nodes c02 and c03. The positive effect on lowering the downstream pressures can be seen in Figure 4.33, without affecting the pressure in d07. This also allows for further reduction of the pressure by lowering the setting to 50 mwc, which is to be seen in the results in Figure 4.34.

Conclusion (PRV location c):

- In the location furthest upstream, the PRV also controls the NEW TOWN district where the hotel in b12 is located (file NT317c). The results in Figure 4.35 show the same problem as in Figure 4.31, namely the limitation of the PRV setting because of its influence in node d07. This can be partly mitigated by connecting nodes b01 and d08, if the routing of the pipe was possible. The results shown in Figure 4.36 indicate pressures after installing a pipe of 370 m length, 100 mm diameter, and a *k* value of 0.25 mm. Moreover, a further increase in the pipe diameters in the NEW TOWN district needs to be implemented to maintain the pressures in this area.

Conclusion (PRV location d):

- The final PRV location was tested starting in node a19 (file NT317d) and with the setting of 60 mwc. The pressures at 06:00 for these conditions are shown in Figure 4.37. It is to be noted that next to still existing pressure problem in node d07, the pressures are also insufficient in the NEW TOWN district; the situation that clearly originates from too low pressure setting.

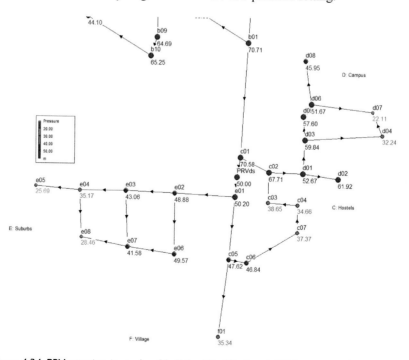

Figure 4.34 PRV starting in node c01, link c02-c03 closed, PRV setting = 50 mwc (file NT317b)

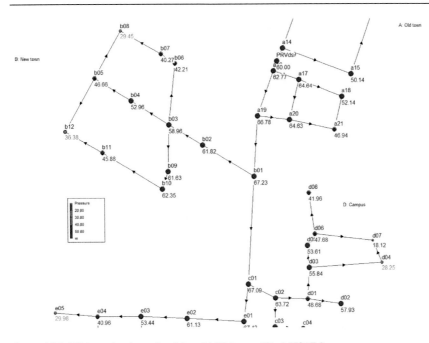

Figure 4.35 PRV starting in node a14, at 06:00 hours (file NT317c)

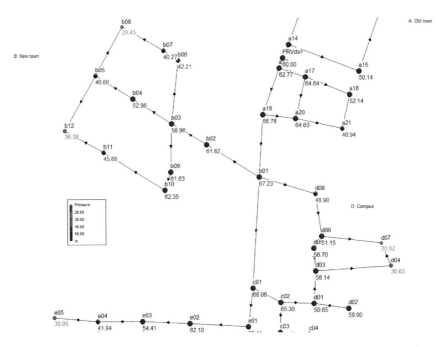

Figure 4.36 PRV starting in node a14, at 06:00 hours, added link b01-d08 (file NT317c)

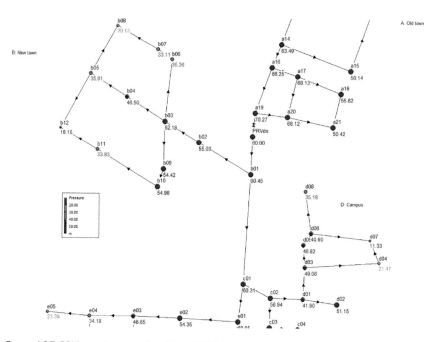

Figure 4.37 PRV starting in node a19, at 06:00 hours (file NT317d)

Comment:

• Despite the pressure problems, the PRV location shown in Figure 4.37 has been adopted for the analyses of the future demand conditions, with the setting increased to 70 mwc, also taking into consideration the fact that the second source is located in the CAMPUS district. Keeping the PRV upstream of this district gives options to control the contributions from the two sources, which is further elaborated in the questions dealing with the pump modelling/ design. Furthermore, the plan for current renovation with this position for the PRV remains the same as shown in Table 4.10.

Reading:
Appendix 8: 'EPANET – Version 2' (information on valve modelling)

4.5.8   Future demand growth (Question 3.2.1)

The network layout from the EPANET file NT317d has been carried forward for the analyses of the future demand situation at the end of the design period, which is after 20 years. The average demand after connecting 3700 consumers in the SUBURBS district is shown in Table 4.11. Applying the annual growth percentages per district, as shown in Table 4.3, with the specific demand increased from 100 to 106 l/c/d, and the uniform leakage demand percentage of 17 %, the average demand after 20 years will be as shown in Table 4.12.

Table 4.11 Structure of the base node demand and leakage – present situation

Node name	Elevation (msl)	Demand pattern	District	NAMETOWN – current situation					
				Population	q (l/c/d)	Q_c (l/s)	Leaks (%)	Q_l (l/s)	Q_d (l/s)
a01	82.48	Domestic	Old town	0	0	0.000	0	0.000	0.000
a02	68.48	Domestic	Old town	868	100	1.005	25	0.335	1.340
a03	65.77	Domestic	Old town	768	100	0.889	25	0.296	1.185
a04	48.90	Domestic	Old town	960	100	1.111	25	0.370	1.481
a05	48.81	Domestic	Old town	656	100	0.759	25	0.253	1.012
a06	42.95	Domestic	Old town	320	100	0.370	25	0.123	0.494
a07	42.07	Domestic	Old town	704	100	0.815	25	0.272	1.086
a08	41.51	Domestic	Old town	368	100	0.426	25	0.142	0.568
a09	41.21	Domestic	Old town	544	100	0.630	25	0.210	0.840
a10	43.83	Domestic	Old town	208	100	0.241	25	0.080	0.321
a11	59.47	Domestic	Old town	304	100	0.352	25	0.117	0.469
a12	56.74	Domestic	Old town	416	100	0.481	25	0.160	0.642
a13	57.12	Domestic	Old town	288	100	0.333	25	0.111	0.444
a14	38.33	Domestic	Old town	976	100	1.130	25	0.377	1.506
a15	40.85	Domestic	Old town	1216	100	1.407	25	0.469	1.877
a16	35.17	Domestic	Old town	704	100	0.815	25	0.272	1.086
a17	30.42	Domestic	Old town	496	100	0.574	25	0.191	0.765
a18	40.40	Domestic	Old town	592	100	0.685	25	0.228	0.914
a19	28.34	Domestic	Old town	1024	100	1.185	25	0.395	1.580
a20	29.91	Domestic	Old town	304	100	0.352	25	0.117	0.469
a21	45.60	Domestic	Old town	528	100	0.611	25	0.204	0.815
b01	24.82	Domestic	New town	960	100	1.111	12	0.152	1.263
b02	27.10	Domestic	New town	544	100	0.630	12	0.086	0.715
b03	28.16	Domestic	New town	1200	100	1.389	12	0.189	1.578
b04	32.41	Domestic	New town	752	100	0.870	12	0.119	0.989
b05	37.71	Domestic	New town	480	100	0.556	12	0.076	0.631
b06	41.81	Domestic	New town	912	100	1.056	12	0.144	1.199
b07	41.78	Domestic	New town	672	100	0.778	12	0.106	0.884
b08	46.80	Domestic	New town	592	100	0.685	12	0.093	0.779
b09	24.71	Domestic	New town	96	100	0.111	12	0.015	0.126
b10	23.74	Domestic	New town	96	100	0.111	12	0.015	0.126
b11	33.35	Domestic	New town	832	100	0.963	12	0.131	1.094
b12	42.72	Domestic	New town	1424	100	1.648	12	0.225	1.873
c01	19.06	Hostels	Hostels	448	100	0.519	15	0.092	0.610
c02	21.47	Hostels	Hostels	112	100	0.130	15	0.023	0.153
c03	20.35	Hostels	Hostels	112	100	0.130	15	0.023	0.153
c04	24.38	Hostels	Hostels	176	100	0.204	15	0.036	0.240
c05	12.50	Hostels	Hostels	272	100	0.315	15	0.056	0.370
c06	13.04	Hostels	Hostels	640	100	0.741	15	0.131	0.871
c07	21.88	Hostels	Hostels	336	100	0.389	15	0.069	0.458
d01	36.06	Campus	Campus	176	100	0.204	15	0.036	0.240
d02	26.27	Campus	Campus	160	100	0.185	15	0.033	0.218
d03	28.55	Campus	Campus	288	100	0.333	15	0.059	0.392
d04	54.35	Campus	Campus	560	100	0.648	15	0.114	0.763
d05	30.01	Campus	Campus	336	100	0.389	15	0.069	0.458
d06	35.63	Campus	Campus	336	100	0.389	15	0.069	0.458
d07	64.33	Campus	Campus	240	100	0.278	15	0.049	0.327
d08	40.20	Campus	Campus	768	100	0.889	15	0.157	1.046
e01	17.91	Domestic	Suburbs	656	100	0.759	20	0.190	0.949

(Continued)

Table 4.11 (Continued)

Node name	Elevation (msl)	Demand pattern	District	NAMETOWN – current situation					
				Population	q (l/c/d)	Q_c (l/s)	Leaks (%)	Q_l (l/s)	Q_d (l/s)
e02	17.77	Domestic	Suburbs	869	100	1.006	20	0.251	1.257
e03	22.99	Domestic	Suburbs	748	100	0.866	20	0.216	1.082
e04	30.10	Domestic	Suburbs	823	100	0.953	20	0.238	1.191
e05	38.88	Domestic	Suburbs	652	100	0.755	20	0.189	0.943
e06	16.13	Domestic	Suburbs	529	100	0.612	20	0.153	0.765
e07	22.96	Domestic	Suburbs	909	100	1.052	20	0.263	1.315
e08	35.39	Domestic	Suburbs	610	100	0.706	20	0.177	0.883
f01	9.66	Domestic	Village	4000	100	4.630	20	1.157	5.787
			TOTAL:	35,560		42.157		9.922	51.080

Table 4.12 Structure of the base node demand and leakage – forecast after 20 years

Node name	Elevation (msl)	Demand pattern	District	NAMETOWN – future situation after 20 years					
				Population	q (l/c/d)	Q_c (l/s)	Leaks (%)	Q_l (l/s)	Q_d (l/s)
a01	82.48	Domestic	Old town	0	0	0.000	0	0.000	0.000
a02	68.48	Domestic	Old town	1124	106	1.379	17	0.282	1.661
a03	65.77	Domestic	Old town	994	106	1.220	17	0.250	1.470
a04	48.90	Domestic	Old town	1243	106	1.525	17	0.312	1.837
a05	48.81	Domestic	Old town	849	106	1.042	17	0.213	1.255
a06	42.95	Domestic	Old town	414	106	0.508	17	0.104	0.612
a07	42.07	Domestic	Old town	912	106	1.118	17	0.229	1.347
a08	41.51	Domestic	Old town	476	106	0.585	17	0.120	0.704
a09	41.21	Domestic	Old town	704	106	0.864	17	0.177	1.041
a10	43.83	Domestic	Old town	269	106	0.330	17	0.068	0.398
a11	59.47	Domestic	Old town	394	106	0.483	17	0.099	0.582
a12	56.74	Domestic	Old town	539	106	0.661	17	0.135	0.796
a13	57.12	Domestic	Old town	373	106	0.457	17	0.094	0.551
a14	38.33	Domestic	Old town	1264	106	1.550	17	0.318	1.868
a15	40.85	Domestic	Old town	1574	106	1.932	17	0.396	2.327
a16	35.17	Domestic	Old town	912	106	1.118	17	0.229	1.347
a17	30.42	Domestic	Old town	642	106	0.788	17	0.161	0.949
a18	40.40	Domestic	Old town	766	106	0.940	17	0.193	1.133
a19	28.34	Domestic	Old town	1326	106	1.627	17	0.333	1.960
a20	29.91	Domestic	Old town	394	106	0.483	17	0.099	0.582
a21	45.60	Domestic	Old town	684	106	0.839	17	0.172	1.011
b01	24.82	Domestic	New town	1427	106	1.750	17	0.358	2.109
b02	27.10	Domestic	New town	808	106	0.992	17	0.203	1.195
b03	28.16	Domestic	New town	1783	106	2.188	17	0.448	2.636
b04	32.41	Domestic	New town	1117	106	1.371	17	0.281	1.652
b05	37.71	Domestic	New town	713	106	0.875	17	0.179	1.054
b06	41.81	Domestic	New town	1355	106	1.663	17	0.341	2.003
b07	41.78	Domestic	New town	999	106	1.225	17	0.251	1.476
b08	46.80	Domestic	New town	880	106	1.079	17	0.221	1.300
b09	24.71	Domestic	New town	143	106	0.175	17	0.036	0.211
b10	23.74	Domestic	New town	143	106	0.175	17	0.036	0.211
b11	33.35	Domestic	New town	1236	106	1.517	17	0.311	1.827

Node name	Elevation (msl)	Demand pattern	District	NAMETOWN – future situation after 20 years					
				Population	q (l/c/d)	Q_c (l/s)	Leaks (%)	Q_l (l/s)	Q_d (l/s)
b12	42.72	Domestic	New town	2116	106	2.596	17	0.532	3.128
c01	19.06	Hostels	Hostels	615	106	0.755	17	0.155	0.910
c02	21.47	Hostels	Hostels	154	106	0.189	17	0.039	0.227
c03	20.35	Hostels	Hostels	154	106	0.189	17	0.039	0.227
c04	24.38	Hostels	Hostels	242	106	0.297	17	0.061	0.357
c05	12.50	Hostels	Hostels	374	106	0.458	17	0.094	0.552
c06	13.04	Hostels	Hostels	879	106	1.079	17	0.221	1.299
c07	21.88	Hostels	Hostels	462	106	0.566	17	0.116	0.682
d01	36.06	Campus	Campus	251	106	0.309	17	0.063	0.372
d02	26.27	Campus	Campus	229	106	0.280	17	0.057	0.338
d03	28.55	Campus	Campus	411	106	0.505	17	0.103	0.608
d04	54.35	Campus	Campus	800	106	0.982	17	0.201	1.183
d05	30.01	Campus	Campus	480	106	0.589	17	0.121	0.710
d06	35.63	Campus	Campus	480	106	0.589	17	0.121	0.710
d07	64.33	Campus	Campus	343	106	0.421	17	0.086	0.507
d08	40.20	Campus	Campus	1097	106	1.346	17	0.276	1.622
e01	17.91	Domestic	Suburbs	884	106	1.084	17	0.222	1.306
e02	17.77	Domestic	Suburbs	1170	106	1.436	17	0.294	1.730
e03	22.99	Domestic	Suburbs	1007	106	1.236	17	0.253	1.489
e04	30.10	Domestic	Suburbs	1108	106	1.360	17	0.279	1.638
e05	38.88	Domestic	Suburbs	878	106	1.077	17	0.221	1.298
e06	16.13	Domestic	Suburbs	712	106	0.874	17	0.179	1.053
e07	22.96	Domestic	Suburbs	1224	106	1.502	17	0.308	1.810
e08	35.39	Domestic	Suburbs	822	106	1.008	17	0.206	1.214
f01	9.66	Domestic	Village	7224	106	8.863	17	1.815	10.679
			TOTAL:	50,575		62.048		12.709	74.756

Clarification:
- The average demand of node b12 shown in both tables does not include the hotel's average demand of 0.694 l/s calculated based on the number of hotel beds. This demand will not grow during the design period because no expansion of the hotel capacity is expected in this period. Hence, the real demand growth in this node is lower than shown in the table (1.873 + 0.694 = 2.567 l/s at present, against 3.128 + 0.694 = 3.822 l/s after 20 years, which is an increase of 48.9 %)

Conclusions:
- The total population growth in the town is from 35,560 to 50,575 inhabitants, which is 42.2 %.
- The total average consumption growth is from 42.157 to 62.048 l/s, which is an increase of 47.2 %. This percentage is higher than the population growth percentage, resulting from the specific demand growth.
- The total average leakage growth is from 9.922 to 12.709 l/s, which is an increase of 28.1 %. This percentage is lower than the consumption growth percentage, resulting from the lower average leakage percentage after the network renovation took place.

- The total average demand growth is from 51.080 to 74.756 l/s, which is an increase of 46.4 %. This percentage is lower than the consumption growth percentage, resulting from the lower average leakage percentage.

4.5.9   Future demand modelling (Question 3.2.2)

A new file named NT322a has been created by introducing the average demand (consumption and leakage) displayed in Table 4.12 to the file NT317d. The hydraulic simulation of this scenario shows the system flow as in Figure 4.38. The results indicate the range of the demands between 67.68 l/s at 00:00/24:00 hours (48.21 l/s at present; hence, the growth of 40.4 %), and 261.29 l/s at 06:00 hours (176.61 l/s at present; hence, the growth of 47.9 %). The red line in the figure shows the

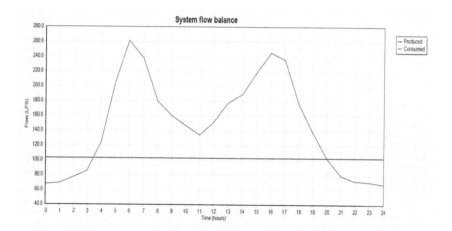

Figure 4.38 NT322a: the future demand variation on the maximum consumption day

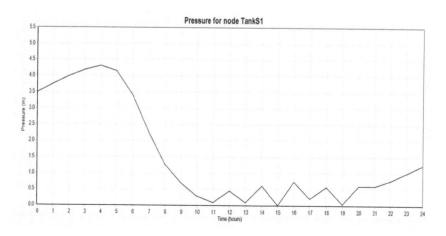

Figure 4.39 NT322a: source tank volume variation for future demand growth (S1)

constant feed of the tank in S1, which originates from the present condition and is still 102.81 l/s. The average system flow for the future demand scenario is 150.14 l/s; thus 46.0% higher that the source feed in the present conditions.

Obviously, if less water is supplied than is demanded, the network will not function properly. This can be seen on the volume variation of the tank in S1, shown in Figure 4.39, where the balancing pattern has completely disappeared.

Clarification:
- In this initial phase of the future demand analyses, the average feed of the source tank is kept at the present demand level; this also points out the limitation of the source which is at maximum capacity (actually, above it). Thus, to mitigate any future demand increase, the second source needs to be activated.
- To verify the extent of the problem, the minimum level in the tank has been reduced to 0.0 m, i.e. the use of the emergency volume is also allowed in what in fact is not a regular situation. Figure 4.39 shows that this volume is also utilised very quickly after the morning demand peak is over.
- When the volume reaches the minimum level, EPANET disconnects the tank from the network. Then follows an interval when the tank is subsequently filled/connected and immediately emptied/disconnected, which is observed in the period between 10:00 and 20:00 hours. Later, the demand drops allowing some water to accumulate in the tank, but not enough to start the demand balancing on the next day.

Comment:
- In the demand-driven mode of calculation, the system is 'supplied' even if there is no water in the tank. This may lead to a confusion while trying to reconcile the pressures in the network with the volume variation in the tank, which can reflect deficiencies in the numerical algorithm.

The snapshot results during the maximum consumption hour (at 06:00) on the maximum consumption day are shown in Figure 4.40 for the pressures, and in Figure 4.41 for the hydraulic gradients.

Conclusions:
- Figure 4.40 shows that the present network layout is not able to convey the future demand without further renovation taking place. The extent of the problem is less than shown in Figure 4.15, but still very serious.
- The most critical situation is in the NEW TOWN district, with a minimum pressure of -26.11 mwc in node b12 (where the hotel is located). Furthermore, negative pressures also appear in the CAMPUS, SUBURBS and VILLAGE districts.
- As is the case for the present demand scenario, the low pressures in general originate from too small pipes generating high friction losses, which is reflected in very high values of hydraulic gradients shown in Figure 4.41; again, generally smaller than those shown in Figure 4.16 but still very high. The highest value of the hydraulic gradient of 116.56 m/km is registered in pipe p66 in the SUBURBS district. The hydraulic gradients in the NEW TOWN district are also very high (p38–73.31 m/km, p34–69.30 m/km).

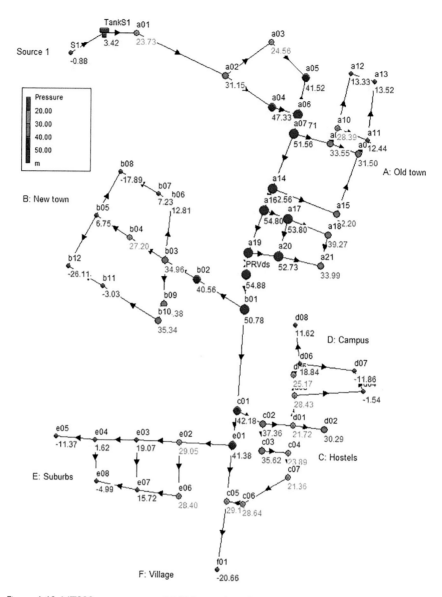

Figure 4.40 NT322a: pressures at 06:00 hours (mwc)

Comments:

• In modelling approaches where the sources are simulated as tanks, given their volume variation, the minimum pressure may occur outside the maximum consumption hour. This is not the case in this example but may be a consequence of an empty tank.

• Demand-driven simulations may show the peculiar results in the case of irregular scenarios. This is why they are more efficient for regular demand scenarios.

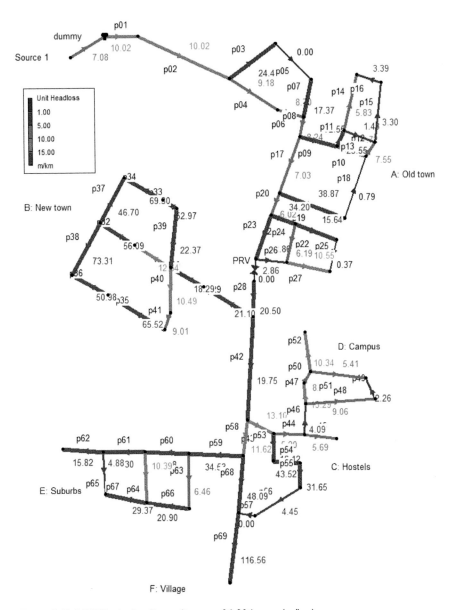

Figure 4.41 NT322a: hydraulic gradients at 06:00 hours (m/km)

The modelling of extreme scenarios will be more accurately represented by the pressure-driven demand mode of calculation.

The overview of the critical nodes and pipes in Figure 4.42, as shown for the present demand scenario in Figure 4.17, indicates the nodes with negative pressure (left) and the pipes with a hydraulic gradient above 10 m/km (right), both in red.

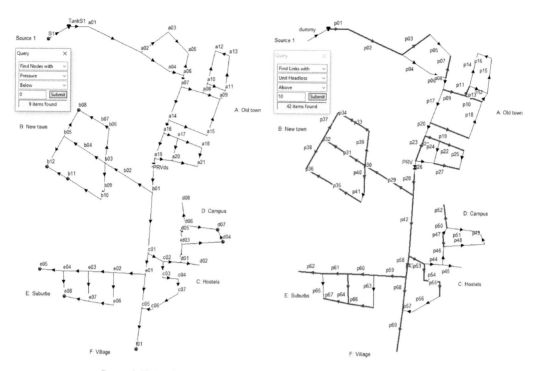

Figure 4.42 NT322a: filtering of the results at 06:00 hours

Conclusion:

• A comparison of figures 4.17 and 4.42 shows 42 nodes with negative pressure in the first figure, and only 9 in the second figure, which confirms the lower level of irregularities in the second network. Nevertheless, comparing the hydraulic gradients, 53 pipes have values above 10 m/km in the first case and 42 in the second case, which is still quite a lot. As it would not make sense to replace the same pipes twice in an interval of just 10 years (indicated as the second phase renovation time scale in Question 4.3.1), these pipes that need to be replaced in present conditions could immediately be upgraded to the diameter that satisfies the future demand conditions, or actually their replacement could be postponed to year 10, assuming that this delay does not significantly negatively affect the pressure in the network. This decision can also be based on financial constraints; this is further elaborated in the answers to Question 4.3.1.

4.5.10 Connection of the second source (Question 3.2.3)

An arbitrary location of the second source has been selected in the CAMPUS district, and the reservoir with a surface water level of 80 msl was connected to node d07 with a pipe that is 375 m long and has a 150 mm diameter. The pipe length and the water elevation in the reservoir have been approximated proportionally to the lengths of the neighbouring pipes and the nodal elevations in the surrounding area. This model modification has been named NT323a and the results of hydraulic simulation are shown in Figure 4.43.

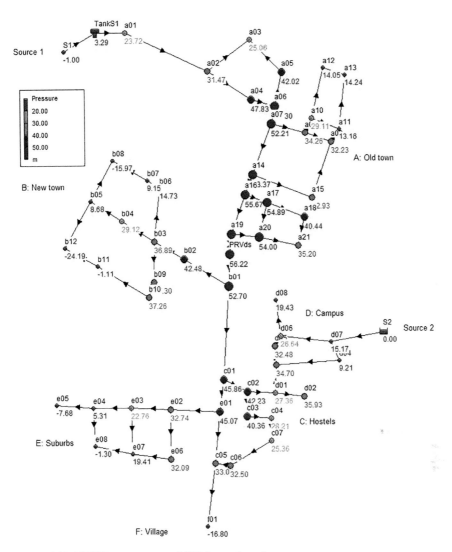

Figure 4.43 NT323a: pressures at 06:00 hours (mwc)

Conclusion:

- A comparison of the results in figures 4.40 and 4.43 does not show significant pressure improvements after connecting the second source. The pressures have improved in the CAMPUS area but not much outside it because the source in S2 has a much lower elevation than the one in S1 and is therefore unable to supply more significant demand. This can be concluded after plotting the flow variation in the connecting pipe, which is shown in Figure 4.44. The results show the maximum supply from S2 of 6.89 l/s (at 06:00 hours), which is less than 3% of the total demand during the maximum consumption hour. Therefore, the second source can be utilised efficiently only with the help of pumping, which will offer more flexibility in balancing the contributions from two sources.

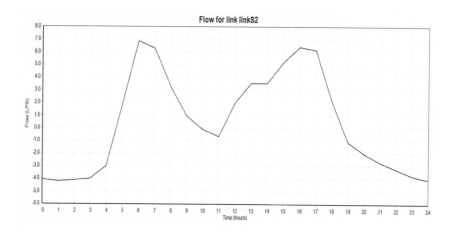

Figure 4.44 NT323a: supply from the reservoir in S2.

Table 4.13 Pipe renovation list, areas A and B (material PE, k = 0.25 mm)

NAMETOWN – *renovated pipes for future demand scenario*

Pipe name	Node from	Node to	New D (mm)	Length (m)	Old D* (mm)
p09	a07	a08	150	256	100
p11	a08	a10	100	111	80
p29	b01	b02	250	376	200 (150)
p30	b02	b03	250	248	200 (150)
p31	b03	b04	200	278	150 (100)
p32	b04	b05	150	270	100 (50)
p33	b06	b07	100	106	80
p34	b07	b08	80	290	50
p35	b10	b11	100	439	80
p36	b11	a12	80	269	50
p37	b08	b05	100	333	50
p38	b05	b12	150	380	80 (50)

*Two values in the column 'Old D' indicate already renovated pipes; the original diameter is shown in brackets.

4.5.11 Renovation plan in the future demand scenario (Question 3.2.4)

Further improvement of the pipe diameters must be analysed before installing the pumping station in the network. As is the case with the pressures, the hydraulic gradients shown in Figure 4.41 for layout NT322a will not be significantly different in layout NT323a.

For easier monitoring, the plan of renovation is illustrated in two stages. Firstly, to improve the situation in areas A and B, twelve pipes have been enlarged as indicated in Table 4.13. Five out of these pipes (p29, p30, p31, p32 and p38) have

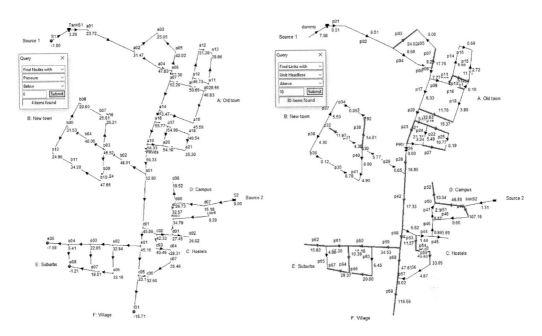

Figure 4.45 NT324a: enlargement of the pipes in areas A and B: the results at 06:00 hours.

Table 4.14 Pipe renovation list, areas C – F (material PE, k = 0.25 mm)

NAMETOWN – *renovated pipes for future demand scenario*

Pipe name	Node from	Node to	New D (mm)	Length (m)	Old D* (mm)
p59	e01	e02	200	361	150 (125)
p60	e02	e03	200	289	150 (80)
p61	e03	e04	150	277	100 (80)
p62	e04	e05	100	266	80 (50)
p66	e06	e07	100	280	50 (25)
p67	e07	e08	80	282	50 (25)
p68	e01	c05	200	366	150
p69	c05	f01	150	452	125

*Two values in the column 'Old D' indicate already renovated pipes; the original diameter is shown in brackets.

already been enlarged and the actual renovation plan in their case will be finalised after forming a complete picture of the final design (the two options are to replace them immediately with a larger diameter, or after 10 years but maintaining the original diameter). The PRV setting is still kept at 70 mwc in this case. The extent of the improvement of this stage is to be seen in Figure 4.45 (layout NT324a).

To improve the situation in the lower part of the network (areas C – F), eight additional pipes have been enlarged as shown in Table 4.14. Six out of these that have already been enlarged (p59, p60, p61, p62, p66 and p67) are added to the list of the five pipes in Table 4.13, for later reconsideration. In this stage the PRV setting has been lowered to 60 mwc, for further pressure optimisation in the lower part of the network. The extent of the improvement of this stage is to be seen in Figure 4.46 (layout NT324b).

Conclusions:

- The results show a gradual improvement of the situation. Only three nodes still show the negative pressure in Figure 4.45 (e05, e08 and f01), while there are no demand nodes with negative pressure in Figure 4.46. However, the minimum pressure criterion of 20 mwc is not satisfied in four demand nodes (e05, d04, d07 and d08). This situation will be mitigated after adding the pumping station.
- The situation with the hydraulic gradients is less favourable but not dramatic. Figure 4.45 shows 30 pipes with a hydraulic gradient above 10 m/km, while there are 21 pipes in this category, shown in Figure 4.46. Continuing the enlargement of pipe diameters would improve this situation but the improvement can also be reached by installing the pumping station, which will change the balance of supply from the two sources with a positive effect on the distribution of flows/ friction losses in the pipes, and volume variation of the balancing tank(s).

Comment:

- Pipe enlargement has implications for the investment costs while additional pumping capacity increases the operational costs. The trade-off analysis between these two requires a more detailed assessment for which optimization tools with sophisticated software applications are available nowadays. This exercise only discusses the hydraulic impact of these two choices.

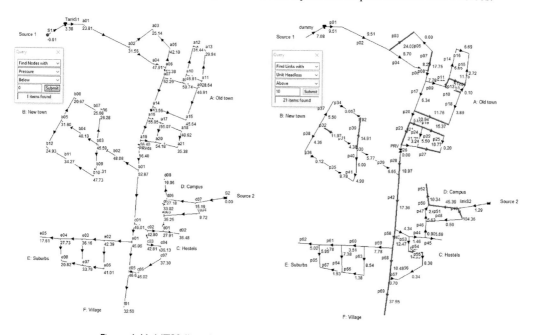

Figure 4.46 NT324b: enlargement of the pipes in areas C – F: the results at 06:00 hours

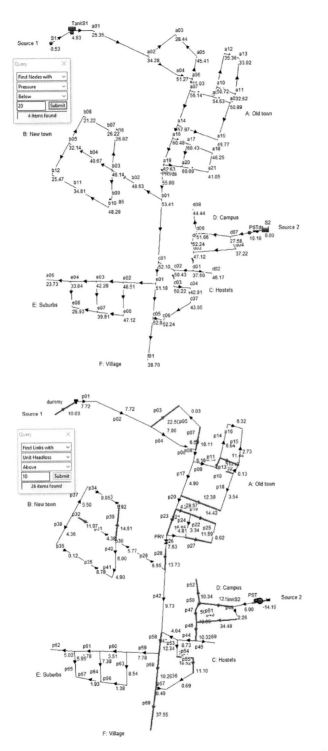

Figure 4.47 NT324c: PST in D, Q_p/H_p = 30 l/s/15 mwc, PRV = 55 mwc: the results at 06:00 hours

Reading:
Volume 1, Section 4.1: 'The planning phase'.

Self-study:
Spreadsheet Lesson 1–7 (Appendix 7).

Further considerations include pumping being modelled with a synthetic pumping curve defined by the duty flow and duty head. Figure 4.47 shows the situation in which a pump with a duty flow of $Q_p = 30$ l/s and a duty head of $H_p = 15$ msl is connected to the second source. For further pressure optimisation in the downstream part of the network, the PRV setting has been reduced to 55 mwc.

Clarification:
- The negative unit headloss in the link of the pump (of -14.16 mwc) actually indicates the pumping head. At 06:00 hours, the pump is therefore operating still relatively close to the value of the duty head.

Conclusions:
- The pumping curve has been selected to be able to bring sufficient pressure in the SUBURBS district. This also raises the pressures in the downstream part of the network in general, which enables the reduction of the PRV setting.
- Because less water is now supplied from the first source, the balancing role of the tank S1 has been largely restored, which can be seen in Figure 4.48; however, this has been achieved with a much reduced emergency volume. The lower supply from S1 has further reduced the friction losses in the upper part of the network, so that the pressures in the NEW TOWN district are not affected by the lower PRV setting; they are actually slightly higher than those shown in Figure 4.46.
- As expected, adding a pump of relatively small size has only marginally affected the hydraulic gradients. The PRV is also not affected by the pump and it is still open during the entire 24-hour simulation day despite the lowered setting.
- Nevertheless, the balance of supply between S1 and S2 has changed in favour of the new source, which is shown in Figure 4.49. The results show the maximum supply from S2 of 32.43 l/s (at 06:00 hours), which is approximately 15 % of the total demand during the maximum consumption hour. This is still slightly low, making the contribution of the source in S1 too high; the average feed of the tank in S1 is 122.77 l/s i.e. well above the available maximum of this source, which is 90 l/s.

In the next iteration, the duty flow and duty head have been increased to 140 l/s and 80 mwc, respectively. This has been done in order to further increase the contribution from the source in S2 and reduce the contribution in S1. The results of this simulation are shown in Figure 4.50.

Conclusions:
- The main objective of the intervention has been reached. The balance of supply between S1 and S2 has significantly changed in favour of the new source, which is shown in Figure 4.51. The results show the supply from S2 of 53.82 l/s at 06:00 hours, which is approximately 21 % of the total demand during the maximum consumption hour.

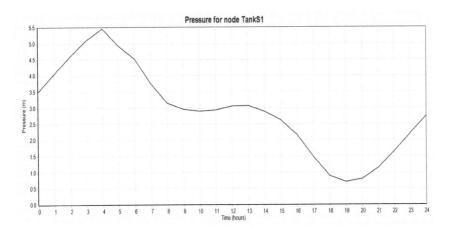

Figure 4.48 NT324c: source tank volume variation in S1 after a small pump is added in S2

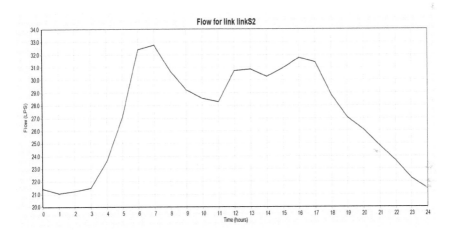

Figure 4.49 NT324c: supply from the reservoir in S2

- At the same time, the average feed of the tank in S1 is lowered to 98.76 l/s i.e. closer to the available maximum of this source, which also gives a further improved balancing curve of the tank in S1, which can be seen in Figure 4.52. However, the emergency volume is still significantly depleted compared to the present demand scenario.
- Nevertheless, the relatively large pump has significantly increased the pressures in the downstream part of the network, and the hydraulic gradients have also been increased, which can be seen in Figure 4.50. Thus, the additional pumping energy has been immediately utilized through the high friction losses, clearly pointing out the limitation caused by the small pipe diameters. Thus, the selection of the pump may also require additional expansion of the network in the vicinity of the new source.

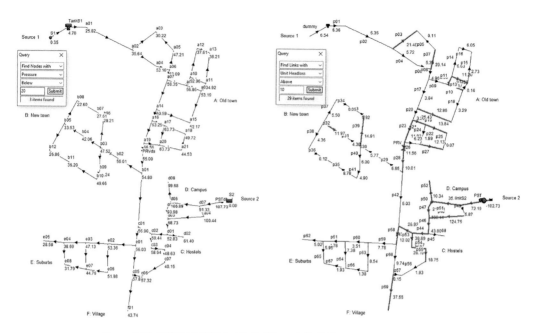

Figure 4.50 NT324d: PST in D, Q_p/H_p = 140 l/s/80 mwc, PRV = 55 mwc: the results at 06:00 hours

- Keeping the same setting of 55 mwc, the PRV operation did not change significantly by installing the large pump.

Comments:
- An intervention that solves one problem may easily create a new one. It is therefore important to be aware of all the implications. All the network components: storage, pumping, and transmission (including pipes and valves), operate dependently, continuously impacting each other.
- Putting a (much) larger pump in the network without further pipe enlargement does not seem to be an efficient plan. The balance of supply from the two sources therefore needs to be solved differently.
- Adding multiple sources does not mean that all of them reach the maximum supply during the maximum consumption hour. Figure 4.51 shows the maximum supply of S2 to be at 12:00 hours (54.91 l/s), while in Figure 4.49, this happens at 07:00 hours (32.78 l/s).

Because the second source is downstream of the PRV and additional pumping is needed to reach the minimum pressure in this part of the network, it is interesting to analyse to what extent the location and the setting of the PRV can influence the balance of supply between the two sources. First of all, we analyse a scenario in which the PRV has been temporarily removed from the network. In this scenario the average supply from S1 has been reduced to 70 l/s and the pump in S2

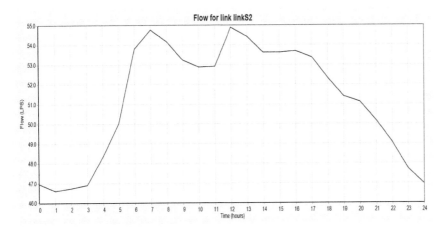

Figure 4.51 NT324d: supply from the reservoir in S2

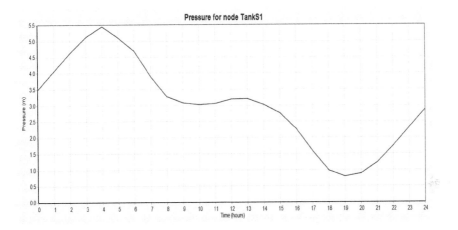

Figure 4.52 NT324d: source tank volume variation in S1 after a large pump is added in S2

is designed with a duty follow and duty head of 15 l/s and 30 mwc, respectively. Moreover, the pump has been modelled as a booster pump, in series with pipe p42, to reduce the pressures in the CAMPUS district. The results of this simulation are shown in Figure 4.53.

Clarification:
* The booster pump has been modelled in series with pipe p42 by adding a dummy node PSTds with the same elevation as node c01. The modelling approach is the same as in the case of the PRV, in order to keep the hydraulic performance (i.e. the friction loss) of pipe p42 intact. Simply replacing this pipe with a booster pump would be wrong.

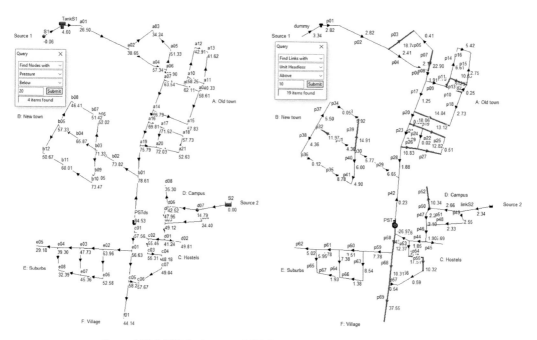

Figure 4.53 NT324e: booster PST, Q_p/H_p = 15 l/s/30 mwc, no PRV: the results at 06:00 hours

Conclusions:

• In the absence of a PRV, the pressure in the downstream part of the network increases again. The exception is node d07, where the pressure is below the minimum of 20 mwc (14.79 mwc). The reason is the low elevation of the source in S2 because this node is actually supplied by gravity from S2.

• The situation with the hydraulic gradients has improved somewhat but there are still 19 pipes with gradients above 10 m/km. This is to be further elaborated in the final design layout.

• Hence, the network can operate without the PRV but the pressure optimisation with the source/booster PST alone will not be an ideal solution, due to the topography in the network.

• In this scenario, this supply from the two sources becomes more even, which can be observed from Figure 4.54. The results show the supply from S2 of 124.34 l/s at 06:00 hours, which is approximately 48 % of the total demand during the maximum consumption hour.

• At the same time, because the average feed of the tank in S1 is lowered to 70 l/s, the emergency volume has been restored, which can be seen in Figure 4.55.

One of the water quality simulation options in EPANET is to check the contribution of each source to each nodal demand at a particular point in time. This is called *source tracing* and gives very useful information about the mixing of water from different sources as a result of a specific hydraulic operation and the location

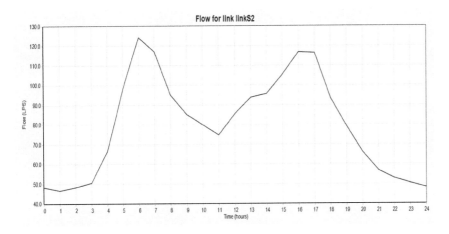

Figure 4.54 NT324e: supply from the reservoir in S2

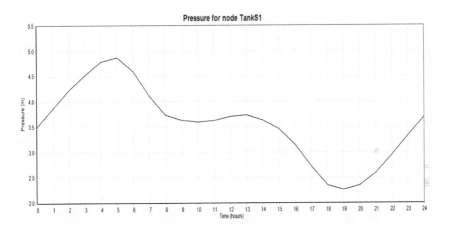

Figure 4.55 NT324e: source tank volume variation in S1 after a large pump is added in S2

of the main network components. In a layout without PRV, Figure 4.56 shows the results for the second source (S2).

Clarifications:
*   The trace option is activated in EPANET by selecting the *Quality Options* editor from the browser, as shown in Figure 4.57. The same editor can be opened in the menu option '**Project>>Analysis Options . . .** ' assuming that it is a default (which is normally the editor of *Hydraulics Options*).
*   The only two fields to modify are (1) for *Parameter* and (2) for *Trace Node*. The other fields are related to the calculation of the WQ residuals, which is not the scope of this exercise. More about this is to be found in the original EPANET manual.

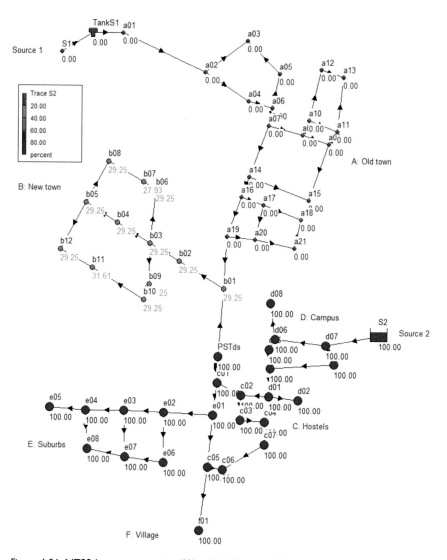

Figure 4.56 NT324e: source tracing (S2) at 06:00 hours (%)

- In the EPANET files discussed with this example, the trace option has been activated in all the files with a source in S2 connected to the network (starting from NT323a).

Conclusion:
- Not surprisingly, all the nodes upstream of the booster PST are supplied 100% from S2. More interesting, the mixing of the water from S1 and S2 (approximately 30%) takes place in the nodes of the NEW TOWN district,

while the OLD TOWN is supplied entirely from S1 (0% from S2). This situation is evidently caused by the selected location and the capacity of the PST.

Comments:
- The results shown in Figure 4.56 do not necessarily reflect the correct situation because they are influenced by the initial values at the beginning of the simulation, which are always 0% in all the nodes. This is therefore why the WQ simulations are often run for period longer than 24 hours (48 or 72 hours) to examine more consistent variation patterns. The duration of EPS can be adjusted in the same browser as shown in Figure 4.57, by selecting the *Times Options* editor. Nevertheless, Figure 4.56 is shown to illustrate a useful WQ feature of EPANET.

Given the positive effects, the concept of the booster pump has been carried forward by reinstalling the PRV in the network for better pressure management. In view of the location of the booster, which looks good from the perspective of the pressure distribution in the downstream part of the network, the PRV has been moved further upstream and installed in series with pipe p08; it has a setting of 45 mwc.

Furthermore, the source in S2 has been modelled in the same way as the one in S1. The modelling approach used here is the same as explained in Section 4.5.6 related to Question 3.1.6, by using the 24-hour supply pattern from the link connected to S2. This has resulted in a tank with a bottom level at 75 msl, a diameter of 22 m, a depth of 5 m and an initial depth of 3.5 m. With this, the total volume of the tank has been designed at approximately 1900 m³. Finally, the booster station in the same location as in the previous scenario has been designed with a duty flow of 12 l/s and a duty head of 15 mwc, resulting in an average feed of the tank in S1 of 67.32 l/s, and in S2 of 82.76 l/s.

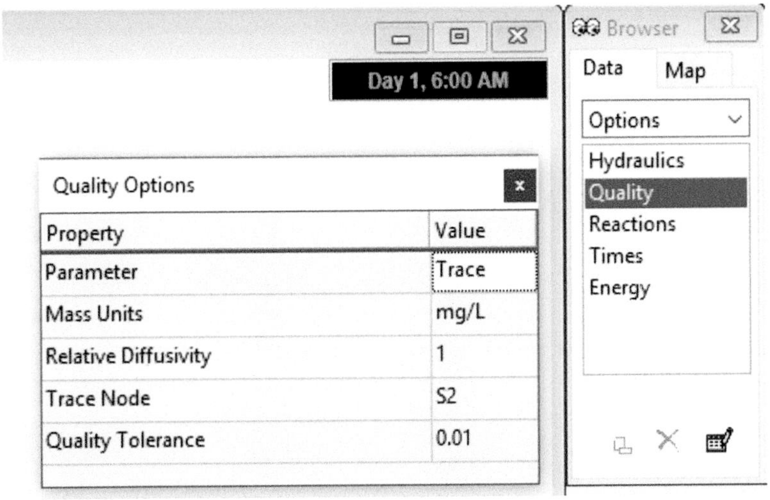

Figure 4.57 Activating the source tracing WQ modelling option (for S2)

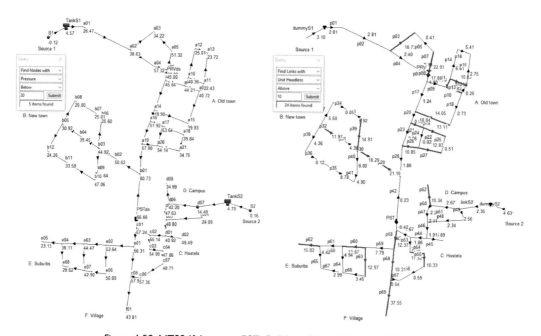

Figure 4.58 NT324f: booster PST, $Q_p/H_p = 12$ l/s/15 mwc, PRV = 45 mwc: the results at 06:00 hours

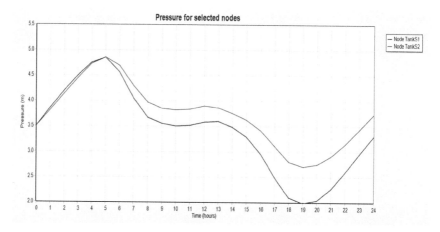

Figure 4.59 NT324f: source tank volume variation in S1 and S2

Thus, the network has been designed to maintain the function of the tank in S1 without any further renovation, and provide more future supply from the source in S2. The results of this scenario are shown in figures 4.58 and 4.59.

Conclusions:

- Except for the pressure in node d07, the results show a fairly balanced pressure distribution with the maximum pressure in node b01 (of 60.73 mwc).

The pressure in node d07 will be considered as an isolated case that can be treated with local pressure boosting. After all, this node is directly connected to S2 and will therefore have a very reliable supply (by gravity) at a somewhat lower pressure in extreme demand conditions. Being strict on the condition of min 20 mwc in this case by further improvement of the distribution network would simply be too costly while the actual benefits are limited.

- Equally, the enlargement of the pipes with hydraulic gradients above 10 m/km are not seen to be of immediate concern, for the following reasons: (1) Figure 4.58 shows the situation in the most extreme hour of supply at the end of the design period; (2) the pipes in question do not significantly affect the pressures; (3) they carry mostly the gravity supply and as such the friction losses reflect the loss of pumping energy only to a lesser extent; (4) most of the pipes form loops with pipes of sufficiently low gradients and therefore this does not significantly affect the reliability of supply; (5) where the pipes are branched or at the end, enlargement of the diameter should exclusively be governed by the pressure criterion because increasing the diameter will not increase the supply reliability.
- Finally, the tank balancing curves indicate a fairly effective design of both tanks.

It is interesting to observe in this network layout that if the booster pump is switched off (for instance, due to an electricity failure), the network will be hydraulically split into the upstream and the downstream parts. However, the hydraulic performance will not be significantly affected, provided the balance of supply is shifted towards S1. The scenario with the PRV setting further reduced to 40 mwc and the supply from S1 close to the maximum capacity i.e. of 85 l/s, and the feed of S2 adjusted accordingly at 65 l/s, is shown in figures 4.60 and 4.61. The booster PST is closed.

Conclusion:
- Under regular supply conditions, the presented alternative looks attractive, having actually two gravity-fed networks from two different sources. However, if any of the sources needs to be temporarily closed or any of the bigger pipes connected to it fails, an emergency supply would be difficult to arrange. Providing a bypass without any pumping in such a situation would eliminate the possibility of balancing the supply of S1 and S2 that would be driven exclusively by the topography i.e. the elevation of the sources. Thus, pumping is still necessary for more flexible pressure management and balance of supply.

Comment:
- Switching the booster pump on in this scenario will unexpectedly generate warning messages as shown in Figure 4.62. The random timing of negative pressures indicates the numerical instability caused by modelling the sources S1 and S2 as negative demand nodes. Thus, sometimes it works, but quite often it does not if the network does not have at least one reservoir (a fixed head point). Once these warning occur, the simplest way to mitigate the

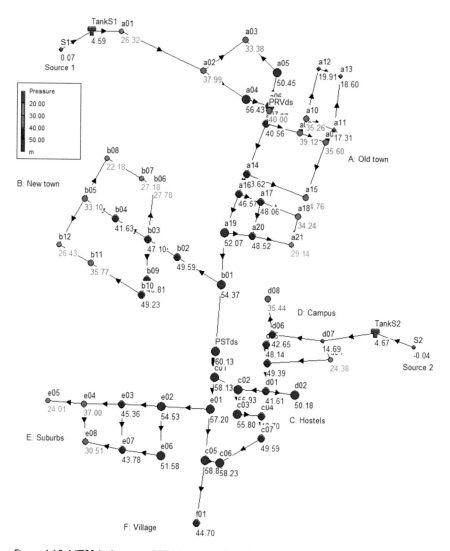

Figure 4.60 NT324g: booster PST closed, PRV = 40 mwc; pressures at 06:00 hours (mwc)

problem is to keep modelling one of the sources, in this case S2, as a reservoir, which was applied in the files NT323a and NT324a-e. Based on the actual feed calculated from S2, the tank volume can be designed with the help of the Spreadsheet Lesson 8–10, instead of being modelled in EPANET.

4.5.12   Selection of the actual pump units (Question 3.3.1)

The design layout carried forward for the selection of the actual pump units from the available KSB ETANORM catalogue is the one presented in the NT324f layout. The summary of the main components of this layout is given in the following bullets:

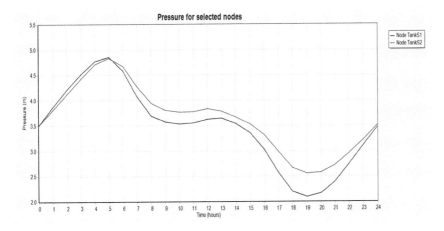

Figure 4.61 NT324g: source tank volume variation in S1 and S2 – booster PST closed

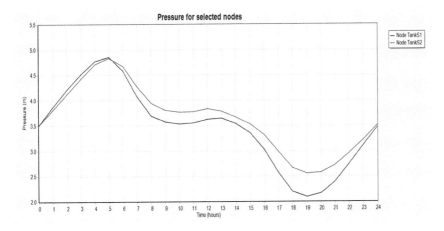

Figure 4.62 NT324g: booster PST open, unstable iterative process

- Modelling of the additional 3700 consumers in the SUBURBS district as elaborated in Section 4.5.1.
- Modelling of the hotel demand with 200 beds in node b12 of the NEW TOWN district as elaborated in Section 4.5.2.
- Pipe renovation as indicated in Table 4.10 for the current demand scenario.

- Modelling of the future demand growth as elaborated in Section 4.5.8, in Table 4.12.
- Pipe renovation as indicated in tables 4.13 and 4.14 for the future demand scenario.
- PRV located at node a06 of pipe p08, with a setting of 45 mwc.
- Average feed of the tank in source S1 of 67.32 l/s.
- Average feed of the tank in source S2 of 82.76 l/s.
- Tank in S1: bottom elevation at 105 msl, $D = 25$ m, $H = 5.5$ m, $H_{ini} = 3.5$ m, $V \approx 2580$ m^3.
- Tank in S2: bottom elevation at 75 msl, $D = 22$ m, $H = 5.0$ m, $H_{ini} = 3.5$ m, $V \approx 1900$ m^3.
- Synthetic pump curve for the booster pump of the duty flow of $Q_p = 12$ l/s, and a duty head of $H_p = 15$ mwc.

As the preferred pump size is based on synthetic curves, the following ETA-NORM (ETABLOC) fixed speed pumps operating at 1450 rpm and 50 Hz have been considered in the KSB documentation available with this exercise (model code/impeller diameter in mm): (1) 40–250/238 (page 49 in the catalogue), (2) 50–200/219 (page 53), (3) 50–250/215 (page 54), (4) 65–200/212 (page 58), and (5) 65–250/220 (page 59). Two pumps in parallel arrangement have been planned and a wider range of flows (and heads), indicated in Figure 4.63, tested with the idea to consider the second pump as a regular one (with smaller pump models), or as a stand-by unit (with bigger pump models). Moreover, for all the pumps under consideration, the efficiency curve has also been included in order to calculate the energy costs for a particular operational mode. The transfer of the pump data to EPANET was carried out manually, which is illustrated in Figure 4.64 for the largest selected model (65–250/220).

Clarifications:
- The format of the pumping curve diagrams in the KSB catalogue enables relatively easy readings of the points where the efficiency intersects with the efficiency values. In the selected example these are the dots with the efficiencies of 58, 68, 73, 74.3, and 73 per cent. The values for Q_{min} and Q_{max} (end of the curve) are extrapolated, based on the standard shape of the efficiency curve. Other manufacturers may opt for a different presentation of the pump and efficiency curves.
- The EPANET curve editor allows the entry of a few dozen points, which is very rarely utilized. Typical pump curves will be accurately described in no more than 10 sets of points.
- Unlike the case with the synthetic curves, the manufacturer ends the pump curve at the point where the operational range has been exhausted. The pump should not be operated beyond the end point because of low efficiency/high energy and high NPSH i.e. the risks of cavitation.

With the prepared network layout, including the actual pumping and efficiency curves (NT331a), the simulation was run using the largest pump considered

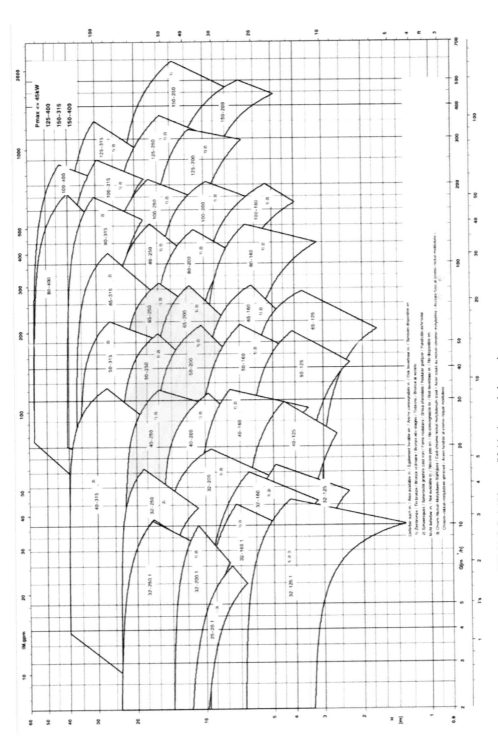

Figure 4.63 KSB pump selection ETABLOC, 1450 rpm, 50 Hz (www.ksb.com)

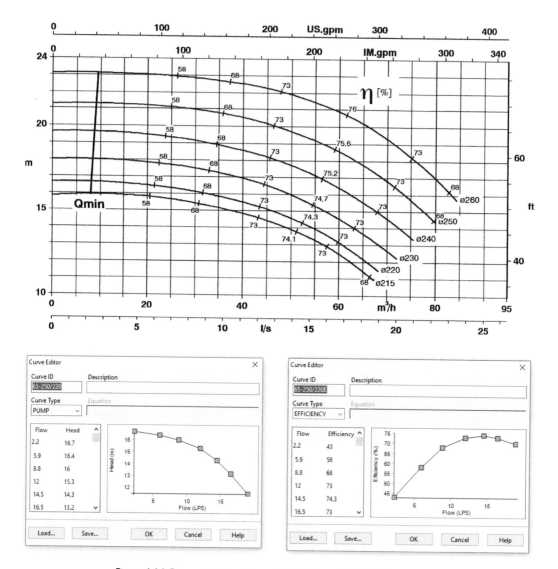

Figure 4.64 Pump data transfer to EPANET, model 65–250, impeller 220 mm

(65–250/220), one unit with the second unit kept closed. The programme returned the warning messages as shown in Figure 4.65.

Conclusions:

*   The warning messages shown in Figure 4.65 signal that the pump operation is beyond the suitable range for most of the day. This is not difficult to observe by comparing the pumping curve in Figure 4.64 with the range of simulated pump flows shown in Figure 4.66. The latter shows the range of flows between 18.11 and 21.76 l/s (with a single unit in operation), while

```
Status Report

Page 1                                    Thu Aug 29 13:41:04 2019

*********************************************************************
*                         E P A N E T                              *
*              Hydraulic and Water Quality                         *
*              Analysis for Pipe Networks                          *
*                    Version 2.00.12                               *
*********************************************************************

Analysis begun Thu Aug 29 13:41:04 2019

WARNING: Pump Pump1 open but exceeds maximum flow at 0:00:00 hrs.

WARNING: Pump Pump1 open but exceeds maximum flow at 1:00:00 hrs.

WARNING: Pump Pump1 open but exceeds maximum flow at 2:00:00 hrs.

WARNING: Pump Pump1 open but exceeds maximum flow at 3:00:00 hrs.

WARNING: Pump Pump1 open but exceeds maximum flow at 4:00:00 hrs.

WARNING: Pump Pump1 open but exceeds maximum flow at 5:00:00 hrs.

WARNING: Pump Pump1 open but exceeds maximum flow at 6:00:00 hrs.

WARNING: Pump Pump1 open but exceeds maximum flow at 7:00:00 hrs.

WARNING: Pump Pump1 open but exceeds maximum flow at 8:00:00 hrs.

WARNING: Pump Pump1 open but exceeds maximum flow at 9:00:00 hrs.

WARNING: Pump Pump1 open but exceeds maximum flow at 10:00:00 hrs.

WARNING: Pump Pump1 open but exceeds maximum flow at 11:00:00 hrs.

WARNING: Pump Pump1 open but exceeds maximum flow at 12:00:00 hrs.

WARNING: Pump Pump1 open but exceeds maximum flow at 13:00:00 hrs.

WARNING: Pump Pump1 open but exceeds maximum flow at 14:00:00 hrs.

WARNING: Pump Pump1 open but exceeds maximum flow at 15:00:00 hrs.

WARNING: Pump Pump1 open but exceeds maximum flow at 16:00:00 hrs.

WARNING: Pump Pump1 open but exceeds maximum flow at 22:00:00 hrs.

WARNING: Pump Pump1 open but exceeds maximum flow at 23:00:00 hrs.

WARNING: Pump Pump1 open but exceeds maximum flow at 24:00:00 hrs.
```

Figure 4.65 NT331a: booster PST open but the pump exceeds the maximum flow

the end point in the pump curve editor indicates the maximum pump flow of 19 l/s. Thus, the maximum pump flow is not exceeded only in the period between 16:00 and 22:00 hours.

- Switching the second unit on will not be helpful because the booster starts pumping even more flow from the source in S2, which will disturb the balancing patterns of the tanks, as can be seen in Figure 4.67. Hence, selecting the model 65–250/220 does not seem to be an ideal alternative.

Comment:

- Based on the fact that the synthetic pump curves are utilized over the unlimited range of flows may overlook the fact that the 'optimal' pump flows from the hydraulic perspective may actually be unfavourable from the perspective

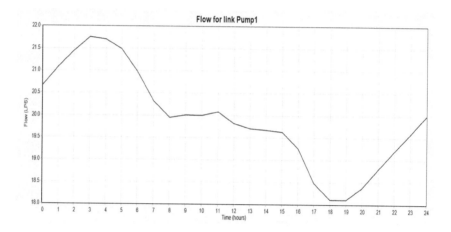

Figure 4.66 NT331a: booster PST flow (single unit in operation)

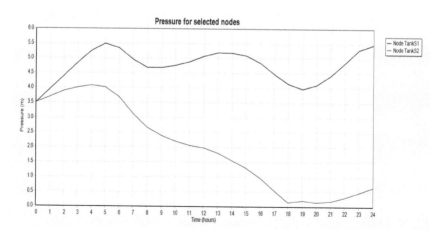

Figure 4.67 NT331a: tank volume variation for both booster pumps switched on

of the pump efficiency/energy consumption and NPSH requirements. It is therefore important to model the pump operation with a synthetic curve as close as possible to the selected values of the duty flow and duty head. This can also be achieved by increasing/reducing the number of pump units in parallel arrangement.

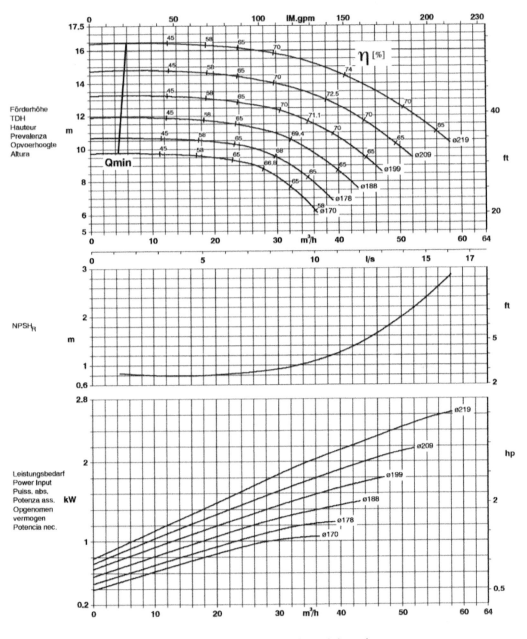

Figure 4.68 Selected pump, model 50–200, impeller 219 mm (www.ksb.com)

After going through further investigations by using other proposed pump models, the final choice in this design has been the model 50–200/219. The full information on this one is given in Figure 4.68. The data for the pump and efficiency curve have been captured in EPANET in the same way as shown in Figure 4.64.

Under regular operation, both pumps are switched on and the results of the computer simulation are shown in figures 4.69–4.73. Minor adaptation in this final scenario is the rounding of the tank feeds to 68 l/s in S1, and 82 l/s in S2 and the increase of the PRV setting to 50 mwc.

Conclusions:
• Under regular operation, Figure 4.70 shows a range of flows between 5.71 l/s and 10.84 l/s per each pump. Comparing this with the curve of the impeller of 219 mm in Figure 4.68 yields the operation with a range of pump efficiencies between 65 % and 74 %, the latter being the maximum possible. This is further converted into an average efficiency of 69.73 % and an average energy consumption of 1.87 kWh, or a total of 89.76 kWh/day for both pumps. This information is available in EPANET by running the menu option '**Report>>Energy**'.
• The pressures in the network have grown somewhat. This results from the higher PRV setting and facilitates the balancing tank operation favourably, which can be seen in Figure 4.69. It is quite possible that by keeping the lower PRV setting while choosing the smaller impeller diameter, say 209 mm, a further pressure optimisation could take place without affecting the tanks. This has however not been done and the design is completed with some additional potential reliability provision in mind. Moreover, both pumps operate full time, while switching them off occasionally could also help in further pressure and energy optimisation. Hence, this final design is still not necessarily the optimum one.

Comment:
• All the illustrated hydraulic simulations show a certain optimum for the selected layout, operation and demand scenario on the maximum consumption day. In reality however, only the layout stays more or less the same, while it is the subject of a much wider range of operation and demand scenarios during the year/the design period. It is therefore impossible to have a network that will satisfy all the criteria on such a wide horizon. Optimisation algorithms available in state-of-the-art network modelling software can help but the final decision is always taken by the design engineers.

4.5.13   Pump motor selection (Question 3.3.2)

The analysis of the operating regimes under Question 3.2.4, using EPANET synthetic pump curves defined by one specified point reflecting the preferred duty flow and duty head, leads to a choice of a number of real pumps that should operate in parallel arrangement and with high efficiency. The primary objective is to enable pump operation in an optimal range of flows/heads and with this reduce the required energy for pumping. This was verified in Question 3.3.1 and the choice was made for the ETANORM model 50–200 with an impeller diameter of 219

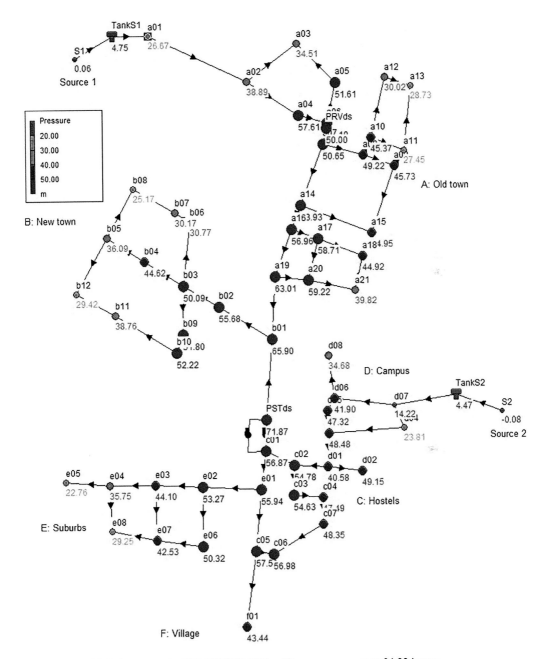

Figure 4.69 NT331b: 2 pump units 500–200/219, PRV = 50 mwc: pressures at 06:00 hours (mwc)

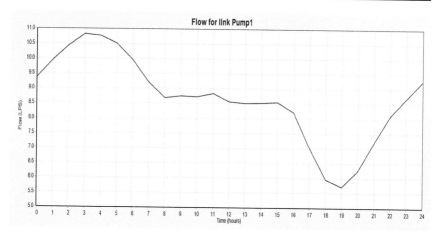

Figure 4.70 NT331b: booster PST flow (single unit operation with both pumps switched on)

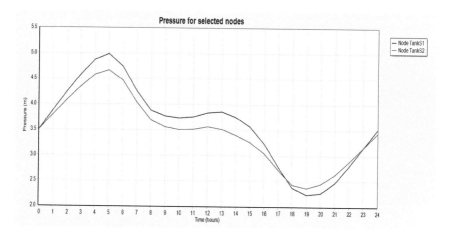

Figure 4.71 NT331b: tank volume variation for both booster pumps switched on

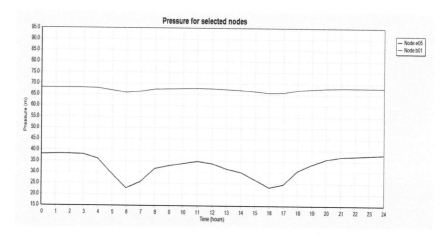

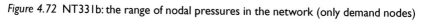

Figure 4.72 NT331b: the range of nodal pressures in the network (only demand nodes)

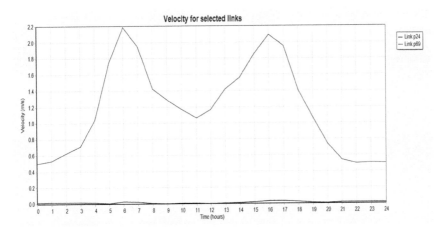

Figure 4.73 NT33Ib: the range of pipe velocities in the network (dummy pipes excluded)

mm, operating at a fixed speed of 1450 rpm and at 50 Hz. Two pumps were simulated in regular operation and by adding a stand-by unit, the booster station will be designed with three equal units connected in parallel arrangement.

Under regular supply conditions, all pumps except the stand-by unit will be in operation during the maximum consumption hour. In the worst-case scenario, the working point of the pump may be shifted to the end of the pump curve as specified by the manufacturer, resulting from the sudden demand increase. The consequence will be an operation at lower efficiency and higher NPSH in which case the pump should be (automatically) switched off beyond this working point. This working point is therefore assumed to reflect the most extreme operating conditions, and the selection of the pump motor logically takes this point into consideration. For the model 50–200/219, the pump curve has been introduced in EPANET reading the maximum pump flow of 16.1 l/s (57.96 m³/h) from the KSB catalogue, which is providing a minimum pump head of 10.4 mwc. The pump efficiency for this operational point has been assumed at 61%. During the emergency situations, all three pumps can be used for a while, resulting in a maximum flow of 48.3 l/s (173.88 m³/h).

The selection of the pump motor is based on the calculation of the pump power. To calculate the required power for the pumps, an allowance of 10% has to be made for unexpected deviation in the pump design. The standard formula is used, as discussed in Volume 1, Section 3.7.3, leading to the following results:

$$N = \frac{\rho g Q H}{0.9\eta} = \frac{1000 \times 9.81 \times 57.96 \times 10.4}{0.9 \times 0.61 \times 3600} = 2.99 \ kW, \text{ for the single unit.}$$

The required capacity of the motor is determined by the efficiency of the motor, usually between 0.92 and 0.95. Assuming the motor efficiency is 95 %:

$$N_m = \frac{2.99}{0.92} = 3.25 \ kW, \text{ for the single unit.}$$

The KSB documentation specifies similar although somewhat lower values. The maximum power input for the model 50–200/219 shown in Figure 4.68 is approximately 2.62 kW, while the selection of motor for this model (the pdf file *Pump and motor selection*, page 11) specifies a maximum motor size of 3.0 kW (for 1450 rpm). Further design is therefore based on this motor. The table on the same page gives the following pump details:

- the inlet diameter, DN_1 (at the suction side of the pump) = 65 mm,
- the outlet diameter, DN_2 (at the pressure side of the pump) = 50 mm,
- the maximum size of foundation plate, l_1 = 1150 mm and b_1 = 450 mm.

The DN dimensions are needed for planning reducers and enlargers (i.e. the determination of the local losses) later when the pipe diameters of the suction and the pressure pipe will be determined. The dimension of the foundation plate will be used as a guideline for estimation of the space between the two neighbouring pumps. The maximum values have been selected without deeper analyses of the specific characteristics of KSB pumps. Other manufacturers can obviously produce different pumps and present their characteristics in different catalogue formats.

4.5.14 Transformer and diesel generator capacity (Question 3.3.3)

If pumps of different size are used in the same pumping station, the motor selection will also differ. The power to be delivered by the transformer is consequently determined by the capacity of the most critical situation, which also includes supply of other electrical installations. The capacity of the transformer is further adjusted by the power factor ($\cos \theta$) that normally takes a value of approximately 0.7 but can be increased to 0.8 using condensers, as discussed in Volume 1, Section 4.2.4.

The transformer capacity, in the above example, can be calculated as follows, assuming conservatively $\cos q = 0.7$:

- motors, $3N_m/\cos \theta = (3 \times 3.0) / 0.7 = 12.86$ kVA
- lighting, $2.0 \text{ kW}/\cos \theta = 2.0 / 0.7 = 2.86$ kVA
- welding corner, $7.0 \text{ kW}/\cos \theta = 7.0 / 0.7 = 10.0$ kVA

Therefore, the total required is 25.72 kVA.

For this exercise, the power required to start a pump can be assumed to be twice the amount required for normal operation. With pumps of different size, sufficient power must be available to start the largest pump unit when other units are in operation.

The required power of the diesel engine can be calculated as follows:

- motors, $N = (1 + 3) \times 3.0 = 12.0$ kW
- lighting, $N = 2.0$ kW

If the efficiency of the diesel engine η_d is assumed to be 0.95, the total power requirement N_d can be calculated as:

$$N_d = (12.0 + 2.0) / 0.95 = 14.74 \, kW$$

The power factor of the generator ($\cos \theta$) is assumed to be 0.7, thus the capacity of the generator, $N_g = 14.74/0.7 = 21.05 \, kVA$.

4.5.15   Pumping station layout (Question 3.3.4)

For a simplified assessment of the hydraulic losses and NPSH requirements, the layout of the pumping station has been taken from the Spreadsheet Lesson 6–6, which is shown in Figure 4.74, where:

- all x components compose the feeder main,
- all y components compose the discharge header,
- all a components indicate the suction pipes,
- all c/g components indicate the gate/butterfly valves,
- all f components indicate the non-return valves,
- all the remaining components (b/d/e/h) indicate reducers/enlargers.

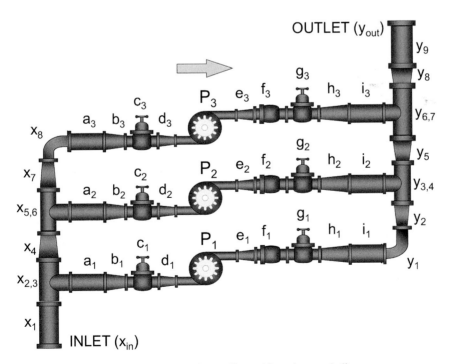

Figure 4.74 Simplified pumping station layout (Spreadsheet Lesson 6–6)

4.5.16 Piping and head losses (Question 3.3.5)

The diameter of the internal pumping station pipe should be calculated based on a recommended range of velocities. For this exercise, the following guideline has been used:

- feeder main (upstream of the pumps), v = 0.6–0.8 m/s
- suction pipe (using a single pump), v = 0.8–1.2 m/s
- pressure pipe (using a single pump), v = 1.5–2.0 m/s
- discharge header (downstream of the pumps) v = 1.2–1.7 m/s

The inlet and outlet diameter of a pump, as detailed in the manufacturer's specifications (DN_1 and DN_2), are normally lower than the suction and discharge station pipe diameters, as higher velocities are required in the pump. Therefore, pipe reducers and enlargers are required at the inlet and outlet of the pump. For example, the inlet and outlet diameters and velocities for the selected ETANORM 50–200 pump unit operating at Q_{max} of 57.96 m³/h will be:

- inlet diameter, DN_1 = 65 mm, v = 4.85 m/s,
- outlet diameter, DN_2 = 50 mm, v = 8.20 m/s,

The pumping station head loss caused by friction and fittings needs to be calculated from the exit of the suction storage tank (for source pumping stations) to the start of the delivery pumping main. This total loss should be within acceptable limits (normally less than 1.0 m). Here, in the absence of a reservoir on the suction side, the starting point is the upstream head at the entrance to the booster pumping station.

For local loss factors, reference is made to Appendix 5 of this book. Local losses should include entry to the suction pipe, all major bends, branches, valves, reducers and enlargers that connect to the pump unit. A complete technical drawing of the pumping station will include a more detailed layout with additional valves, bends, and metering equipment.

The calculation of the total head loss is given in Table 4.15 for the friction losses, and in Table 4.16 for the minor losses. Both tables shows the flows/velocities in each component assuming extreme conditions: all three pumps operating at maximum capacity. Furthermore, the downstream velocities have been applied in the calculation of minor losses for all the valves and fittings shown in Table 4.16. In Figure 4.74, x_{in} and y_{out} indicate the entry and the exit minor loss; the first appearing at the connection to the reservoir, and the second if the discharge takes place into the atmosphere. These conditions are not met in this case and both losses have been neglected in the case of the booster station.

The total head loss can be observed along the route of any of the three pumps, namely:

- route P_1: x_1-x_2-x_3-a_1-b_1-c_1-d_1-P_1-e_1-f_1-g_1-h_1-i_1-y_1-y_2-y_3-y_4-y_5-y_6-y_7-y_8-y_9,
- route P_2: x_1-x_2-x_3-x_4-x_5-x_6-a_2-b_2-c_2-d_2-P_2-e_2-f_2-g_2-h_2-i_2-y_3-y_4-y_5-y_6-y_7-y_8-y_9, and
- route P_3: x_1-x_2-x_3-x_4-x_5-x_6-x_7-x_8-a_3-b_3-c_3-d_3-P_3-e_3-f_3-g_3-h_3-i_3-y_6-y_7-y_8-y_9.

Summarising the head losses from tables 4.15 and 4.16 to compose the total head loss from the inlet to the outlet of the pumping station, the values per each of the three routes are given in Table 4.17.

Clarifications:
- The selection of minor loss factors is based on the information given in Appendix 5. The coefficients for the bends have been taken from Table 5.1, the values for the reducers/enlargers from Table 5.3, those for the T branches from Table 5.4 and those for the valves from the table shown in Figure 5.9.
- In the absence of a more complete and more detailed layout description some approximations have been introduced, such as in the case of the angle α, in the case of enlargers. Also in the case of T branches the minor loss was calculated only for the diverted flow Q_2 (coefficient ζ_1), neglecting the minor

Table 4.15 Friction loss calculation, 3 pumps in operation at Q_{max} (16.1 l/s, each)

Section	Location	L (m)	D (mm)	k (mm)	v range (m/s)	v (m/s)	hf (mwc)
x_1	Feeder main	6	300	0.1	0.6–0.8	0.68	0.009
x_2	Feeder main	4	300	0.1	0.6–0.8	0.68	0.006
x_5	Feeder main	4	250	0.1	0.6–0.8	0.66	0.007
$a_{1,2,3}$	Suction pipe, $P_{1,2,3}$	3	150	0.1	0.8–1.2	0.91	0.018
$i_{1,2,3}$	Pressure pipe, $P_{1,2,3}$	3	100	0.1	1.5–2.0	2.05	0.138
y_3	Discharge header	4	125	0.1	1.2–1.7	1.31	0.059
y_6	Discharge header	4	175	0.1	1.2–1.7	1.34	0.041
y_9	Discharge header	6	200	0.1	1.2–1.7	1.54	0.068

Table 4.16 Minor loss calculation, 3 pumps in operation at Q_{max} (16.1 l/s, each)

Section	Component	d/D	ζ (-)	v (m/s)	hm (mwc)
x_3	T branch, $Q_2/Q_3 = 0.33$ (see Table 5.4)	–	0.89	0.91	0.038
x_4	Reducer 300/250 mm (Table 5.3)	0.83	0.02	0.66	0.000
x_6	T branch, $Q_2/Q_3 = 0.50$	–	0.92	0.91	0.039
x_7	Reducer 250/150 mm	0.60	0.10	0.91	0.004
x_8	Pipe bend 90°, 150 mm, R = 300 mm (Table 5.1)	–	0.15	0.91	0.006
$b_{1,2,3}$	Reducer 150/125 mm	0.83	0.02	1.31	0.002
$c_{1,2,3}$	Flat gate valve 125 mm (Figure 5.9)	–	0.30	1.31	0.026
$d_{1,2,3}$	Reducer 125/65 mm	0.52	0.20	4.85	0.240
$e_{1,2,3}$	Enlarger 50/80 mm	0.63	0.10	3.20	0.052
$f_{1,2,3}$	Non-return (foot) valve 80 mm	–	0.80	3.20	0.418
$g_{1,2,3}$	Flat gate valve 80 mm	–	0.35	3.20	0.183
$h_{1,2,3}$	Enlarger 80/100 mm	0.80	0.03	2.05	0.006
y_1	Pipe bend 90°, 100 mm, R = 300 mm	–	0.12	2.05	0.026
y_2	Enlarger 100/125 mm	0.80	0.03	1.31	0.003
y_4	T branch, $Q_2/Q_3 = 0.50$	–	0.28	1.31	0.025
y_5	Enlarger 125/175 mm	0.71	0.07	1.34	0.006
y_7	T branch, $Q_2/Q_3 = 0.33$	–	0.02	1.34	0.002
y_8	Enlarger 175/200 mm	0.88	0.01	1.54	0.001

Table 4.17 Total head loss per route (mwc)

Inlet to P_1	Inlet to P_2	Inlet to P_3
0.338	0.384	0.395
P_1 to outlet	P_2 to outlet	P_3 to outlet
1.026	0.998	0.910
Inlet to outlet, P_1	Inlet to outlet, P_2	Inlet to outlet, P_3
1.364	1.382	1.305

loss caused by the coefficient that, as Table 5.4 shows, occasionally takes a negative value.

- All the results shown in tables 4.15–4.17 are derived from Spreadsheet Lesson 6–6. The velocities in the tables are automatically calculated taking the continuity equation for the specified flows and the selected pipe diameters into consideration.

Conclusions:

- The routes from the inlet to each pump and further to the outlet show different values in a logical trend. The head loss upstream of the pump gradually grows from P_1 to P_3 because the route to P_3 is the longest and most complex of all three. Equally, the head loss downstream of the pump gradually reduces from P_1 to P_3 because the route from P_3 is the shortest and least complex of all three.
- Following the integrity of the hydraulic grade lines, the sum of the upstream and downstream head losses should yield the same total head loss from the inlet to the outlet of the pumping station, regardless of the selected route. However, as can be seen in Table 4.17, this is not the case and minor deviations apply, mostly due to a still inconsistent selection of the minor loss factors. Nonetheless, the differences are not detrimental and the largest values with some rounding can be taken for further consideration.
- The total head loss for the maximum flow and all three pumps in operation can be assumed at 1.4 mwc. This is above the guideline of 1.0 mwc but is still considered acceptable in view of the fact that this operational scenario is reserved for extreme i.e. rare events. Additional considerations using Spreadsheet Lesson 6–6 will show the drop of the total head loss to approximately 1.2 mwc if the standby pump is switched off (two others still working at 16.1 l/s), and further to approximately 0.6 mwc when two pumps operate at the maximum flow on the maximum consumption day (regular operation of 10.84 l/s per pump registered in the simulation of NT331b, total 21.68 l/s).

Comments:

- The calculations clearly illustrate that the friction losses are far smaller compared to the minor losses in the design of pumping station layouts. This is not a surprise given the use of relatively short pipe lengths, and smooth pipe materials (mostly steel).
- Despite fairly elaborate consideration of each location/component, the minor loss calculation is still provisional owing to the empirical character of the

minor loss factors available in the literature. It is therefore essential to follow the recommendations of the manufacturers of pumps and accessories when analysing the head losses in pumping stations. Making assumptions and simplifications does not necessarily lead to significant errors.

- As is the case in distribution networks, the recommended range of velocity is difficult to satisfy in all demand scenarios and will also depend on the selected pipe diameters available in the market.

Reading:
Volume 1, Section 3.2.2: 'Minor losses',
Volume 1, Section 3.7.3: 'Pumped systems',
Volume 1, Section 4.2.4: 'Pumping station design',
Appendix 5: 'Minor Loss Factors',
Appendix 7: 'Spreadsheet Hydraulic Lessons – Overview'.

4.5.17 NPSH calculation (Question 3.3.6)

The head loss from the pump inlet needs to be calculated and the pump located at a sufficient elevation, to ensure the available NPSH is greater than the required NPSH of the pump specified by the manufacturer. The NPSH available can be calculated using the formula elaborated in Volume 1, Chapter 4 (Equation 4.14):

$$NPSH_{available} = h_{atm} - h_{vp} - \Delta H \pm \Delta Z - fs$$

Following the information in tables 4.4 and 4.5, the difference in atmospheric pressure, h_{atm}, and the vapour pressure, h_{vp}, can be assumed at 10 mwc in view of the fact that the network is located at an altitude of up to 100 msl and the assumption of the water temperature of 10–12 °C is reasonable; in this respect, Equation 4.13 is also applicable.

The total head loss, ΔH, between the inlet and the pump will be taken from Table 4.17 for route P3 that is the most critical from the NPSH perspective because of the largest value (0.395 mwc), which is rounded at 0.4 mwc.

The elevation difference, ΔZ, is the static head between the pump axis and the minimum suction level (m) at the inlet; the value becomes negative if the pump axis is located above the suction level. In the case of source pumping stations, the minimum suction level is in fact the minimum water level of the source reservoir. In the case of booster stations, the minimum head at the inlet of the pumping station needs to be taken into consideration. This value depends on the overall hydraulic picture in the vicinity of the pumping station.

In EPANET file NT331b, the upstream and downstream node of the booster station are set at 19.06 msl. A fair assumption can therefore be made with the pump axis located at 20 msl. The simulation of this demand scenario shows the maximum pump flow at 03:00 hours (10.84 l/s per pump) at the upstream head in node c01 of 78.57 msl. Thus at 03:00 hours, ΔZ of 78.57–20 ≈ 60 m. The minimum head of 74.95 m appears in node c01 at 17:00 hours. Based on this scenario, it is unlikely to expect that in any extreme scenario, the static head would become lower than 40–50 m.

From the pump specification curves shown in Figure 4.68, it can be seen that the required NPSH for the unit ETANORM 50–200/219 at the flow Q_{max} of 58 m³/h is approximately 2.85 mwc. The NPSH available must be greater than the NPSH required to avoid cavitation, as specified by the manufacturer. This therefore specifically requires a maximum allowed value of the static head, as shown in Equation 4.15 in Volume 1, Chapter 4:

$$\Delta Z = h_{atm} - h_{vp} - \Delta H - NPSH_{required} - fs$$

Using the calculated/adopted values:

$$\Delta Z = 10.20-0.12-0.4-2.85-1.5 = 5.33 \text{ mwc}$$

This means that the maximum allowed elevation of the pump impeller centreline is 5.33 m *above* the minimum suction/inlet head (or water level in the reservoir, if existing). The EPANET simulation shows that the heads in node c01 are clearly much higher than the pump elevation. Thus, the available NPSH will be in all circumstances significantly higher than the required one, creating no concerns about possible cavitation.

Reading:
Volume 1, Section 4.2.4: 'Pumping station design',

4.5.18 Cost calculation of investments (Question 4.3.1)

The investment costs include the pipe replacement for the major part, as well as the construction of the reservoirs, the booster station, and installation of the PRV. As specified in the assignment introduction, the design period is to be 20 years with some renovation taking place immediately, and a final layout that will satisfy the demand at the end of the design period constructed halfway through the design period i.e. after 10 years. Without further simulations (that could possibly modify this plan), in order to add additional safety to the design, the eleven pipes that were only to be enlarged twice within 10 years will be immediately enlarged to the diameter suitable for the demand at the end of the design period (p29, p30, p31, p32, p38, p59, p60, p61, p62, p66 and p67). The final plan of pipe renovation shown in tables 4.18 and 4.19 has been compiled from tables 4.10, 4.13 and 4.14.

Using the prices indicated in Table 4.5 in the assignment introduction, the investment costs of the final layouts are summarised in Table 4.20

Clarification:
• The installed capacity of the pumping stations has been assumed with all three units (including stand-by) in operation. The price is calculated for the capacities converted from l/s into m³/h.

Assuming that 100% of the investment needed at the beginning of the design period will be borrowed immediately, and the entire sum for the second phase in

Table 4.18 Pipe renovation list, beginning of the design period

NAMETOWN – renovated pipes for current demand scenario

Pipe name	Node from	Node to	New D (mm)	Length (m)	Old D (mm)
p01	S1	a01	400	220	300
p02	a01	a02	400	657	300
p04	a02	a04	400	371	300
p06	a04	a06	400	180	300
p08	a06	a07	400	125	300
p14	a10	a12	80	369	50
p15	a11	a13	80	386	50
p17	a07	a14	400	390	300
p20	a14	a16	400	153	300
p23	a16	a19	300	295	250
p28	a19	b01	300	372	250
p29	b01	b02	250	376	150
p30	b02	b03	250	248	150
p31	b03	b04	200	278	100
p32	b04	b05	150	270	50
p38	b05	b12	150	380	50
p42	b01	c01	250	727	200
p43	c01	c02	150	184	125
p44	c02	d01	150	201	100
p45	d01	d02	50	214	25
p46	d01	d03	150	195	100
p47	d03	d05	100	136	80
p49	d04	d07	50	148	25
p50	d05	d06	100	85	80
p52	d06	d08	80	256	50
p58	c01	e01	250	168	200
p59	e01	e02	200	361	125
p60	e02	e03	200	289	80
p61	e03	e04	150	277	80
p62	e04	e05	100	266	50
p63	e02	e06	100	354	50
p64	e03	e07	100	325	50
p65	e04	e08	80	272	50
p66	e06	e07	100	280	25
p67	e07	e08	80	282	25

Table 4.19 Pipe renovation list, planned in year 10

NAMETOWN – renovated pipes for future demand scenario

Pipe name	Node from	Node to	New D (mm)	Length (m)	Old D (mm)
p09	a07	a08	150	256	100
p11	a08	a10	100	111	80
p33	b06	b07	100	106	80
p34	b07	b08	80	290	50
p35	b10	b11	100	439	80
p36	b11	a12	80	269	50
p37	b08	b05	100	333	50
p68	e01	c05	200	366	150
p69	c05	f01	150	452	125

Table 4.20 Investment costs, rounded in Euros (EUR)

Pipe diameter	Beginning of the design period		Investment at year 10	
	Total length (m)	Total price (EUR)	Total length (m)	Total price (EUR)
D = 50 mm	362	16,290	–	–
D = 80 mm	1565	112,680	559	40,248
D = 100 mm	1446	130,140	989	89,010
D = 150 mm	1507	203,445	708	95,590
D = 200 mm	928	167,040	366	65,880
D = 250 mm	1519	341,775	–	–
D = 300 mm	667	180,090	–	–
D = 400 mm	2096	754,560	–	–
D = 500 mm	–	–	–	–
D = 600 mm	–	–	–	–
Total pipes	10,090	1,906,020	2622	290,728
Pumping stations	Installed capacity (m³/h)	Total price (EUR)	Installed capacity (m³/h)	Total price (EUR)
Source 1	–	–	–	–
Source 2	–	–	3 × 57.96 ≈ 180	318,558
Total PST	–	–	180	318,558
Reservoirs	Total volume (m³)	Total price (EUR)	Total volume (m³)	Total price (EUR)
Source 1	2580	687,000	–	–
Source 2	–	–	1900	585,000
Total reservoirs	2580	687,000	1900	585,000
Valves	Diameter (mm)	Total price (EUR)	Diameter (mm)	Total price (EUR)
PRV	400	13,000	–	–
Total investment	–	2,606,020	–	1,194,286

year 10 when the extension takes place, the total present value of the investments, rounded in Euros is (with the loan interest of 8%):

$$2,606,020 + 1,194,286/1.08^{10} = EUR\ 3,159,206$$

If the re-payment of the loan starts immediately, the annual instalments to be paid over a period of 20 years are:

$$3,159,206 \times (0.08 \times 1.08^{20})/(1.08^{20} - 1) = EUR\ 321,772$$

Reading:
- See Volume 1, Section 4.1.2: 'Economic aspects'.

4.5.19 Cost calculation of operation and maintenance (Question 4.3.2)

In the assignment description, the annual O&M costs of the existing infrastructure have been assessed at EUR 100,000 at the beginning of the design period, and EUR 130,000 at the end of the design period. For the added/replaced infrastructure, the operation and maintenance costs are calculated as a percentage of the raw investment costs (Table 4.6). The system operates at the beginning of the design period still as a gravity system, with the following additional costs, rounded in Euros:

Distribution system (including PRV): $0.005 \times 1,919,020 = $ EUR 9595
Pumping stations: $0.020 \times 0 = $ EUR 0
Reservoirs: $0.008 \times 687,000 = $ EUR 5496
TOTAL: EUR 115,091 per year.

The additional O&M costs at the end of the design period are, rounded in Euros:

Distribution system: $0.005 \times 290,728 = $ EUR 1454
Pumping stations: $0.020 \times 318,558 = $ EUR 6371
Reservoirs: $0.008 \times 585,000 = $ EUR 4680
TOTAL: EUR 12,505 per year.

On top of this, the cost of pumping energy is EUR 13.45 per day (EPANET simulation of NT331b), which on an annual basis is rounded at EUR 4909 at the end of the design period. Hence, the total O&M costs amount to 115,091 + 30,000 + 12,505 + 4909 = EUR 162,505.

4.5.20 Calculation of total costs (Question 4.3.3)

At the beginning of the design period, source S1 supplies the average flow of Q_{avg} = 102.81 l/s on the maximum consumption day (EPANET simulation of NT317d). Assuming that this consumption equals two times the consumption on the average consumption day (selected IN = 2.0), the total annual production of water at the beginning of the design period can be assessed at:

$$102.81 \times 3.6 \times 24 \times 365/2.0 = 1,621,108 \text{ m}^3 \text{ per year.}$$

The production cost of this water is specified at EUR 0.45 per m³. Evenly distributing the annual O&M costs of EUR 115,091 and the loan repayment of EUR 321,772 adds an additional EUR 0.16 per m³ to the cost, making it EUR 0.61 per m³. The total annual price of water supplied is then $1,621,108 \times 0.61 = $ EUR 988,876.

At the end of the design period, sources S1 and S2 supply the average flows of Q_{avg} = 68 and 82 l/s, respectively; both on the maximum consumption day

(EPANET simulation of NT331b). With the same assumption of the higher consumption compared to the average consumption day, the total annual production of water at the end of the design period can be assessed at:

$(68 + 82) \times 3.6 \times 24 \times 365/2.0 = 2,365,200$ m³ per year.

The production cost of this water is specified at EUR 0.60 per m³. Evenly distributing the annual O&M costs of EUR 162,505 and the loan repayment of EUR 321,772 adds an additional EUR 0.12 to the cost per m³, making it EUR 0.72 per m³. The total annual price of water supplied in this case is then $2,365,200 \times 0.72 =$ EUR 1,702,944.

Comments:
- The presented case is simplified in many aspects and is only one possible ('good enough') solution; there could be more.
- This shows very clearly that satisfying the design criteria for multiple demand scenarios is virtually impossible. All decisions are therefore made taking into consideration an average situation. Distribution networks live their (design) life just as humans: they grow and develop, but they also age.
- More computer simulations could have been run to test the demand and pressures around the point of renovation in year 10 to verify the plan shown in tables 4.18 and 4.19. This has been avoided due to repetitive work of the same nature that does not necessarily contribute to the educational value of the exercise.
- Further considerations could include the reliability assessment. It is clear that the network in its main structure still largely has a branched configuration, which may create concerns in the case of the failure of any of the main pipes. This is deliberately left out of the scope, being part of the advanced programme on water transport and distribution. For the same reason, the water quality simulations, except for the source tracing, have not been included in the assessment of the network performance.

Minor loss factors

The general minor loss formula is:

$$h_m = \frac{8\xi}{\pi^2 g D^4} Q^2 = \frac{\xi}{12.1 D^4} Q^2$$

where:

h_m	=	Minor loss (mwc).
Q	=	Pipe flow (m³/s).
ξ	=	Minor loss factor (-)
g	=	Gravity, g = 9.81 m/s²
D	=	Pipe diameter (m). D is the downstream diameter if the cross-section changes, unless stated differently.

5.1 Bends and elbows

By coupling two 90°-bends or elbows together, the ξ-value for a single bend/elbow should not be doubled but multiplied by the value shown in Figure 5.2.

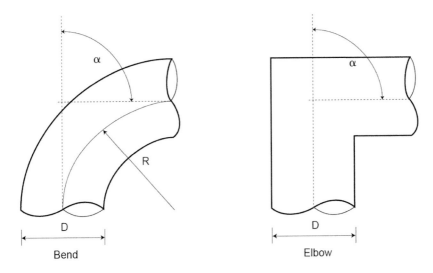

Bend

Elbow

Figure 5.1 Geometry of bends and elbows (Table 5.1)

Table 5.1 Minor loss factors for bends and elbows

Deflection angle, α (°)	15	30	45	60	90
Bends, R/D = 1	0.05	0.09	0.13	0.16	0.21
Bends, R/D = 2	0.04	0.07	0.10	0.12	0.15
Bends, R/D = 3	0.03	0.05	0.08	0.09	0.12
Bends, R/D = 4	0.02	0.04	0.06	0.07	0.09
Elbows	0.06	0.13	0.25	0.50	1.20

1.4	1.6	1.8

Figure 5.2 Multipliers for coupled 90°-bends

5.2 Enlargements and reducers

The minor loss factor is calculated as:

$$\xi = C \left(\frac{A_2}{A_1} - 1 \right)^2$$

where:

C   = Factor dependant on the deflection angle α.

$A_{1(2)}$ = Cross-section area up or down-stream of the obstruction.

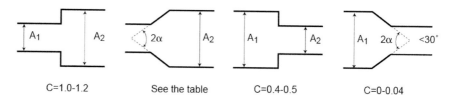

| C=1.0-1.2 | See the table | C=0.4-0.5 | C=0-0.04 |

Figure 5.3 Geometry of pipe enlargements and reducers (Table 5.2)

Table 5.2 Minor loss factor C at a gradual change of the cross section

Deflection angle α (°)	10	20	30	40	50	60	90
C-factor	0.2	0.4	0.7	0.9	1.0	1.1	1.1

For the same shapes, KSB[1] (1990) recommends the following straightforward ξ-values (see Figure 5.4):

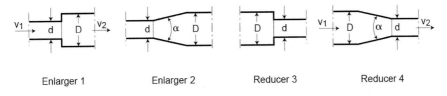

| Enlarger 1 | Enlarger 2 | Reducer 3 | Reducer 4 |

Figure 5.4 Geometry of pipe enlargements and reducers (Table 5.3)

Table 5.3 Minor loss factor ξ, at a gradual change of the cross section (KSB, 1990)

Type	d/D = 0.5	0.6	0.7	0.8	0.9
Enlarger 1	0.56	0.41	0.26	0.13	0.04
Enlarger 2, α = 8°	0.07	0.05	0.03	0.02	0.01
Enlarger 2, α =15°	0.15	0.11	0.07	0.03	0.01
Enlarger 2, α =20°	0.23	0.17	0.11	0.05	0.02
Reducer 3	4.80	2.01	0.88	0.34	0.11
Reducer 4, 20° < α < 40°	0.21	0.10	0.05	0.02	0.01

5.3   Branches

The recommended values for ξ are shown in Figure 5.5 and Table 5.3 (KSB, 1990).

Knee piece

	α = 45°		α = 60°		α = 90°	
surface	ζ	surface	ζ	surface	ζ
smooth	0.25	smooth	0.50	smooth	0.15
rough	0.35	rough	0.70	rough	0.30

Combinations with 90° knee pieces

ζ = 2.5

ζ = 3

ζ = 5

T pieces (subdivision of flow)

With sharp edges
ζ = 1.3

Rounded with
straight bottom
ζ = 0.7

Spherical with
inward rounded
neck
ζ = 0.7

Spherical

ζ = 2.5 to 4.9

Figure 5.5 Minor loss factors for various types of branches

1 KSB. 1990. *Centrifugal Pump Design*. KSB.

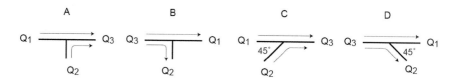

Figure 5.6 Various types of branches (Table 5.4)

Table 5.4 Minor loss factor ξ, at various T-branches (KSB, 1990)

Type	$Q_2/Q_3 =$	0.2	0.4	0.6	0.8	1.0
A	ξ_2 ⌐	−0.4	0.08	0.47	0.72	0.91
	ξ_1 ⌐	0.17	0.30	0.41	0.51	−
B	$\xi_2 \approx$	0.88	0.89	0.95	1.10	1.28
	$\xi_1 \approx$	−0.08	−0.05	0.07	0.21	−
C	$\xi_2 \approx$	−0.38	0	0.22	0.37	0.37
	$\xi_1 \approx$	0.17	0.19	0.09	−0.17	−
D	$\xi_2 \approx$	0.68	0.50	0.38	0.35	0.48
	$\xi_1 \approx$	0.06	0.04	0.07	0.20	−

5.4   Inlets and outlets

The recommended values for ξ are shown in Figure 5.7 (KSB, 1990).

Inlet edge sharp or chamfered

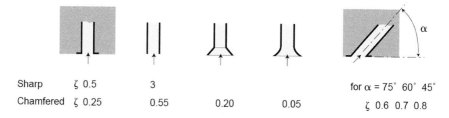

Sharp	ζ 0.5	3		for α = 75° 60° 45°
Chamfered	ζ 0.25	0.55	0.20	0.05	ζ 0.6 0.7 0.8

Figure 5.7 Minor loss factors for various types of inlets and outlets

$\xi = 1$ for downstream of a straight pipe with an approximately uniform velocity distribution in the outlet cross-section.

$\xi = 2$ in the case of a very unequal velocity distribution, e.g. immediately downstream of an elbow or a valve, etc.

5.5   Flow meters

The recommended ξ-values for the Venturi meters and orifice plates are shown in Table 5.5 (KSB, 1990).

Figure 5.8 Geometry of Venturi tube and standard orifice plate (Table 5.5)

Table 5.5 Minor loss factor ξ, for Venturi tube and orifice plate (KSB, 1990)

Diameter ratio d/D	0.30	0.40	0.50	0.60	0.70	0.80
Aperture ratio m = (d/D)²	0.09	0.16	0.25	0.36	0.49	0.64
Short Venturi tube	21	6	2	0.7	0.3	0.2
Standard orifice plate	300	85	30	12	4.5	2

For volumetric water meters $\xi \approx 10$, as an approximation.

5.6   Valves

The recommended ξ-values for various types of valves as shown below are displayed in Figure 5.9 (KSB, 1990).

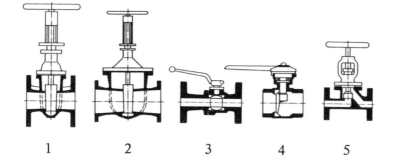
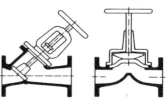
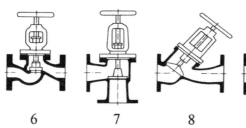

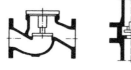

Figure 5.9 Minor loss factors for various types of valves

Loss coefficient ζ for DN =

Type of valve/fitting		Design³	15	20	25	32	40	50	65	80	100	125	150	200	250	300	400	500	600	800	1000	Remarks
Shut-off valves																						
flat gate valves (dE = DN)	min	1	0.1																		0.1	for dE < DN cf. footnote 1)
	max		0.65	0.6	0.55	0.5	0.5	0.45	0.4	0.35	0.3										0.3	
round-body gate valves (dE = DN)	min	2						0.25	0.24	0.23	0.22	0.21	0.19	0.18	0.17	0.16	0.15	0.13	0.12	0.11	0.1	
	max							0.32	0.31	0.30	0.28	0.26	0.23	0.23	0.22	0.20	0.19	0.18	0.16	0.15	0.14	
cocks (dE = DN)	min	3	0.10		0.09	0.09	0.08	0.08	0.07	0.07	0.06	0.05	0.05	0.04	0.03	0.03	0.02					for dE < DN ζ = 0.4 to 1.1
	max		0.15														0.15					
swing-type valves PN ≥ 2.5		4					0.90	0.76	0.60	0.50	0.42	0.36	0.30	0.25	0.20	0.16	0.13	0.10	0.08	0.06	0.05	
PN ≤ 40													1.50	1.20	1.00	0.92	0.83	0.76	0.71	0.67	0.63	
valves, forged	min	5																				ζ = 2 to 3 possible for optimized valve
	max																					
valves, cast	min	6	3.0		6.0			6.0														
	max		6.0		6.8			6.8														
angle valves	min	7	2.0														2.0					
	max		3.1				3.4	3.8	4.1	4.4	4.7	5.0	5.3	5.7	6.0	6.3	6.6					
slanted-seat valves	min	8	1.5													1.5						
	max		2.6													2.6						
full-bore valves	min	9	0.6														0.6					
	max		1.6														1.6					
diaphragm valves	min	10	0.8							0.8				0.8								
	max		2.2							2.2				2.2								
Backflow preventors																						
non-return valves, straight-seat	min	11	3.0										3.0	3.0								
	max		6.0										6.0	6.0								
non-return valves, axial	min	12	3.2					3.2	3.7	5.0	7.3											
	max		3.4	3.4	3.5	3.6	3.8	4.2	5.0	6.4	8.2											
non-return valves, axially expanded	min	13										4.3				4.3						
	max											4.6				4.6						
non-return valves, slanted seat	min	14	2.5	2.4	2.2	2.1	2.0	1.9	1.7	1.6	1.5		1.5			1.5						
	max		3.0													3.0						
foot valves	min	15						1.0	0.9	0.8	0.7	0.6	0.5	0.4	0.4	0.4	(7.0)	(6.1)	(5.5)	(4.5)	(4.0)	( ) in groups
	max							3.0								3.0						
swing-type check valves	min	16	0.5				0.5	0.4								0.4	0.3				0.3	swing-type valves without levers and weights 2)
	max		2.4	2.3	2.3	2.2	2.1	2.0	1.9	1.8	1.8	1.7	1.6	1.5	1.5	1.4	1.3	1.2	1.2	1.1	1.0	
hydrostops v = 4 m/s		17						0.9			3.0	3.0	3.0	2.5	2.5	1.2	2.2					
v = 3 m/s								1.8			4.0	4.5	4.5	4.0	4.0	1.8	3.4					
v = 2 m/s								5.0			6.0	8.0	8.0	7.5	6.5	6.0	7.0					
filters		18					2.8									2.8						in clean condition
screens		19					1.0									1.0						

1) If the narrowest shut-off diameter dE is smaller than the nominal diameter DN, the loss coefficient ζ must be increased by $(DN/dE)^x$, with x = 5 to 6
2) In the case of partial opening, i.e. low flow velocities, the loss coefficients increase
3) Designs: cf. page 15

Appendix 6

Hydraulic tables
(Darcy-Weisbach/Colebrook-White)

k(mm) =   0.01   T(°C) =   10

D	S =	0.0005		S =	0.001		S =	0.002		S =	0.003	
(mm)	v(m/s)	Q(l/s)	Q(m3/h)	v(m/s)	Q(l/s)	Q(m3/h)	v(m/s)	Q(l/s)	Q(m3/h)	v(m/s)	Q(l/s)	Q(m3/h)
50	0.11	0.2	0.8	0.17	0.3	1.2	0.25	0.5	1.8	0.31	0.6	2.2
80	0.16	0.8	2.9	0.24	1.2	4.3	0.35	1.8	6.3	0.44	2.2	7.9
100	0.19	1.5	5.3	0.28	2.2	7.8	0.41	3.2	11.5	0.51	4.0	14.5
125	0.22	2.7	9.6	0.32	4.0	14.3	0.48	5.8	21.1	0.60	7.3	26.4
150	0.25	4.4	15.8	0.37	6.5	23.3	0.54	9.5	34.3	0.68	12.0	43.1
200	0.30	9.5	34.2	0.45	14.0	50.4	0.66	20.6	74.2	0.82	25.8	92.9
250	0.35	17.3	62.3	0.52	25.5	91.7	0.76	37.4	134.6	0.95	46.7	168.3
300	0.40	28.2	101.5	0.59	41.4	149.1	0.86	60.7	218.6	1.07	75.9	273.2
350	0.44	42.6	153.3	0.65	62.5	224.9	0.95	91.5	329.4	1.19	114.3	411.3
400	0.48	60.8	218.8	0.71	89.1	320.9	1.04	130.4	469.5	1.30	162.8	585.9
450	0.52	83.2	299.5	0.77	121.9	438.8	1.12	178.2	641.5	1.40	222.3	800.3
500	0.56	110.1	396.5	0.82	161.2	580.4	1.20	235.5	848.0	1.50	293.7	1057.5
600	0.63	178.8	643.6	0.92	261.4	941.1	1.35	381.5	1373.4	1.68	475.4	1711.6
700	0.70	269.1	968.8	1.02	393.1	1415.2	1.49	573.2	2063.4	1.86	714.0	2570.2
800	0.76	383.3	1380.0	1.11	559.5	2014.4	1.62	815.2	2934.7	2.02	1014.9	3653.8
900	0.82	523.5	1884.7	1.20	763.7	2749.3	1.75	1111.9	4002.7	2.18	1383.8	4981.5
1000	0.88	691.7	2490.2	1.28	1008.4	3630.3	1.87	1467.3	5282.3	2.32	1825.5	6571.8
1100	0.94	889.8	3203.2	1.36	1296.5	4667.3	1.98	1885.4	6787.6	2.47	2345.0	8441.9
1200	0.99	1119.5	4030.3	1.44	1630.5	5869.7	2.10	2370.0	8532.1	2.61	2946.9	10608.8
1300	1.04	1382.7	4977.9	1.52	2012.9	7246.6	2.20	2924.7	10529.1	2.74	3635.7	13088.5
1400	1.09	1681.1	6052.0	1.59	2446.3	8806.8	2.31	3553.0	12791.0	2.87	4415.7	15896.5
1500	1.14	2016.3	7258.6	1.66	2933.0	10558.8	2.41	4258.3	15330.0	2.99	5291.1	19048.0
1600	1.19	2389.8	8603.4	1.73	3475.2	12510.8	2.51	5043.9	18158.1	3.12	6266.0	22557.6

D	S=	0.004		S=	0.005		S=	0.006		S=	0.007	
(mm)	v(m/s)	Q(l/s)	Q(m3/h)	v(m/s)	Q(l/s)	Q(m3/h)	v(m/s)	Q(l/s)	Q(m3/h)	v(m/s)	Q(l/s)	Q(m3/h)
50	0.37	0.7	2.6	0.42	0.8	3.0	0.47	0.9	3.3	0.51	1.0	3.6
80	0.52	2.6	9.3	0.59	2.9	10.6	0.65	3.3	11.7	0.71	3.6	12.8
100	0.60	4.7	17.0	0.68	5.4	19.3	0.76	5.9	21.4	0.82	6.5	23.3
125	0.70	8.6	31.0	0.79	9.8	35.1	0.88	10.8	38.8	0.96	11.8	42.3
150	0.79	14.0	50.5	0.90	15.9	57.2	0.99	17.6	63.2	1.08	19.1	68.8
200	0.96	30.2	108.8	1.09	34.2	123.1	1.20	37.8	136.0	1.31	41.1	148.0
250	1.12	54.7	197.1	1.26	61.9	222.7	1.39	68.4	246.1	1.52	74.4	267.8
300	1.26	88.8	319.8	1.42	100.4	361.3	1.57	110.9	399.1	1.71	120.6	434.1
350	1.39	133.7	481.3	1.57	151.0	543.5	1.73	166.7	600.2	1.88	181.3	652.7
400	1.52	190.4	685.4	1.71	215.0	773.8	1.89	237.3	854.3	2.05	258.0	928.8
450	1.63	260.0	935.9	1.85	293.4	1056.4	2.04	323.9	1166.0	2.21	352.1	1267.4
500	1.75	343.4	1236.3	1.97	387.5	1395.1	2.18	427.7	1539.7	2.37	464.8	1673.3
600	1.96	555.6	2000.1	2.22	626.7	2256.3	2.45	691.5	2489.3	2.66	751.3	2704.7
700	2.17	834.0	3002.2	2.44	940.5	3385.8	2.70	1037.4	3734.7	2.93	1126.9	4057.0
800	2.36	1185.2	4266.5	2.66	1336.2	4810.5	2.93	1473.6	5305.1	3.18	1600.5	5761.9
900	2.54	1615.4	5815.4	2.86	1820.9	6555.3	3.16	2007.8	7228.0	3.43	2180.4	7849.3
1000	2.71	2130.5	7669.9	3.06	2401.2	8644.2	3.37	2647.2	9529.8	3.66	2874.3	10347.6
1100	2.88	2736.2	9850.4	3.24	3083.3	11099.9	3.58	3398.7	12235.3	3.88	3689.9	13283.7
1200	3.04	3437.9	12376.4	3.42	3873.4	13944.1	3.77	4269.0	15368.6	4.10	4634.4	16683.7
1300	3.19	4240.7	15266.5	3.60	4777.2	17197.9	3.97	5264.6	18952.6	4.31	5714.6	20572.4
1400	3.35	5149.7	18538.8	3.77	5800.4	20881.5	4.15	6391.6	23009.6	4.51	6937.2	24974.0
1500	3.49	6169.6	22210.7	3.93	6948.5	25014.5	4.33	7655.9	27561.2	4.70	8308.8	29911.7
1600	3.63	7305.4	26299.3	4.09	8226.7	29616.0	4.51	9063.4	32628.2	4.89	9835.6	35408.2

$$k(mm) = 0.01 \qquad T(°C) = 10$$

D	S=	0.008		S=	0.009		S=	0.010		S=	0.012	
(mm)	v(m/s)	Q(l/s)	Q(m3/h)	v(m/s)	Q(l/s)	Q(m3/h)	v(m/s)	Q(l/s)	Q(m3/h)	v(m/s)	Q(l/s)	Q(m3/h)
50	0.55	1.1	3.9	0.59	1.2	4.1	0.62	1.2	4.4	0.69	1.4	4.9
80	0.76	3.8	13.8	0.81	4.1	14.7	0.86	4.3	15.6	0.95	4.8	17.3
100	0.89	7.0	25.1	0.95	7.4	26.8	1.00	7.9	28.4	1.11	8.7	31.4
125	1.03	12.7	45.5	1.10	13.5	48.6	1.17	14.3	51.5	1.29	15.8	57.0
150	1.16	20.6	74.1	1.24	22.0	79.0	1.32	23.3	83.8	1.46	25.7	92.6
200	1.41	44.2	159.3	1.50	47.2	169.9	1.59	50.0	180.0	1.76	55.2	198.8
250	1.63	80.0	288.0	1.74	85.3	307.1	1.84	90.3	325.2	2.03	99.8	359.1
300	1.83	129.7	466.8	1.96	138.2	497.6	2.07	146.4	526.9	2.29	161.6	581.6
350	2.03	194.9	701.7	2.16	207.8	747.9	2.29	219.9	791.8	2.52	242.7	873.8
400	2.21	277.3	998.4	2.35	295.6	1064.0	2.49	312.9	1126.3	2.75	345.2	1242.7
450	2.38	378.4	1362.2	2.54	403.2	1451.6	2.68	426.8	1536.4	2.96	470.8	1694.8
500	2.54	499.5	1798.3	2.71	532.2	1916.0	2.87	563.3	2027.8	3.16	621.2	2236.4
600	2.86	807.2	2906.0	3.04	859.9	3095.8	3.22	909.9	3275.8	3.55	1003.3	3611.8
700	3.15	1210.6	4358.2	3.35	1289.5	4642.0	3.54	1364.2	4911.3	3.91	1503.8	5413.8
800	3.42	1719.1	6188.8	3.64	1830.8	6590.9	3.85	1936.8	6972.4	4.25	2134.5	7684.2
900	3.68	2341.6	8429.7	3.92	2493.5	8976.5	4.15	2637.5	9495.0	4.57	2906.3	10462.6
1000	3.93	3086.5	11111.4	4.18	3286.4	11831.0	4.43	3475.9	12513.3	4.88	3829.5	13786.2
1100	4.17	3961.9	14262.9	4.44	4218.1	15185.1	4.69	4461.0	16059.5	5.17	4914.1	17690.8
1200	4.40	4975.5	17911.8	4.68	5296.8	19068.5	4.95	5601.4	20165.0	5.46	6169.6	22210.6
1300	4.62	6134.7	22085.0	4.92	6530.4	23509.4	5.20	6905.5	24859.7	5.73	7605.1	27378.3
1400	4.84	7446.7	26808.1	5.15	7926.4	28535.2	5.44	8381.2	30172.3	6.00	9229.4	33225.8
1500	5.05	8918.4	32106.2	5.37	9492.3	34172.4	5.68	10036.4	36131.0	6.25	11051.0	39783.6
1600	5.25	10556.5	38003.4	5.59	11235.2	40446.8	5.91	11878.6	42762.8	6.50	13078.2	47081.7

D	S=	0.014		S=	0.016		S=	0.018		S=	0.020	
(mm)	v(m/s)	Q(l/s)	Q(m3/h)	v(m/s)	Q(l/s)	Q(m3/h)	v(m/s)	Q(l/s)	Q(m3/h)	v(m/s)	Q(l/s)	Q(m3/h)
50	0.75	1.5	5.3	0.81	1.6	5.7	0.87	1.7	6.1	0.92	1.8	6.5
80	1.04	5.2	18.8	1.12	5.6	20.3	1.20	6.0	21.6	1.27	6.4	22.9
100	1.21	9.5	34.2	1.30	10.2	36.8	1.39	10.9	39.3	1.47	11.6	41.6
125	1.40	17.2	62.0	1.51	18.5	66.7	1.61	19.8	71.2	1.71	20.9	75.4
150	1.58	28.0	100.7	1.70	30.1	108.4	1.82	32.1	115.6	1.92	34.0	122.4
200	1.91	60.1	216.2	2.06	64.6	232.5	2.19	68.9	247.9	2.32	72.9	262.5
250	2.21	108.5	390.5	2.38	116.6	419.8	2.53	124.3	447.4	2.68	131.6	473.6
300	2.48	175.6	632.2	2.67	188.7	679.5	2.85	201.1	724.1	3.01	212.9	766.4
350	2.74	263.8	949.6	2.95	283.5	1020.4	3.14	302.0	1087.2	3.32	319.6	1150.5
400	2.98	375.1	1350.2	3.21	403.0	1450.7	3.42	429.3	1545.4	3.61	454.2	1635.2
450	3.22	511.4	1841.1	3.45	549.4	1977.9	3.68	585.2	2106.7	3.89	619.1	2228.9
500	3.44	674.8	2429.2	3.69	724.8	2609.3	3.93	771.9	2778.9	4.16	816.6	2939.8
600	3.85	1089.5	3922.1	4.14	1170.0	4212.0	4.41	1245.8	4485.0	4.66	1317.8	4743.9
700	4.24	1632.7	5877.7	4.56	1753.1	6311.0	4.85	1866.4	6719.1	5.13	1973.9	7105.9
800	4.61	2317.0	8341.3	4.95	2487.5	8954.9	5.27	2648.0	9532.6	5.57	2800.1	10080.3
900	4.96	3154.3	11355.6	5.32	3385.9	12189.3	5.67	3604.0	12974.2	5.99	3810.6	13718.2
1000	5.29	4155.8	14961.0	5.68	4460.4	16057.5	6.04	4747.2	17089.9	6.39	5019.0	18068.3
1100	5.61	5332.2	19196.0	6.02	5722.5	20600.9	6.41	6089.8	21923.4	6.77	6438.0	23176.7
1200	5.92	6693.8	24097.8	6.35	7183.1	25859.0	6.76	7643.6	27516.9	7.14	8079.9	29087.8
1300	6.22	8250.5	29701.8	6.67	8852.8	31869.9	7.10	9419.6	33910.6	7.50	9956.8	35844.3
1400	6.50	10011.7	36042.3	6.98	10741.8	38670.4	7.42	11428.8	41143.8	7.85	12079.8	43487.3
1500	6.78	11986.8	43152.6	7.28	12860.0	46295.8	7.74	13681.7	49254.0	8.18	14460.2	52056.6
1600	7.05	14184.7	51064.8	7.57	15216.9	54780.8	8.05	16188.3	58277.8	8.51	17108.5	61590.8

k(mm) =   0.05   T(°C) =   10

D	S =	0.0005		S =	0.001		S =	0.002		S =	0.003	
(mm)	v(m/s)	Q(l/s)	Q(m3/h)	v(m/s)	Q(l/s)	Q(m3/h)	v(m/s)	Q(l/s)	Q(m3/h)	v(m/s)	Q(l/s)	Q(m3/h)
50	0.11	0.2	0.8	0.16	0.3	1.2	0.24	0.5	1.7	0.31	0.6	2.2
80	0.16	0.8	2.8	0.23	1.2	4.2	0.34	1.7	6.2	0.43	2.1	7.7
100	0.18	1.4	5.2	0.27	2.1	7.7	0.40	3.1	11.3	0.50	3.9	14.1
125	0.21	2.6	9.5	0.32	3.9	14.0	0.46	5.7	20.5	0.58	7.1	25.6
150	0.24	4.3	15.5	0.36	6.3	22.8	0.53	9.3	33.4	0.66	11.6	41.7
200	0.30	9.3	33.7	0.44	13.7	49.3	0.64	20.0	72.1	0.79	25.0	89.8
250	0.35	17.0	61.2	0.51	24.9	89.5	0.74	36.3	130.6	0.92	45.1	162.5
300	0.39	27.7	99.6	0.57	40.4	145.5	0.83	58.8	211.9	1.04	73.2	263.5
350	0.43	41.7	150.3	0.63	60.9	219.2	0.92	88.5	318.8	1.14	110.0	396.1
400	0.47	59.6	214.4	0.69	86.8	312.5	1.00	126.1	453.8	1.25	156.6	563.6
450	0.51	81.5	293.3	0.75	118.6	427.0	1.08	172.1	619.6	1.34	213.6	769.0
500	0.55	107.8	388.0	0.80	156.8	564.4	1.16	227.3	818.2	1.44	282.0	1015.2
600	0.62	174.8	629.1	0.90	253.9	913.9	1.30	367.5	1323.2	1.61	455.7	1640.4
700	0.68	262.8	946.1	0.99	381.3	1372.8	1.43	551.5	1985.3	1.78	683.3	2459.7
800	0.74	374.0	1346.5	1.08	542.2	1951.9	1.56	783.4	2820.2	1.93	970.1	3492.3
900	0.80	510.4	1837.6	1.16	739.3	2661.5	1.68	1067.3	3842.4	2.08	1321.1	4756.1
1000	0.86	673.9	2426.2	1.24	975.4	3511.4	1.79	1407.2	5065.9	2.22	1741.1	6268.0
1100	0.91	866.3	3118.7	1.32	1253.0	4510.9	1.90	1806.6	6503.8	2.35	2234.5	8044.3
1200	0.96	1089.3	3921.6	1.39	1574.7	5668.8	2.01	2269.1	8168.9	2.48	2805.7	10100.7
1300	1.01	1344.7	4840.8	1.46	1942.7	6993.8	2.11	2798.1	10073.2	2.61	3458.9	12451.9
1400	1.06	1633.9	5882.1	1.53	2359.5	8494.2	2.21	3396.8	12228.6	2.73	4197.9	15112.6
1500	1.11	1958.7	7051.2	1.60	2827.2	10178.0	2.30	4068.5	14646.5	2.84	5026.8	18096.5
1600	1.15	2320.4	8353.5	1.67	3348.0	12052.9	2.40	4816.1	17337.9	2.96	5949.3	21417.4

D	S=	0.004		S=	0.005		S=	0.006		S=	0.007	
(mm)	v(m/s)	Q(l/s)	Q(m3/h)	v(m/s)	Q(l/s)	Q(m3/h)	v(m/s)	Q(l/s)	Q(m3/h)	v(m/s)	Q(l/s)	Q(m3/h)
50	0.36	0.7	2.6	0.41	0.8	2.9	0.45	0.9	3.2	0.49	1.0	3.5
80	0.50	2.5	9.1	0.57	2.8	10.3	0.63	3.1	11.3	0.68	3.4	12.3
100	0.58	4.6	16.5	0.66	5.2	18.7	0.73	5.7	20.6	0.79	6.2	22.4
125	0.68	8.3	30.0	0.77	9.4	33.9	0.85	10.4	37.4	0.92	11.3	40.7
150	0.77	13.6	48.8	0.87	15.3	55.1	0.96	16.9	60.8	1.04	18.3	66.0
200	0.93	29.1	104.9	1.05	32.9	118.3	1.15	36.2	130.5	1.25	39.4	141.7
250	1.07	52.7	189.7	1.21	59.4	213.7	1.33	65.4	235.5	1.45	71.0	255.7
300	1.21	85.4	307.3	1.36	96.1	346.1	1.50	105.9	381.3	1.63	114.9	413.8
350	1.33	128.3	461.8	1.50	144.4	519.9	1.65	159.1	572.6	1.79	172.6	621.2
400	1.45	182.4	656.8	1.63	205.3	739.2	1.80	226.1	814.0	1.95	245.2	882.9
450	1.56	248.8	895.8	1.76	280.0	1008.0	1.94	308.2	1109.7	2.10	334.3	1203.4
500	1.67	328.4	1182.2	1.88	369.4	1329.9	2.07	406.6	1463.7	2.25	440.9	1587.1
600	1.88	530.3	1909.2	2.11	596.4	2146.9	2.32	656.2	2362.3	2.52	711.3	2560.7
700	2.07	794.9	2861.7	2.32	893.6	3216.9	2.55	983.0	3538.7	2.77	1065.3	3835.1
800	2.24	1128.2	4061.5	2.52	1267.9	4564.4	2.77	1394.4	5020.0	3.01	1511.0	5439.6
900	2.41	1536.0	5529.5	2.71	1725.8	6212.8	2.98	1897.7	6831.7	3.23	2056.0	7401.6
1000	2.58	2023.7	7285.4	2.89	2273.3	8184.1	3.18	2499.4	8997.9	3.45	2707.6	9747.3
1100	2.73	2596.7	9347.9	3.07	2916.4	10499.2	3.37	3206.0	11541.6	3.65	3472.6	12501.5
1200	2.88	3259.7	11735.1	3.24	3660.6	13178.2	3.56	4023.6	14485.0	3.85	4357.8	15688.0
1300	3.03	4017.8	14464.1	3.40	4511.3	16240.6	3.74	4958.1	17849.0	4.05	5369.4	19329.8
1400	3.17	4875.5	17551.7	3.56	5473.6	19704.9	3.91	6015.1	21654.4	4.23	6513.6	23448.9
1500	3.30	5837.2	21014.0	3.71	6552.6	23589.3	4.07	7200.2	25920.7	4.41	7796.3	28066.7
1600	3.44	6907.4	24866.7	3.86	7753.1	27911.3	4.24	8518.7	30667.3	4.59	9223.3	33204.0

k(mm) = 0.05   T(°C) = 10

D	S=	0.008		S=	0.009		S=	0.010		S=	0.012	
(mm)	v(m/s)	Q(l/s)	Q(m3/h)	v(m/s)	Q(l/s)	Q(m3/h)	v(m/s)	Q(l/s)	Q(m3/h)	v(m/s)	Q(l/s)	Q(m3/h)
50	0.53	1.0	3.8	0.57	1.1	4.0	0.60	1.2	4.2	0.66	1.3	4.7
80	0.73	3.7	13.3	0.78	3.9	14.2	0.83	4.2	15.0	0.91	4.6	16.5
100	0.85	6.7	24.1	0.91	7.1	25.7	0.96	7.6	27.2	1.06	8.3	30.0
125	0.99	12.1	43.7	1.05	12.9	46.5	1.12	13.7	49.3	1.23	15.1	54.3
150	1.12	19.7	70.9	1.19	21.0	75.6	1.26	22.2	80.0	1.39	24.5	88.1
200	1.35	42.3	152.1	1.43	45.0	162.0	1.51	47.6	171.3	1.67	52.4	188.7
250	1.55	76.2	274.5	1.65	81.2	292.2	1.75	85.8	308.9	1.92	94.5	340.1
300	1.75	123.4	444.1	1.86	131.3	472.6	1.96	138.8	499.6	2.16	152.7	549.9
350	1.92	185.1	666.5	2.05	197.0	709.2	2.16	208.2	749.6	2.38	229.1	824.8
400	2.09	263.1	947.1	2.23	279.9	1007.6	2.35	295.8	1064.8	2.59	325.4	1171.4
450	2.25	358.5	1290.8	2.40	381.4	1373.0	2.53	403.0	1450.8	2.79	443.3	1595.7
500	2.41	472.8	1702.1	2.56	502.9	1810.3	2.71	531.3	1912.7	2.98	584.3	2103.5
600	2.70	762.7	2745.7	2.87	811.0	2919.6	3.03	856.7	3084.3	3.33	941.9	3390.9
700	2.97	1142.1	4111.4	3.16	1214.2	4371.2	3.33	1282.5	4617.1	3.66	1409.7	5075.0
800	3.22	1619.6	5830.6	3.43	1721.7	6198.2	3.62	1818.4	6546.2	3.98	1998.3	7194.0
900	3.46	2203.5	7932.7	3.68	2342.2	8431.9	3.89	2473.5	8904.4	4.27	2717.8	9784.1
1000	3.69	2901.5	10445.5	3.93	3083.9	11101.9	4.15	3256.4	11723.1	4.56	3577.6	12879.4
1100	3.92	3721.0	13395.8	4.16	3954.5	14236.3	4.39	4175.5	15031.8	4.83	4586.8	16512.5
1200	4.13	4669.1	16808.8	4.39	4961.7	17862.2	4.63	5238.7	18859.2	5.09	5754.1	20714.7
1300	4.33	5752.6	20709.2	4.61	6112.7	22005.6	4.86	6453.5	23232.5	5.34	7087.8	25515.9
1400	4.53	6977.9	25120.5	4.82	7414.3	26691.6	5.08	7827.3	28178.3	5.58	8595.9	30945.2
1500	4.73	8351.6	30065.7	5.02	8873.4	31944.4	5.30	9367.3	33722.1	5.82	10286.3	37030.5
1600	4.91	9879.7	35566.9	5.22	10496.5	37787.5	5.51	11080.2	39888.7	6.05	12166.4	43799.1

D	S=	0.014		S=	0.016		S=	0.018		S=	0.020	
(mm)	v(m/s)	Q(l/s)	Q(m3/h)	v(m/s)	Q(l/s)	Q(m3/h)	v(m/s)	Q(l/s)	Q(m3/h)	v(m/s)	Q(l/s)	Q(m3/h)
50	0.72	1.4	5.1	0.78	1.5	5.5	0.83	1.6	5.9	0.88	1.7	6.2
80	0.99	5.0	18.0	1.07	5.4	19.3	1.14	5.7	20.6	1.20	6.1	21.8
100	1.15	9.1	32.6	1.24	9.7	35.0	1.32	10.4	37.3	1.40	11.0	39.5
125	1.34	16.4	59.0	1.43	17.6	63.3	1.53	18.7	67.4	1.61	19.8	71.3
150	1.50	26.6	95.7	1.61	28.5	102.7	1.72	30.4	109.3	1.82	32.1	115.6
200	1.81	56.9	204.8	1.94	61.0	219.7	2.07	64.9	233.8	2.19	68.6	247.1
250	2.09	102.5	368.9	2.24	109.9	395.7	2.38	116.9	421.0	2.52	123.6	444.9
300	2.34	165.6	596.2	2.51	177.6	639.4	2.67	188.9	680.0	2.82	199.6	718.5
350	2.58	248.4	894.1	2.77	266.3	958.7	2.94	283.2	1019.5	3.11	299.2	1077.0
400	2.81	352.7	1269.6	3.01	378.1	1361.1	3.20	402.0	1447.2	3.38	424.6	1528.6
450	3.02	480.3	1729.2	3.24	514.9	1853.6	3.44	547.4	1970.6	3.64	578.1	2081.3
500	3.22	633.1	2279.1	3.46	678.6	2442.8	3.67	721.3	2596.6	3.88	761.7	2742.3
600	3.61	1020.3	3673.2	3.87	1093.4	3936.2	4.11	1162.1	4183.5	4.34	1227.1	4417.4
700	3.97	1526.8	5496.5	4.25	1635.9	5889.1	4.52	1738.4	6258.2	4.77	1835.4	6607.4
800	4.31	2164.0	7790.3	4.61	2318.3	8345.8	4.90	2463.3	8867.8	5.17	2600.5	9361.8
900	4.63	2942.7	10593.7	4.95	3152.2	11347.8	5.26	3349.0	12056.5	5.56	3535.3	12727.1
1000	4.93	3873.2	13943.6	5.28	4148.6	14934.8	5.61	4407.3	15866.3	5.92	4652.1	16747.6
1100	5.22	4965.3	17875.2	5.60	5317.9	19144.4	5.94	5649.2	20337.0	6.27	5962.6	21465.4
1200	5.51	6228.4	22422.3	5.90	6678.6	24012.6	6.26	7085.2	25506.9	6.61	7478.0	26920.7
1300	5.78	7671.4	27617.2	6.19	8215.0	29574.1	6.57	8725.8	31412.7	6.94	9209.0	33152.4
1400	6.04	9303.1	33491.3	6.47	9961.8	35862.3	6.87	10580.6	38090.1	7.25	11166.1	40197.9
1500	6.30	11131.9	40074.8	6.74	11919.4	42909.7	7.16	12659.2	45573.3	7.56	13359.3	48093.3
1600	6.55	13165.9	47397.1	7.01	14096.6	50747.6	7.45	14971.0	53895.6	7.86	15798.3	56873.8

k(mm) =   0.1   T(°C) =   10

D	S =	0.0005		S =	0.001		S =	0.002		S =	0.003	
(mm)	v(m/s)	Q(l/s)	Q(m3/h)	v(m/s)	Q(l/s)	Q(m3/h)	v(m/s)	Q(l/s)	Q(m3/h)	v(m/s)	Q(l/s)	Q(m3/h)
50	0.11	0.2	0.8	0.16	0.3	1.1	0.24	0.5	1.7	0.30	0.6	2.1
80	0.15	0.8	2.8	0.23	1.1	4.1	0.33	1.7	6.0	0.42	2.1	7.5
100	0.18	1.4	5.1	0.27	2.1	7.5	0.39	3.1	11.0	0.48	3.8	13.7
125	0.21	2.6	9.3	0.31	3.8	13.7	0.45	5.6	20.0	0.56	6.9	24.9
150	0.24	4.2	15.3	0.35	6.2	22.3	0.51	9.0	32.5	0.64	11.2	40.5
200	0.29	9.2	33.0	0.43	13.4	48.2	0.62	19.5	70.0	0.77	24.2	87.0
250	0.34	16.7	60.0	0.49	24.3	87.4	0.72	35.2	126.7	0.89	43.7	157.2
300	0.38	27.1	97.6	0.56	39.4	141.9	0.81	57.1	205.4	1.00	70.7	254.6
350	0.42	40.9	147.2	0.62	59.3	213.6	0.89	85.8	308.8	1.10	106.2	382.5
400	0.46	58.3	209.9	0.67	84.5	304.3	0.97	122.1	439.4	1.20	151.1	543.9
450	0.50	79.7	286.9	0.73	115.4	415.5	1.05	166.5	599.6	1.30	206.0	741.8
500	0.54	105.4	379.3	0.78	152.5	548.9	1.12	219.8	791.5	1.38	271.9	978.7
600	0.60	170.7	614.6	0.87	246.7	888.1	1.26	355.2	1278.8	1.55	439.0	1580.3
700	0.67	256.5	923.5	0.96	370.3	1333.1	1.38	532.6	1917.5	1.71	657.8	2368.2
800	0.73	364.9	1313.6	1.05	526.2	1894.3	1.50	756.2	2722.3	1.86	933.5	3360.6
900	0.78	497.7	1791.7	1.13	717.1	2581.5	1.62	1029.8	3707.2	2.00	1270.7	4574.7
1000	0.84	656.8	2364.4	1.20	945.6	3404.3	1.73	1357.1	4885.6	2.13	1674.1	6026.7
1100	0.89	843.9	3037.9	1.28	1214.3	4371.4	1.83	1741.7	6269.9	2.26	2147.8	7732.1
1200	0.94	1060.7	3818.5	1.35	1525.4	5491.4	1.93	2186.8	7872.5	2.38	2696.1	9705.9
1300	0.99	1308.8	4711.7	1.42	1881.3	6772.6	2.03	2695.8	9704.8	2.50	3322.8	11962.2
1400	1.03	1589.8	5723.3	1.48	2284.1	8222.9	2.13	3271.7	11778.3	2.62	4031.9	14514.8
1500	1.08	1905.2	6858.7	1.55	2736.1	9849.9	2.22	3917.7	14103.6	2.73	4827.0	17377.2
1600	1.12	2256.4	8123.0	1.61	3239.2	11661.3	2.31	4636.5	16691.5	2.84	5711.7	20562.2

D	S=	0.004		S=	0.005		S=	0.006		S=	0.007	
(mm)	v(m/s)	Q(l/s)	Q(m3/h)	v(m/s)	Q(l/s)	Q(m3/h)	v(m/s)	Q(l/s)	Q(m3/h)	v(m/s)	Q(l/s)	Q(m3/h)
50	0.35	0.7	2.5	0.40	0.8	2.8	0.44	0.9	3.1	0.48	0.9	3.4
80	0.49	2.4	8.8	0.55	2.8	9.9	0.61	3.0	11.0	0.66	3.3	11.9
100	0.57	4.4	16.0	0.64	5.0	18.0	0.70	5.5	19.9	0.76	6.0	21.6
125	0.66	8.1	29.0	0.74	9.1	32.7	0.82	10.0	36.1	0.89	10.9	39.1
150	0.74	13.1	47.2	0.84	14.8	53.2	0.92	16.3	58.6	1.00	17.6	63.5
200	0.90	28.1	101.3	1.01	31.7	114.0	1.11	34.9	125.5	1.20	37.8	136.1
250	1.04	50.8	183.0	1.16	57.2	205.8	1.28	62.9	226.4	1.39	68.2	245.5
300	1.16	82.3	296.2	1.31	92.5	333.0	1.44	101.7	366.3	1.56	110.3	397.0
350	1.28	123.6	444.8	1.44	138.8	499.9	1.59	152.7	549.7	1.72	165.4	595.6
400	1.40	175.6	632.3	1.57	197.3	710.4	1.73	217.0	781.0	1.87	235.0	846.1
450	1.51	239.5	862.0	1.69	268.9	968.2	1.86	295.6	1064.3	2.01	320.2	1152.8
500	1.61	315.9	1137.1	1.81	354.7	1276.9	1.99	389.8	1403.4	2.15	422.2	1519.9
600	1.80	509.8	1835.1	2.02	572.2	2060.0	2.22	628.8	2263.6	2.41	680.8	2450.9
700	1.98	763.6	2749.1	2.23	857.0	3085.1	2.45	941.5	3389.3	2.65	1019.2	3669.2
800	2.16	1083.3	3899.9	2.42	1215.5	4375.7	2.66	1335.1	4806.2	2.87	1445.1	5202.4
900	2.32	1474.3	5307.5	2.60	1653.8	5953.9	2.86	1816.3	6538.7	3.09	1965.8	7076.9
1000	2.47	1941.8	6990.5	2.77	2178.0	7840.6	3.05	2391.6	8609.8	3.30	2588.2	9317.5
1100	2.62	2490.8	8967.0	2.94	2793.3	10056.1	3.23	3067.0	11041.3	3.49	3318.9	11947.9
1200	2.76	3126.1	11254.0	3.10	3505.4	12619.3	3.40	3848.5	13854.4	3.68	4164.1	14990.8
1300	2.90	3852.3	13868.1	3.25	4319.1	15548.8	3.57	4741.5	17069.3	3.86	5130.0	18468.0
1400	3.04	4673.7	16825.2	3.40	5239.6	18862.5	3.74	5751.5	20705.3	4.04	6222.4	22400.7
1500	3.17	5594.6	20140.7	3.55	6271.5	22577.4	3.90	6883.8	24781.6	4.21	7447.0	26809.3
1600	3.29	6619.3	23829.6	3.69	7419.6	26710.4	4.05	8143.4	29316.2	4.38	8809.3	31713.4

k(mm) =   0.1   T(°C) =   10

D	S=	0.008		S=	0.009		S=	0.010		S=	0.012	
(mm)	v(m/s)	Q(l/s)	Q(m3/h)	v(m/s)	Q(l/s)	Q(m3/h)	v(m/s)	Q(l/s)	Q(m3/h)	v(m/s)	Q(l/s)	Q(m3/h)
50	0.51	1.0	3.6	0.55	1.1	3.9	0.58	1.1	4.1	0.64	1.3	4.5
80	0.71	3.6	12.8	0.75	3.8	13.6	0.80	4.0	14.4	0.88	4.4	15.9
100	0.82	6.4	23.2	0.87	6.9	24.7	0.92	7.3	26.1	1.02	8.0	28.8
125	0.95	11.7	42.0	1.01	12.4	44.7	1.07	13.1	47.3	1.18	14.5	52.0
150	1.07	18.9	68.2	1.14	20.1	72.5	1.21	21.3	76.7	1.33	23.4	84.4
200	1.29	40.6	146.0	1.37	43.1	155.3	1.45	45.6	164.1	1.60	50.1	180.5
250	1.49	73.1	263.2	1.58	77.7	279.9	1.67	82.1	295.7	1.84	90.3	325.0
300	1.67	118.2	425.5	1.78	125.7	452.4	1.88	132.7	477.8	2.06	145.9	525.2
350	1.84	177.3	638.4	1.96	188.5	678.6	2.07	199.1	716.6	2.27	218.7	787.5
400	2.00	251.9	906.7	2.13	267.7	963.7	2.25	282.7	1017.6	2.47	310.5	1118.0
450	2.16	343.1	1235.3	2.29	364.6	1312.7	2.42	385.0	1386.0	2.66	422.9	1522.5
500	2.30	452.3	1628.4	2.45	480.7	1730.4	2.58	507.5	1826.9	2.84	557.3	2006.4
600	2.58	729.3	2625.4	2.74	774.8	2789.4	2.89	817.9	2944.6	3.18	898.1	3233.3
700	2.84	1091.6	3929.8	3.01	1159.7	4174.8	3.18	1224.0	4406.5	3.49	1343.8	4837.7
800	3.08	1547.6	5571.3	3.27	1643.9	5918.0	3.45	1735.0	6246.0	3.79	1904.5	6856.2
900	3.31	2105.0	7578.0	3.51	2235.8	8048.9	3.71	2359.5	8494.4	4.07	2589.8	9323.2
1000	3.53	2771.2	9976.5	3.75	2943.2	10595.6	3.95	3105.9	11181.3	4.34	3408.6	12271.1
1100	3.74	3553.3	12792.0	3.97	3773.6	13585.0	4.19	3982.0	14335.3	4.60	4369.7	15731.1
1200	3.94	4458.0	16048.8	4.19	4734.1	17042.9	4.42	4995.4	17983.3	4.85	5481.3	19732.8
1300	4.14	5491.8	19770.4	4.39	5831.7	20994.0	4.64	6153.2	22151.5	5.09	6751.4	24304.9
1400	4.33	6660.9	23979.3	4.59	7072.8	25462.2	4.85	7462.5	26865.1	5.32	8187.5	29475.0
1500	4.51	7971.4	28697.2	4.79	8464.1	30470.7	5.05	8930.1	32148.5	5.54	9797.2	35269.8
1600	4.69	9429.2	33945.3	4.98	10011.6	36041.9	5.25	10562.6	38025.4	5.76	11587.5	41715.2

D	S=	0.014		S=	0.016		S=	0.018		S=	0.020	
(mm)	v(m/s)	Q(l/s)	Q(m3/h)	v(m/s)	Q(l/s)	Q(m3/h)	v(m/s)	Q(l/s)	Q(m3/h)	v(m/s)	Q(l/s)	Q(m3/h)
50	0.69	1.4	4.9	0.75	1.5	5.3	0.79	1.6	5.6	0.84	1.6	5.9
80	0.95	4.8	17.2	1.02	5.1	18.5	1.09	5.5	19.7	1.15	5.8	20.8
100	1.10	8.7	31.2	1.18	9.3	33.5	1.26	9.9	35.6	1.33	10.5	37.7
125	1.28	15.7	56.4	1.37	16.8	60.5	1.46	17.9	64.4	1.54	18.9	68.0
150	1.44	25.4	91.5	1.54	27.2	98.1	1.64	29.0	104.3	1.73	30.6	110.2
200	1.73	54.3	195.6	1.85	58.2	209.6	1.97	61.9	222.8	2.08	65.4	235.3
250	1.99	97.8	352.1	2.14	104.8	377.3	2.27	111.4	401.0	2.40	117.6	423.4
300	2.24	158.0	568.8	2.39	169.3	609.3	2.54	179.9	647.5	2.69	189.9	683.6
350	2.46	236.8	852.6	2.64	253.7	913.3	2.80	269.5	970.4	2.96	284.5	1024.3
400	2.68	336.2	1210.3	2.87	360.1	1296.3	3.04	382.5	1377.1	3.21	403.8	1453.6
450	2.88	457.8	1648.0	3.08	490.3	1764.9	3.27	520.8	1874.8	3.46	549.6	1978.7
500	3.07	603.2	2171.7	3.29	646.0	2325.5	3.49	686.1	2470.1	3.69	724.1	2606.8
600	3.44	971.9	3498.9	3.68	1040.6	3746.2	3.91	1105.2	3978.6	4.12	1166.2	4198.5
700	3.78	1454.0	5234.4	4.04	1556.6	5603.8	4.30	1653.0	5950.8	4.53	1744.2	6279.1
800	4.10	2060.5	7417.6	4.39	2205.7	7940.3	4.66	2342.1	8431.4	4.92	2471.1	8895.9
900	4.40	2801.6	10085.7	4.71	2998.8	10795.5	5.00	3184.0	11462.4	5.28	3359.2	12093.2
1000	4.69	3687.1	13273.6	5.02	3946.4	14206.9	5.33	4189.9	15083.7	5.63	4420.3	15913.1
1100	4.97	4726.4	17015.1	5.32	5058.5	18210.4	5.65	5370.4	19333.3	5.96	5665.4	20395.6
1200	5.24	5928.4	21342.1	5.61	6344.5	22840.3	5.96	6735.5	24247.7	6.28	7105.3	25579.0
1300	5.50	7301.6	26285.7	5.89	7813.8	28129.7	6.25	8295.0	29861.9	6.59	8750.1	31500.5
1400	5.75	8854.3	31875.6	6.16	9475.1	34110.5	6.53	10058.3	36209.8	6.89	10609.9	38195.5
1500	6.00	10594.6	38140.7	6.42	11337.0	40813.4	6.81	12034.4	43323.9	7.18	12694.1	45698.7
1600	6.23	12530.3	45109.0	6.67	13407.9	48268.4	7.08	14232.3	51236.1	7.47	15012.0	54043.3

k(mm) =   0.5   T(°C) =   10

D	S =	0.0005		S =	0.001		S =	0.002		S =	0.003	
(mm)	v(m/s)	Q(l/s)	Q(m3/h)	v(m/s)	Q(l/s)	Q(m3/h)	v(m/s)	Q(l/s)	Q(m3/h)	v(m/s)	Q(l/s)	Q(m3/h)
50	0.10	0.2	0.7	0.15	0.3	1.0	0.21	0.4	1.5	0.26	0.5	1.8
80	0.14	0.7	2.5	0.20	1.0	3.7	0.29	1.5	5.3	0.36	1.8	6.6
100	0.16	1.3	4.6	0.24	1.9	6.7	0.34	2.7	9.7	0.42	3.3	11.9
125	0.19	2.4	8.5	0.28	3.4	12.2	0.40	4.9	17.6	0.49	6.0	21.7
150	0.22	3.8	13.8	0.31	5.5	19.9	0.45	7.9	28.6	0.55	9.8	35.2
200	0.26	8.3	29.9	0.38	11.9	42.9	0.54	17.1	61.4	0.67	21.0	75.7
250	0.31	15.0	54.2	0.44	21.6	77.7	0.63	30.9	111.1	0.77	38.0	136.7
300	0.35	24.4	88.0	0.50	35.0	126.0	0.71	50.0	180.0	0.87	61.5	221.5
350	0.38	36.8	132.5	0.55	52.7	189.6	0.78	75.2	270.5	0.96	92.4	332.7
400	0.42	52.4	188.7	0.60	75.0	269.9	0.85	106.9	384.9	1.05	131.4	473.2
450	0.45	71.6	257.8	0.64	102.3	368.4	0.92	145.9	525.1	1.13	179.3	645.4
500	0.48	94.6	340.6	0.69	135.1	486.4	0.98	192.5	693.0	1.20	236.6	851.7
600	0.54	153.1	551.2	0.77	218.5	786.5	1.10	311.0	1119.7	1.35	382.1	1375.6
700	0.60	229.9	827.5	0.85	327.8	1179.9	1.21	466.4	1679.0	1.49	572.8	2062.1
800	0.65	326.7	1176.0	0.93	465.6	1676.0	1.32	662.2	2383.8	1.62	813.1	2927.2
900	0.70	445.3	1603.0	1.00	634.3	2283.5	1.42	901.9	3246.7	1.74	1107.2	3986.1
1000	0.75	587.3	2114.3	1.06	836.3	3010.6	1.51	1188.7	4279.2	1.86	1459.2	5253.0
1100	0.79	754.3	2715.5	1.13	1073.7	3865.2	1.61	1525.7	5492.5	1.97	1872.7	6741.6
1200	0.84	947.7	3411.8	1.19	1348.6	4855.0	1.69	1915.9	6897.4	2.08	2351.4	8465.1
1300	0.88	1169.1	4208.6	1.25	1663.1	5987.2	1.78	2362.3	8504.2	2.18	2898.9	10436.1
1400	0.92	1419.6	5110.7	1.31	2019.1	7268.8	1.86	2867.4	10322.7	2.29	3518.5	12666.8
1500	0.96	1700.8	6122.9	1.37	2418.5	8706.7	1.94	3434.1	12362.8	2.38	4213.6	15168.9
1600	1.00	2013.9	7250.0	1.42	2863.2	10307.4	2.02	4064.9	14633.6	2.48	4987.2	17954.1

D	S=	0.004		S=	0.005		S=	0.006		S=	0.007	
(mm)	v(m/s)	Q(l/s)	Q(m3/h)	v(m/s)	Q(l/s)	Q(m3/h)	v(m/s)	Q(l/s)	Q(m3/h)	v(m/s)	Q(l/s)	Q(m3/h)
50	0.30	0.6	2.1	0.34	0.7	2.4	0.38	0.7	2.7	0.41	0.8	2.9
80	0.42	2.1	7.6	0.47	2.4	8.6	0.52	2.6	9.4	0.56	2.8	10.2
100	0.49	3.8	13.9	0.55	4.3	15.5	0.60	4.7	17.1	0.65	5.1	18.5
125	0.57	7.0	25.1	0.64	7.8	28.2	0.70	8.6	31.0	0.76	9.3	33.5
150	0.64	11.3	40.8	0.72	12.7	45.8	0.79	14.0	50.3	0.86	15.1	54.4
200	0.78	24.4	87.7	0.87	27.3	98.3	0.95	30.0	107.9	1.03	32.4	116.7
250	0.90	44.0	158.4	1.00	49.3	177.4	1.10	54.1	194.7	1.19	58.5	210.5
300	1.01	71.2	256.4	1.13	79.8	287.2	1.24	87.5	315.1	1.34	94.6	340.7
350	1.11	107.0	385.2	1.25	119.8	431.4	1.37	131.4	473.2	1.48	142.1	511.6
400	1.21	152.1	547.7	1.36	170.4	613.3	1.49	186.9	672.7	1.61	202.0	727.3
450	1.30	207.5	746.9	1.46	232.3	836.3	1.60	254.8	917.2	1.73	275.4	991.5
500	1.39	273.7	985.5	1.56	306.5	1103.4	1.71	336.1	1210.0	1.85	363.3	1308.0
600	1.56	442.0	1591.4	1.75	494.9	1781.5	1.92	542.6	1953.4	2.07	586.5	2111.5
700	1.72	662.6	2385.2	1.93	741.6	2669.9	2.11	813.1	2927.3	2.28	878.9	3164.0
800	1.87	940.4	3385.4	2.09	1052.5	3789.2	2.30	1153.9	4154.2	2.48	1247.2	4489.9
900	2.01	1280.4	4609.5	2.25	1433.0	5158.9	2.47	1571.0	5655.6	2.67	1697.9	6112.3
1000	2.15	1687.2	6074.1	2.40	1888.2	6797.5	2.64	2069.9	7451.6	2.85	2237.0	8053.2
1100	2.28	2165.2	7794.8	2.55	2423.0	8722.8	2.79	2656.1	9561.8	3.02	2870.4	10333.4
1200	2.40	2718.6	9786.9	2.69	3042.1	10951.6	2.95	3334.6	12004.6	3.19	3603.6	12972.9
1300	2.52	3351.4	12065.0	2.83	3750.1	13500.2	3.10	4110.5	14797.8	3.35	4442.0	15991.1
1400	2.64	4067.5	14643.1	2.96	4551.2	16384.4	3.24	4988.5	17958.8	3.50	5390.7	19406.6
1500	2.76	4870.8	17534.9	3.08	5449.9	19619.5	3.38	5973.4	21504.2	3.65	6454.8	23237.4
1600	2.87	5764.9	20753.6	3.21	6450.1	23220.2	3.52	7069.5	25450.2	3.80	7639.2	27501.0

k(mm) =   0.5   T(°C) =   10

D	S=	0.008		S=	0.009		S=	0.010		S=	0.012	
(mm)	v(m/s)	Q(l/s)	Q(m3/h)	v(m/s)	Q(l/s)	Q(m3/h)	v(m/s)	Q(l/s)	Q(m3/h)	v(m/s)	Q(l/s)	Q(m3/h)
50	0.44	0.9	3.1	0.46	0.9	3.3	0.49	1.0	3.5	0.54	1.1	3.8
80	0.60	3.0	10.9	0.64	3.2	11.6	0.68	3.4	12.2	0.74	3.7	13.4
100	0.70	5.5	19.8	0.74	5.8	21.0	0.78	6.2	22.2	0.86	6.8	24.4
125	0.81	10.0	35.9	0.86	10.6	38.1	0.91	11.2	40.2	1.00	12.3	44.1
150	0.92	16.2	58.2	0.97	17.2	61.8	1.03	18.1	65.2	1.13	19.9	71.6
200	1.10	34.7	124.9	1.17	36.8	132.6	1.24	38.8	139.8	1.36	42.6	153.4
250	1.27	62.6	225.3	1.35	66.4	239.2	1.43	70.1	252.3	1.57	76.9	276.7
300	1.43	101.3	364.6	1.52	107.5	387.0	1.60	113.4	408.2	1.76	124.3	447.6
350	1.58	152.1	547.4	1.68	161.4	581.0	1.77	170.2	612.8	1.94	186.7	672.0
400	1.72	216.1	778.1	1.83	229.4	825.8	1.93	241.9	871.0	2.11	265.3	955.0
450	1.85	294.7	1060.8	1.97	312.7	1125.8	2.07	329.8	1187.3	2.27	361.6	1301.7
500	1.98	388.7	1399.3	2.10	412.5	1485.0	2.22	435.0	1566.1	2.43	476.9	1716.9
600	2.22	627.4	2258.7	2.35	665.8	2396.9	2.48	702.1	2527.6	2.72	769.7	2770.8
700	2.44	940.1	3384.3	2.59	997.6	3591.2	2.73	1051.9	3787.0	3.00	1153.1	4151.1
800	2.65	1334.0	4802.3	2.82	1415.5	5095.8	2.97	1492.6	5373.4	3.25	1636.0	5889.7
900	2.85	1816.0	6537.5	3.03	1926.9	6936.8	3.19	2031.8	7314.5	3.50	2226.9	8017.0
1000	3.05	2392.5	8613.1	3.23	2538.6	9139.0	3.41	2676.8	9636.4	3.74	2933.8	10561.6
1100	3.23	3069.9	11051.5	3.43	3257.2	11726.1	3.61	3434.5	12364.1	3.96	3764.1	13550.7
1200	3.41	3853.9	13874.2	3.62	4089.1	14720.8	3.81	4311.5	15521.5	4.18	4725.2	17010.8
1300	3.58	4750.5	17101.8	3.80	5040.3	18145.0	4.00	5314.4	19131.7	4.39	5824.2	20967.0
1400	3.75	5765.0	20754.2	3.97	6116.6	22019.9	4.19	6449.2	23217.0	4.59	7067.7	25443.7
1500	3.91	6903.0	24850.6	4.14	7323.8	26365.8	4.37	7721.9	27799.0	4.79	8462.4	30464.6
1600	4.06	8169.4	29409.9	4.31	8667.4	31202.8	4.55	9138.5	32898.5	4.98	10014.6	36052.5

D	S=	0.014		S=	0.016		S=	0.018		S=	0.020	
(mm)	v(m/s)	Q(l/s)	Q(m3/h)	v(m/s)	Q(l/s)	Q(m3/h)	v(m/s)	Q(l/s)	Q(m3/h)	v(m/s)	Q(l/s)	Q(m3/h)
50	0.58	1.1	4.1	0.62	1.2	4.4	0.66	1.3	4.7	0.70	1.4	5.0
80	0.80	4.0	14.5	0.86	4.3	15.6	0.91	4.6	16.5	0.96	4.8	17.4
100	0.93	7.3	26.3	1.00	7.8	28.2	1.06	8.3	29.9	1.12	8.8	31.6
125	1.08	13.3	47.7	1.16	14.2	51.1	1.23	15.1	54.2	1.29	15.9	57.2
150	1.22	21.5	77.4	1.30	23.0	82.8	1.38	24.4	87.9	1.46	25.8	92.8
200	1.47	46.1	165.9	1.57	49.3	177.5	1.67	52.3	188.4	1.76	55.2	198.7
250	1.69	83.1	299.1	1.81	88.9	320.0	1.92	94.3	339.6	2.03	99.5	358.2
300	1.90	134.4	483.9	2.03	143.8	517.7	2.16	152.6	549.4	2.28	160.9	579.3
350	2.10	201.8	726.3	2.24	215.8	777.0	2.38	229.0	824.5	2.51	241.5	869.5
400	2.28	286.7	1032.2	2.44	306.7	1104.1	2.59	325.4	1171.6	2.73	343.2	1235.4
450	2.46	390.8	1406.9	2.63	418.0	1504.8	2.79	443.5	1596.8	2.94	467.7	1683.8
500	2.63	515.4	1855.6	2.81	551.3	1984.7	2.98	585.0	2105.9	3.14	616.8	2220.6
600	2.94	831.8	2994.4	3.15	889.6	3202.6	3.34	943.9	3398.1	3.52	995.3	3583.0
700	3.24	1246.1	4485.9	3.46	1332.6	4797.5	3.67	1414.0	5090.2	3.87	1490.9	5367.1
800	3.52	1767.9	6364.5	3.76	1890.7	6806.5	3.99	2006.0	7221.6	4.21	2115.1	7614.2
900	3.78	2406.4	8663.0	4.05	2573.4	9264.4	4.29	2730.3	9829.2	4.53	2878.7	10363.4
1000	4.04	3170.1	11412.3	4.32	3390.1	12204.3	4.58	3596.7	12948.1	4.83	3792.1	13651.6
1100	4.28	4067.2	14642.0	4.58	4349.4	15657.7	4.86	4614.4	16611.8	5.12	4865.0	17514.1
1200	4.51	5105.7	18380.3	4.83	5459.8	19655.1	5.12	5792.3	20852.4	5.40	6106.9	21984.9
1300	4.74	6293.0	22654.7	5.07	6729.3	24225.6	5.38	7139.2	25701.0	5.67	7526.8	27096.5
1400	4.96	7636.5	27491.3	5.30	8165.9	29397.3	5.63	8663.2	31187.4	5.93	9133.5	32880.5
1500	5.17	9143.3	32915.9	5.53	9777.1	35197.5	5.87	10372.4	37340.5	6.19	10935.4	39367.4
1600	5.38	10820.3	38953.0	5.75	11570.2	41652.7	6.10	12274.5	44188.3	6.44	12940.7	46586.6

$$k(mm) = \quad 1 \quad T(°C) = \quad 10$$

D	S =	0.0005		S =	0.001		S =	0.002		S =	0.003	
(mm)	v(m/s)	Q(l/s)	Q(m3/h)	v(m/s)	Q(l/s)	Q(m3/h)	v(m/s)	Q(l/s)	Q(m3/h)	v(m/s)	Q(l/s)	Q(m3/h)
50	0.09	0.2	0.7	0.13	0.3	0.9	0.19	0.4	1.4	0.24	0.5	1.7
80	0.13	0.7	2.3	0.19	0.9	3.4	0.27	1.3	4.8	0.33	1.7	6.0
100	0.15	1.2	4.3	0.22	1.7	6.2	0.31	2.5	8.8	0.38	3.0	10.9
125	0.18	2.2	7.8	0.25	3.1	11.2	0.36	4.5	16.1	0.45	5.5	19.8
150	0.20	3.6	12.8	0.29	5.1	18.3	0.41	7.3	26.2	0.51	8.9	32.2
200	0.24	7.7	27.7	0.35	11.0	39.6	0.50	15.7	56.4	0.61	19.3	69.3
250	0.28	14.0	50.3	0.41	19.9	71.7	0.58	28.4	102.1	0.71	34.8	125.4
300	0.32	22.7	81.7	0.46	32.3	116.5	0.65	46.0	165.7	0.80	56.5	203.5
350	0.36	34.2	123.1	0.51	48.7	175.3	0.72	69.2	249.3	0.88	85.0	306.0
400	0.39	48.7	175.4	0.55	69.4	249.7	0.78	98.6	354.9	0.96	121.0	435.7
450	0.42	66.6	239.6	0.60	94.7	341.1	0.85	134.6	484.6	1.04	165.2	594.7
500	0.45	88.0	316.7	0.64	125.2	450.6	0.91	177.8	640.0	1.11	218.2	785.4
600	0.50	142.5	512.8	0.72	202.6	729.2	1.02	287.6	1035.3	1.25	352.8	1270.2
700	0.56	214.0	770.4	0.79	304.1	1094.8	1.12	431.6	1553.9	1.38	529.5	1906.2
800	0.61	304.3	1095.4	0.86	432.3	1556.2	1.22	613.4	2208.1	1.50	752.3	2708.5
900	0.65	414.9	1493.8	0.93	589.3	2121.5	1.31	836.0	3009.6	1.61	1025.3	3691.1
1000	0.70	547.5	1971.0	0.99	777.4	2798.6	1.40	1102.6	3969.3	1.72	1352.2	4867.8
1100	0.74	703.4	2532.3	1.05	998.6	3594.8	1.49	1416.1	5097.8	1.83	1736.5	6251.3
1200	0.78	884.1	3182.7	1.11	1254.8	4517.3	1.57	1779.2	6405.2	1.93	2181.6	7853.9
1300	0.82	1090.9	3927.2	1.17	1548.1	5573.1	1.65	2194.8	7901.1	2.03	2691.0	9687.7
1400	0.86	1325.1	4770.4	1.22	1880.2	6768.6	1.73	2665.3	9595.1	2.12	3267.8	11764.1
1500	0.90	1588.0	5716.8	1.27	2252.9	8110.4	1.81	3193.4	11496.1	2.22	3915.1	14094.2
1600	0.94	1880.8	6770.8	1.33	2668.0	9604.7	1.88	3781.4	13613.0	2.31	4635.8	16688.9

D	S=	0.004		S=	0.005		S=	0.006		S=	0.007	
(mm)	v(m/s)	Q(l/s)	Q(m3/h)	v(m/s)	Q(l/s)	Q(m3/h)	v(m/s)	Q(l/s)	Q(m3/h)	v(m/s)	Q(l/s)	Q(m3/h)
50	0.27	0.5	1.9	0.31	0.6	2.2	0.34	0.7	2.4	0.37	0.7	2.6
80	0.38	1.9	6.9	0.43	2.2	7.7	0.47	2.4	8.5	0.51	2.6	9.2
100	0.45	3.5	12.6	0.50	3.9	14.1	0.55	4.3	15.5	0.59	4.7	16.7
125	0.52	6.4	22.9	0.58	7.1	25.6	0.64	7.8	28.1	0.69	8.5	30.4
150	0.59	10.4	37.3	0.66	11.6	41.7	0.72	12.7	45.8	0.78	13.7	49.5
200	0.71	22.3	80.2	0.79	24.9	89.8	0.87	27.4	98.5	0.94	29.6	106.5
250	0.82	40.3	145.1	0.92	45.1	162.4	1.01	49.5	178.1	1.09	53.5	192.5
300	0.92	65.4	235.3	1.04	73.2	263.4	1.13	80.2	288.8	1.23	86.7	312.1
350	1.02	98.3	353.9	1.14	110.0	396.1	1.25	120.6	434.2	1.35	130.4	469.3
400	1.11	139.9	503.8	1.25	156.6	563.8	1.37	171.7	618.0	1.48	185.5	667.9
450	1.20	191.0	687.6	1.34	213.7	769.5	1.47	234.3	843.4	1.59	253.2	911.5
500	1.28	252.2	908.0	1.44	282.2	1016.0	1.58	309.4	1113.7	1.70	334.3	1203.5
600	1.44	407.9	1468.3	1.61	456.3	1642.8	1.77	500.2	1800.6	1.91	540.5	1945.7
700	1.59	612.0	2203.3	1.78	684.7	2465.0	1.95	750.4	2701.6	2.11	810.9	2919.2
800	1.73	869.5	3130.3	1.94	972.8	3501.9	2.12	1066.1	3837.9	2.29	1151.9	4146.9
900	1.86	1184.9	4265.7	2.08	1325.5	4772.0	2.28	1452.7	5229.7	2.47	1569.6	5650.6
1000	1.99	1562.6	5625.2	2.23	1747.9	6292.6	2.44	1915.5	6896.0	2.64	2069.7	7450.8
1100	2.11	2006.6	7223.7	2.36	2244.6	8080.4	2.59	2459.7	8855.0	2.80	2657.6	9567.3
1200	2.23	2520.9	9075.3	2.49	2819.8	10151.4	2.73	3090.1	11124.3	2.95	3338.6	12018.9
1300	2.34	3109.4	11193.9	2.62	3478.0	12520.9	2.87	3811.3	13720.6	3.10	4117.7	14823.9
1400	2.45	3775.7	13592.7	2.74	4223.3	15203.8	3.01	4627.9	16660.3	3.25	4999.9	17999.8
1500	2.56	4523.5	16284.6	2.86	5059.6	18214.4	3.14	5544.2	19959.1	3.39	5989.9	21563.6
1600	2.66	5356.1	19282.1	2.98	5990.8	21566.8	3.26	6564.5	23632.4	3.53	7092.2	25531.8

k(mm) =   1   T(°C) =   10

D	S=	0.008		S=	0.009		S=	0.010		S=	0.012	
(mm)	v(m/s)	Q(l/s)	Q(m3/h)	v(m/s)	Q(l/s)	Q(m3/h)	v(m/s)	Q(l/s)	Q(m3/h)	v(m/s)	Q(l/s)	Q(m3/h)
50	0.39	0.8	2.8	0.42	0.8	2.9	0.44	0.9	3.1	0.48	0.9	3.4
80	0.54	2.7	9.8	0.58	2.9	10.5	0.61	3.1	11.0	0.67	3.4	12.1
100	0.63	5.0	17.9	0.67	5.3	19.0	0.71	5.6	20.1	0.78	6.1	22.0
125	0.74	9.0	32.6	0.78	9.6	34.6	0.82	10.1	36.4	0.90	11.1	40.0
150	0.83	14.7	53.0	0.88	15.6	56.2	0.93	16.5	59.3	1.02	18.1	65.0
200	1.01	31.6	113.9	1.07	33.6	120.9	1.13	35.4	127.5	1.24	38.8	139.7
250	1.17	57.2	205.9	1.24	60.7	218.5	1.30	64.0	230.4	1.43	70.2	252.6
300	1.31	92.7	333.8	1.39	98.4	354.2	1.47	103.8	373.5	1.61	113.7	409.4
350	1.45	139.4	501.9	1.54	147.9	532.6	1.62	156.0	561.6	1.78	171.0	615.5
400	1.58	198.4	714.3	1.68	210.5	757.9	1.77	222.0	799.2	1.94	243.3	875.9
450	1.70	270.8	974.8	1.81	287.3	1034.3	1.90	302.9	1090.6	2.09	332.0	1195.2
500	1.82	357.5	1287.1	1.93	379.3	1365.6	2.04	400.0	1439.9	2.23	438.3	1578.0
600	2.04	578.0	2080.8	2.17	613.2	2207.6	2.29	646.6	2327.6	2.51	708.5	2550.7
700	2.25	867.1	3121.7	2.39	920.0	3311.9	2.52	970.0	3491.8	2.76	1062.9	3826.5
800	2.45	1231.8	4434.5	2.60	1306.8	4704.6	2.74	1377.8	4960.1	3.00	1509.8	5435.3
900	2.64	1678.4	6042.3	2.80	1780.6	6410.3	2.95	1877.3	6758.3	3.23	2057.1	7405.6
1000	2.82	2213.1	7967.2	2.99	2347.9	8452.3	3.15	2475.3	8911.1	3.45	2712.3	9764.4
1100	2.99	2841.7	10230.3	3.17	3014.7	10853.0	3.34	3178.3	11442.0	3.66	3482.6	12537.8
1200	3.16	3569.9	12851.7	3.35	3787.2	13633.8	3.53	3992.7	14373.5	3.87	4374.8	15749.4
1300	3.32	4403.0	15850.8	3.52	4670.9	16815.3	3.71	4924.3	17727.5	4.07	5395.6	19424.3
1400	3.47	5346.3	19246.5	3.68	5671.5	20417.5	3.88	5979.2	21525.0	4.26	6551.4	23584.9
1500	3.62	6404.7	23057.0	3.84	6794.3	24459.6	4.05	7162.8	25786.2	4.44	7848.2	28253.7
1600	3.77	7583.3	27299.8	4.00	8044.5	28960.4	4.22	8480.8	30531.0	4.62	9292.3	33452.1

D	S=	0.014		S=	0.016		S=	0.018		S=	0.020	
(mm)	v(m/s)	Q(l/s)	Q(m3/h)	v(m/s)	Q(l/s)	Q(m3/h)	v(m/s)	Q(l/s)	Q(m3/h)	v(m/s)	Q(l/s)	Q(m3/h)
50	0.52	1.0	3.7	0.56	1.1	3.9	0.59	1.2	4.2	0.63	1.2	4.4
80	0.72	3.6	13.1	0.77	3.9	14.0	0.82	4.1	14.9	0.87	4.4	15.7
100	0.84	6.6	23.8	0.90	7.1	25.5	0.96	7.5	27.0	1.01	7.9	28.5
125	0.98	12.0	43.2	1.05	12.8	46.2	1.11	13.6	49.0	1.17	14.4	51.7
150	1.10	19.5	70.3	1.18	20.9	75.1	1.25	22.2	79.7	1.32	23.4	84.1
200	1.34	41.9	151.0	1.43	44.9	161.5	1.52	47.6	171.4	1.60	50.2	180.7
250	1.54	75.8	273.0	1.65	81.1	292.0	1.75	86.1	309.8	1.85	90.7	326.6
300	1.74	122.9	442.4	1.86	131.4	473.2	1.97	139.5	502.0	2.08	147.0	529.3
350	1.92	184.8	665.1	2.05	197.6	711.3	2.18	209.6	754.7	2.30	221.0	795.7
400	2.09	262.9	946.4	2.24	281.1	1012.1	2.37	298.3	1073.8	2.50	314.5	1132.1
450	2.26	358.7	1291.4	2.41	383.6	1381.0	2.56	407.0	1465.1	2.70	429.1	1544.7
500	2.41	473.6	1705.0	2.58	506.4	1823.2	2.74	537.3	1934.2	2.88	566.5	2039.3
600	2.71	765.5	2755.9	2.90	818.6	2947.0	3.07	868.4	3126.4	3.24	915.6	3296.0
700	2.98	1148.4	4134.2	3.19	1227.9	4420.6	3.38	1302.7	4689.6	3.57	1373.4	4944.1
800	3.25	1631.2	5872.3	3.47	1744.2	6279.0	3.68	1850.3	6661.0	3.88	1950.6	7022.3
900	3.49	2222.5	8000.8	3.74	2376.4	8554.9	3.96	2520.9	9075.3	4.18	2657.6	9567.4
1000	3.73	2930.3	10549.1	3.99	3133.2	11279.4	4.23	3323.7	11965.4	4.46	3504.0	12614.2
1100	3.96	3762.4	13544.8	4.23	4022.9	14482.4	4.49	4267.5	15363.1	4.73	4498.9	16196.0
1200	4.18	4726.3	17014.7	4.47	5053.4	18192.4	4.74	5360.7	19298.5	5.00	5651.3	20344.7
1300	4.39	5829.0	20984.6	4.70	6232.5	22436.9	4.98	6611.4	23800.9	5.25	6969.7	25091.0
1400	4.60	7077.6	25479.2	4.92	7567.3	27242.4	5.21	8027.3	28898.4	5.50	8462.4	30464.7
1500	4.80	8478.5	30522.7	5.13	9065.2	32634.7	5.44	9616.2	34618.4	5.74	10137.4	36494.5
1600	4.99	10038.5	36138.5	5.34	10733.0	38638.8	5.66	11385.3	40987.2	5.97	12002.3	43208.4

k(mm) =   5   T(°C) =   10

D	S =	0.0005		S =	0.001		S =	0.002		S =	0.003	
(mm)	v(m/s)	Q(l/s)	Q(m3/h)	v(m/s)	Q(l/s)	Q(m3/h)	v(m/s)	Q(l/s)	Q(m3/h)	v(m/s)	Q(l/s)	Q(m3/h)
50	0.07	0.1	0.5	0.10	0.2	0.7	0.14	0.3	1.0	0.17	0.3	1.2
80	0.10	0.5	1.8	0.14	0.7	2.5	0.20	1.0	3.6	0.24	1.2	4.4
100	0.12	0.9	3.3	0.16	1.3	4.6	0.23	1.8	6.6	0.28	2.2	8.1
125	0.14	1.7	6.0	0.19	2.4	8.5	0.27	3.4	12.1	0.34	4.1	14.8
150	0.15	2.7	9.9	0.22	3.9	14.0	0.31	5.5	19.8	0.38	6.8	24.3
200	0.19	6.0	21.5	0.27	8.5	30.5	0.38	12.0	43.3	0.47	14.7	53.1
250	0.22	10.9	39.3	0.32	15.5	55.8	0.45	22.0	79.0	0.55	26.9	96.9
300	0.25	17.9	64.3	0.36	25.3	91.1	0.51	35.9	129.1	0.62	44.0	158.2
350	0.28	27.0	97.3	0.40	38.3	137.9	0.56	54.2	195.3	0.69	66.5	239.3
400	0.31	38.7	139.2	0.44	54.8	197.2	0.62	77.6	279.3	0.76	95.1	342.3
450	0.33	53.0	190.8	0.47	75.1	270.3	0.67	106.3	382.8	0.82	130.3	469.1
500	0.36	70.3	252.9	0.51	99.5	358.3	0.72	140.9	507.3	0.88	172.7	621.7
600	0.40	114.3	411.5	0.57	161.9	582.9	0.81	229.2	825.2	0.99	280.9	1011.1
700	0.45	172.4	620.6	0.63	244.1	878.9	0.90	345.6	1244.1	1.10	423.4	1524.4
800	0.49	246.0	885.4	0.69	348.3	1253.8	0.98	493.0	1774.6	1.20	604.0	2174.3
900	0.53	336.4	1210.9	0.75	476.2	1714.5	1.06	674.1	2426.6	1.30	825.8	2973.0
1000	0.57	444.9	1601.8	0.80	629.9	2267.7	1.14	891.5	3209.4	1.39	1092.2	3932.1
1100	0.60	572.9	2062.5	0.85	811.0	2919.8	1.21	1147.8	4132.1	1.48	1406.2	5062.4
1200	0.64	721.5	2597.5	0.90	1021.4	3676.9	1.28	1445.4	5203.4	1.57	1770.8	6374.8
1300	0.67	891.9	3210.8	0.95	1262.5	4544.9	1.35	1786.5	6431.6	1.65	2188.7	7879.3
1400	0.70	1085.2	3906.6	1.00	1536.0	5529.5	1.41	2173.5	7824.7	1.73	2662.7	9585.9
1500	0.74	1302.4	4688.7	1.04	1843.4	6636.3	1.48	2608.5	9390.7	1.81	3195.6	11504.2
1600	0.77	1544.7	5561.0	1.09	2186.3	7870.7	1.54	3093.6	11137.1	1.88	3789.9	13643.5

D	S=	0.004		S=	0.005		S=	0.006		S=	0.007	
(mm)	v(m/s)	Q(l/s)	Q(m3/h)	v(m/s)	Q(l/s)	Q(m3/h)	v(m/s)	Q(l/s)	Q(m3/h)	v(m/s)	Q(l/s)	Q(m3/h)
50	0.19	0.4	1.4	0.22	0.4	1.5	0.24	0.5	1.7	0.26	0.5	1.8
80	0.28	1.4	5.0	0.31	1.6	5.6	0.34	1.7	6.2	0.37	1.9	6.7
100	0.33	2.6	9.3	0.37	2.9	10.4	0.40	3.2	11.4	0.44	3.4	12.3
125	0.39	4.8	17.1	0.43	5.3	19.2	0.48	5.8	21.0	0.51	6.3	22.7
150	0.44	7.8	28.1	0.49	8.7	31.5	0.54	9.6	34.5	0.59	10.3	37.2
200	0.54	17.0	61.3	0.61	19.0	68.6	0.66	20.9	75.1	0.72	22.5	81.2
250	0.63	31.1	111.9	0.71	34.8	125.2	0.78	38.1	137.2	0.84	41.2	148.2
300	0.72	50.8	182.8	0.80	56.8	204.4	0.88	62.2	224.0	0.95	67.2	242.0
350	0.80	76.8	276.5	0.89	85.9	309.2	0.98	94.1	338.8	1.06	101.7	366.0
400	0.87	109.8	395.4	0.98	122.8	442.2	1.07	134.6	484.5	1.16	145.4	523.4
450	0.95	150.5	541.9	1.06	168.3	606.0	1.16	184.4	663.9	1.25	199.2	717.2
500	1.02	199.5	718.1	1.14	223.1	803.0	1.24	244.4	879.8	1.34	264.0	950.4
600	1.15	324.4	1167.9	1.28	362.8	1306.0	1.41	397.5	1430.8	1.52	429.3	1545.6
700	1.27	489.1	1760.7	1.42	546.9	1968.8	1.56	599.2	2157.0	1.68	647.2	2330.1
800	1.39	697.6	2511.3	1.55	780.0	2808.2	1.70	854.6	3076.5	1.84	923.2	3323.3
900	1.50	953.8	3433.7	1.68	1066.5	3839.6	1.84	1168.5	4206.5	1.98	1262.2	4543.9
1000	1.61	1261.5	4541.3	1.80	1410.6	5078.0	1.97	1545.3	5563.2	2.13	1669.3	6009.4
1100	1.71	1624.1	5846.7	1.91	1816.0	6537.6	2.09	1989.5	7162.3	2.26	2149.1	7736.8
1200	1.81	2045.1	7362.3	2.02	2286.7	8232.3	2.22	2505.2	9018.8	2.39	2706.1	9742.1
1300	1.90	2527.7	9099.8	2.13	2826.4	10175.0	2.33	3096.4	11147.1	2.52	3344.7	12041.1
1400	2.00	3075.2	11070.6	2.23	3438.5	12378.7	2.45	3767.0	13561.3	2.64	4069.1	14648.8
1500	2.09	3690.5	13286.0	2.34	4126.6	14855.7	2.56	4520.8	16274.9	2.76	4883.3	17580.0
1600	2.18	4376.8	15756.5	2.43	4893.9	17618.1	2.67	5361.4	19301.1	2.88	5791.3	20848.8

k(mm) =   5   T(°C) =   10

D	S=	0.008		S=	0.009		S=	0.010		S=	0.012	
(mm)	v(m/s)	Q(l/s)	Q(m3/h)	v(m/s)	Q(l/s)	Q(m3/h)	v(m/s)	Q(l/s)	Q(m3/h)	v(m/s)	Q(l/s)	Q(m3/h)
50	0.28	0.5	2.0	0.29	0.6	2.1	0.31	0.6	2.2	0.34	0.7	2.4
80	0.40	2.0	7.2	0.42	2.1	7.6	0.44	2.2	8.0	0.48	2.4	8.8
100	0.47	3.7	13.2	0.49	3.9	14.0	0.52	4.1	14.8	0.57	4.5	16.2
125	0.55	6.7	24.3	0.58	7.1	25.7	0.61	7.5	27.1	0.67	8.3	29.7
150	0.63	11.1	39.8	0.66	11.7	42.2	0.70	12.4	44.5	0.77	13.6	48.8
200	0.77	24.1	86.8	0.81	25.6	92.1	0.86	27.0	97.1	0.94	29.5	106.3
250	0.90	44.0	158.4	0.95	46.7	168.1	1.00	49.2	177.2	1.10	53.9	194.1
300	1.02	71.9	258.7	1.08	76.2	274.4	1.14	80.4	289.3	1.25	88.1	317.0
350	1.13	108.7	391.3	1.20	115.3	415.1	1.26	121.5	437.6	1.38	133.2	479.4
400	1.24	155.4	559.6	1.31	164.9	593.6	1.38	173.8	625.7	1.52	190.4	685.6
450	1.34	213.0	766.8	1.42	225.9	813.4	1.50	238.2	857.5	1.64	261.0	939.4
500	1.44	282.2	1016.1	1.52	299.4	1077.8	1.61	315.6	1136.2	1.76	345.8	1244.8
600	1.62	459.0	1652.5	1.72	486.9	1752.9	1.82	513.3	1847.8	1.99	562.3	2024.4
700	1.80	692.0	2491.2	1.91	734.0	2642.5	2.01	773.8	2785.6	2.20	847.7	3051.7
800	1.96	987.0	3553.1	2.08	1046.9	3768.8	2.20	1103.6	3972.9	2.41	1209.0	4352.6
900	2.12	1349.4	4858.0	2.25	1431.4	5153.0	2.37	1508.9	5432.0	2.60	1653.0	5950.9
1000	2.27	1784.7	6424.8	2.41	1893.0	6814.9	2.54	1995.5	7183.8	2.78	2186.1	7870.0
1100	2.42	2297.6	8271.4	2.56	2437.1	8773.6	2.70	2569.1	9248.6	2.96	2814.5	10132.0
1200	2.56	2893.1	10415.3	2.71	3068.8	11047.7	2.86	3234.9	11645.7	3.13	3543.9	12758.1
1300	2.69	3575.9	12873.2	2.86	3793.0	13654.7	3.01	3998.3	14393.8	3.30	4380.2	15768.6
1400	2.83	4350.3	15661.1	3.00	4614.4	16611.8	3.16	4864.2	17511.0	3.46	5328.7	19183.4
1500	2.95	5220.8	18794.7	3.13	5537.7	19935.6	3.30	5837.4	21014.7	3.62	6394.9	23021.7
1600	3.08	6191.5	22289.4	3.27	6567.3	23642.4	3.44	6922.8	24922.1	3.77	7583.9	27302.2

D	S=	0.014		S=	0.016		S=	0.018		S=	0.020	
(mm)	v(m/s)	Q(l/s)	Q(m3/h)	v(m/s)	Q(l/s)	Q(m3/h)	v(m/s)	Q(l/s)	Q(m3/h)	v(m/s)	Q(l/s)	Q(m3/h)
50	0.37	0.7	2.6	0.39	0.8	2.8	0.41	0.8	2.9	0.44	0.9	3.1
80	0.52	2.6	9.5	0.56	2.8	10.1	0.59	3.0	10.8	0.63	3.1	11.3
100	0.62	4.9	17.5	0.66	5.2	18.7	0.70	5.5	19.8	0.74	5.8	20.9
125	0.73	8.9	32.1	0.78	9.5	34.3	0.82	10.1	36.4	0.87	10.7	38.4
150	0.83	14.6	52.7	0.89	15.7	56.4	0.94	16.6	59.8	0.99	17.5	63.0
200	1.02	31.9	114.9	1.09	34.1	122.8	1.15	36.2	130.3	1.21	38.2	137.4
250	1.19	58.3	209.7	1.27	62.3	224.2	1.35	66.1	237.8	1.42	69.6	250.7
300	1.35	95.1	342.4	1.44	101.7	366.1	1.53	107.9	388.3	1.61	113.7	409.4
350	1.50	143.9	517.9	1.60	153.8	553.7	1.70	163.1	587.3	1.79	172.0	619.1
400	1.64	205.7	740.6	1.75	219.9	791.8	1.86	233.3	839.8	1.96	245.9	885.3
450	1.77	281.9	1014.8	1.89	301.4	1084.9	2.01	319.7	1150.8	2.12	337.0	1213.1
500	1.90	373.5	1344.6	2.03	399.3	1437.6	2.16	423.6	1524.9	2.27	446.5	1607.4
600	2.15	607.4	2186.7	2.30	649.4	2337.9	2.44	688.8	2479.8	2.57	726.1	2614.1
700	2.38	915.7	3296.5	2.54	979.0	3524.3	2.70	1038.4	3738.2	2.84	1094.6	3940.6
800	2.60	1306.0	4701.5	2.78	1396.2	5026.4	2.95	1481.0	5331.5	3.11	1561.1	5620.1
900	2.81	1785.6	6428.1	3.00	1909.0	6872.2	3.18	2024.8	7289.4	3.36	2134.4	7683.9
1000	3.01	2361.4	8501.1	3.21	2524.6	9088.5	3.41	2677.8	9640.1	3.59	2822.8	10161.9
1100	3.20	3040.1	10944.4	3.42	3250.2	11700.6	3.63	3447.4	12410.8	3.82	3634.0	13082.5
1200	3.38	3828.0	13781.0	3.62	4092.5	14733.1	3.84	4340.9	15627.3	4.05	4575.9	16473.1
1300	3.56	4731.3	17032.8	3.81	5058.2	18209.6	4.04	5365.2	19314.8	4.26	5655.6	20360.1
1400	3.74	5755.9	20721.4	4.00	6153.6	22152.9	4.24	6527.1	23497.4	4.47	6880.3	24769.1
1500	3.91	6907.6	24867.4	4.18	7384.8	26585.3	4.43	7833.0	28198.8	4.67	8256.9	29724.8
1600	4.07	8191.9	29491.0	4.36	8757.8	31528.2	4.62	9289.3	33441.6	4.87	9792.1	35251.4

k(mm) =   0.01   T(°C) =   20

D	S =	0.0005		S =	0.001		S =	0.002		S =	0.003	
(mm)	v(m/s)	Q(l/s)	Q(m3/h)	v(m/s)	Q(l/s)	Q(m3/h)	v(m/s)	Q(l/s)	Q(m3/h)	v(m/s)	Q(l/s)	Q(m3/h)
50	0.12	0.2	0.8	0.17	0.3	1.2	0.26	0.5	1.8	0.33	0.6	2.3
80	0.16	0.8	3.0	0.24	1.2	4.4	0.36	1.8	6.5	0.45	2.3	8.2
100	0.19	1.5	5.5	0.29	2.2	8.1	0.42	3.3	11.9	0.53	4.2	15.0
125	0.23	2.8	10.0	0.33	4.1	14.7	0.49	6.0	21.7	0.62	7.6	27.2
150	0.26	4.5	16.3	0.38	6.7	24.0	0.56	9.8	35.4	0.70	12.3	44.3
200	0.31	9.8	35.3	0.46	14.4	52.0	0.67	21.2	76.3	0.84	26.5	95.4
250	0.36	17.8	64.2	0.53	26.2	94.3	0.78	38.4	138.2	0.98	48.0	172.6
300	0.41	29.0	104.5	0.60	42.6	153.3	0.88	62.3	224.3	1.10	77.8	280.0
350	0.46	43.8	157.6	0.67	64.2	231.0	0.97	93.8	337.7	1.22	117.0	421.3
400	0.50	62.5	224.9	0.73	91.5	329.3	1.06	133.6	481.0	1.33	166.6	599.7
450	0.54	85.5	307.7	0.79	125.0	450.0	1.15	182.5	656.9	1.43	227.4	818.7
500	0.58	113.1	407.0	0.84	165.3	594.9	1.23	241.1	867.8	1.53	300.3	1081.2
600	0.65	183.4	660.1	0.95	267.7	963.9	1.38	390.1	1404.4	1.72	485.7	1748.5
700	0.72	275.8	993.0	1.05	402.4	1448.5	1.52	585.7	2108.6	1.89	728.9	2623.9
800	0.78	392.7	1413.6	1.14	572.3	2060.4	1.66	832.5	2997.1	2.06	1035.5	3727.9
900	0.84	536.0	1929.6	1.23	780.8	2810.7	1.78	1134.9	4085.8	2.22	1411.1	5080.0
1000	0.90	707.9	2548.4	1.31	1030.5	3709.8	1.91	1497.1	5389.6	2.37	1860.7	6698.7
1100	0.96	910.2	3276.7	1.39	1324.3	4767.6	2.02	1923.0	6922.7	2.51	2389.3	8601.6
1200	1.01	1144.8	4121.4	1.47	1664.9	5993.7	2.14	2416.4	8698.9	2.65	3001.5	10805.5
1300	1.06	1413.5	5088.7	1.55	2054.8	7397.3	2.25	2980.9	10731.4	2.79	3701.9	13326.9
1400	1.12	1718.0	6184.9	1.62	2496.5	8987.3	2.35	3620.2	13032.8	2.92	4494.8	16181.2
1500	1.17	2060.0	7416.0	1.69	2992.3	10772.3	2.45	4337.7	15615.6	3.05	5384.4	19383.8
1600	1.21	2441.0	8787.7	1.76	3544.6	12760.6	2.55	5136.6	18491.7	3.17	6374.8	22949.3

D	S=	0.004		S=	0.005		S=	0.006		S=	0.007	
(mm)	v(m/s)	Q(l/s)	Q(m3/h)	v(m/s)	Q(l/s)	Q(m3/h)	v(m/s)	Q(l/s)	Q(m3/h)	v(m/s)	Q(l/s)	Q(m3/h)
50	0.38	0.8	2.7	0.44	0.9	3.1	0.48	0.9	3.4	0.53	1.0	3.7
80	0.53	2.7	9.6	0.60	3.0	10.9	0.67	3.4	12.1	0.73	3.7	13.2
100	0.62	4.9	17.6	0.70	5.5	19.9	0.78	6.1	22.0	0.85	6.7	24.0
125	0.72	8.9	31.9	0.82	10.0	36.1	0.90	11.1	39.9	0.98	12.1	43.5
150	0.82	14.4	51.9	0.92	16.3	58.7	1.02	18.0	64.9	1.11	19.6	70.7
200	0.99	31.0	111.7	1.12	35.1	126.2	1.23	38.7	139.5	1.34	42.1	151.7
250	1.14	56.1	202.1	1.29	63.4	228.2	1.43	70.0	252.0	1.55	76.1	274.1
300	1.29	91.0	327.6	1.45	102.7	369.9	1.60	113.4	408.3	1.74	123.3	443.9
350	1.42	136.8	492.6	1.61	154.4	556.0	1.77	170.5	613.6	1.93	185.3	667.0
400	1.55	194.7	701.0	1.75	219.7	791.0	1.93	242.5	872.9	2.10	263.5	948.5
450	1.67	265.7	956.7	1.88	299.8	1079.2	2.08	330.7	1190.7	2.26	359.3	1293.7
500	1.79	350.9	1263.1	2.02	395.7	1424.6	2.22	436.5	1571.4	2.42	474.2	1707.1
600	2.01	567.2	2041.7	2.26	639.4	2302.0	2.49	705.1	2538.5	2.71	765.8	2757.0
700	2.21	850.8	3062.7	2.49	958.9	3452.0	2.75	1057.2	3805.9	2.98	1147.9	4132.6
800	2.40	1208.3	4349.9	2.71	1361.6	4901.7	2.99	1500.8	5403.0	3.24	1629.4	5865.9
900	2.59	1646.1	5926.0	2.92	1854.5	6676.2	3.21	2043.8	7357.8	3.49	2218.6	7986.9
1000	2.76	2170.1	7812.4	3.11	2444.4	8799.7	3.43	2693.5	9696.5	3.72	2923.9	10524.2
1100	2.93	2786.0	10029.4	3.30	3137.5	11295.0	3.64	3456.8	12444.4	3.95	3751.4	13505.1
1200	3.09	3499.1	12596.8	3.48	3940.1	14184.2	3.84	4340.4	15625.6	4.16	4709.9	16955.6
1300	3.25	4314.8	15533.3	3.66	4857.9	17488.3	4.03	5350.9	19263.2	4.37	5805.8	20900.9
1400	3.40	5238.1	18857.1	3.83	5896.6	21227.7	4.22	6494.4	23379.7	4.58	7045.9	25365.2
1500	3.55	6273.8	22585.8	4.00	7061.7	25422.2	4.40	7776.9	27996.8	4.77	8436.6	30371.8
1600	3.69	7426.8	26736.6	4.16	8358.6	30090.9	4.58	9204.3	33135.4	4.97	9984.3	35943.6

k(mm) = 0.01 T(°C) = 20

D	S=	0.008		S=	0.009		S=	0.010		S=	0.012	
(mm)	v(m/s)	Q(l/s)	Q(m3/h)	v(m/s)	Q(l/s)	Q(m3/h)	v(m/s)	Q(l/s)	Q(m3/h)	v(m/s)	Q(l/s)	Q(m3/h)
50	0.57	1.1	4.0	0.61	1.2	4.3	0.64	1.3	4.5	0.71	1.4	5.0
80	0.78	3.9	14.2	0.84	4.2	15.2	0.89	4.5	16.1	0.98	4.9	17.8
100	0.91	7.2	25.8	0.97	7.6	27.5	1.03	8.1	29.2	1.14	9.0	32.3
125	1.06	13.0	46.8	1.13	13.9	49.9	1.20	14.7	52.9	1.32	16.2	58.4
150	1.19	21.1	76.0	1.27	22.5	81.1	1.35	23.9	85.9	1.49	26.4	94.9
200	1.44	45.3	163.2	1.54	48.3	174.0	1.63	51.2	184.3	1.80	56.5	203.4
250	1.67	81.9	294.7	1.78	87.3	314.1	1.88	92.4	332.6	2.08	102.0	367.0
300	1.88	132.6	477.2	2.00	141.3	508.6	2.12	149.5	538.3	2.33	165.0	593.9
350	2.07	199.1	716.8	2.21	212.2	763.8	2.33	224.5	808.4	2.57	247.7	891.6
400	2.25	283.1	1019.2	2.40	301.6	1085.9	2.54	319.2	1149.1	2.80	352.0	1267.2
450	2.43	386.1	1389.9	2.59	411.3	1480.6	2.74	435.2	1566.6	3.02	479.8	1727.2
500	2.59	509.4	1833.9	2.76	542.6	1953.3	2.92	574.1	2066.6	3.22	632.8	2278.1
600	2.91	822.5	2961.1	3.10	876.0	3153.4	3.28	926.6	3335.8	3.61	1021.1	3676.1
700	3.20	1232.7	4437.8	3.41	1312.6	4725.2	3.61	1388.3	4997.8	3.97	1529.5	5506.3
800	3.48	1749.5	6298.1	3.71	1862.5	6705.2	3.92	1969.8	7091.1	4.32	2169.7	7811.0
900	3.74	2381.7	8574.3	3.99	2535.4	9127.5	4.21	2681.1	9651.8	4.64	2952.7	10629.9
1000	4.00	3138.0	11297.0	4.25	3340.2	12024.6	4.50	3531.7	12714.2	4.95	3889.0	14000.4
1100	4.24	4026.5	14495.3	4.51	4285.4	15427.5	4.77	4530.8	16311.0	5.25	4988.5	17958.6
1200	4.47	5054.8	18197.1	4.76	5379.4	19365.8	5.03	5687.1	20473.5	5.54	6260.7	22538.7
1300	4.69	6230.4	22429.4	5.00	6630.1	23868.2	5.28	7008.8	25231.7	5.81	7714.9	27773.7
1400	4.91	7560.6	27218.0	5.23	8045.0	28962.1	5.52	8504.1	30614.7	6.08	9359.9	33695.6
1500	5.12	9052.2	32588.1	5.45	9631.7	34674.1	5.76	10180.7	36650.7	6.34	11204.2	40335.2
1600	5.33	10712.2	38563.9	5.67	11397.3	41030.1	5.99	12046.3	43366.8	6.59	13256.2	47722.3

D	S=	0.014		S=	0.016		S=	0.018		S=	0.020	
(mm)	v(m/s)	Q(l/s)	Q(m3/h)	v(m/s)	Q(l/s)	Q(m3/h)	v(m/s)	Q(l/s)	Q(m3/h)	v(m/s)	Q(l/s)	Q(m3/h)
50	0.78	1.5	5.5	0.84	1.6	5.9	0.89	1.8	6.3	0.95	1.9	6.7
80	1.07	5.4	19.4	1.15	5.8	20.8	1.23	6.2	22.2	1.30	6.5	23.6
100	1.24	9.8	35.1	1.34	10.5	37.8	1.43	11.2	40.3	1.51	11.9	42.7
125	1.44	17.7	63.6	1.55	19.0	68.4	1.65	20.3	72.9	1.75	21.5	77.3
150	1.62	28.7	103.2	1.74	30.8	111.0	1.86	32.9	118.3	1.97	34.8	125.3
200	1.96	61.4	221.2	2.10	66.0	237.7	2.24	70.4	253.4	2.37	74.5	268.2
250	2.26	110.8	398.9	2.43	119.1	428.7	2.58	126.9	456.7	2.74	134.3	483.3
300	2.54	179.2	645.3	2.72	192.6	693.3	2.90	205.1	738.5	3.07	217.1	781.4
350	2.80	269.0	968.5	3.00	289.0	1040.3	3.20	307.8	1108.0	3.38	325.6	1172.2
400	3.04	382.3	1376.2	3.27	410.6	1478.0	3.48	437.2	1573.9	3.68	462.5	1664.9
450	3.28	521.0	1875.5	3.52	559.4	2014.0	3.75	595.7	2144.4	3.96	630.0	2268.1
500	3.50	687.0	2473.3	3.76	737.7	2655.6	4.00	785.4	2827.3	4.23	830.6	2990.1
600	3.92	1108.4	3990.1	4.21	1189.8	4283.3	4.48	1266.5	4559.4	4.74	1339.2	4821.0
700	4.31	1659.9	5975.5	4.63	1781.5	6413.5	4.93	1896.1	6825.8	5.21	2004.6	7216.5
800	4.68	2354.2	8475.2	5.03	2526.4	9095.0	5.35	2688.5	9678.4	5.65	2842.0	10231.3
900	5.04	3203.3	11532.0	5.40	3437.2	12373.8	5.75	3657.3	13166.1	6.08	3865.8	13916.8
1000	5.37	4218.5	15186.7	5.76	4526.0	16293.4	6.13	4815.3	17335.0	6.48	5089.4	18321.7
1100	5.69	5410.5	19477.9	6.11	5804.3	20895.4	6.50	6174.8	22229.1	6.87	6525.7	23492.7
1200	6.00	6789.7	24443.0	6.44	7283.2	26219.3	6.85	7747.4	27890.8	7.24	8187.2	29474.1
1300	6.30	8366.0	30117.5	6.76	8973.2	32303.6	7.19	9544.6	34360.5	7.60	10085.8	36308.7
1400	6.59	10148.9	36536.0	7.07	10884.7	39185.1	7.52	11577.0	41677.3	7.95	12232.7	44037.9
1500	6.87	12147.7	43731.7	7.37	13027.6	46899.3	7.84	13855.3	49879.2	8.28	14639.3	52701.6
1600	7.15	14371.4	51737.0	7.66	15411.4	55481.0	8.15	16389.7	59002.9	8.61	17316.2	62338.4

$$k(mm) = \quad 0.05 \quad T(°C) = \quad 20$$

D	S =	0.0005		S =	0.001		S =	0.002		S =	0.003	
(mm)	v(m/s)	Q(l/s)	Q(m3/h)	v(m/s)	Q(l/s)	Q(m3/h)	v(m/s)	Q(l/s)	Q(m3/h)	v(m/s)	Q(l/s)	Q(m3/h)
50	0.11	0.2	0.8	0.17	0.3	1.2	0.25	0.5	1.8	0.32	0.6	2.2
80	0.16	0.8	2.9	0.24	1.2	4.3	0.35	1.8	6.4	0.44	2.2	8.0
100	0.19	1.5	5.4	0.28	2.2	7.9	0.41	3.2	11.6	0.51	4.0	14.5
125	0.22	2.7	9.8	0.33	4.0	14.4	0.48	5.8	21.1	0.59	7.3	26.3
150	0.25	4.4	16.0	0.37	6.5	23.5	0.54	9.5	34.3	0.67	11.9	42.7
200	0.31	9.6	34.6	0.45	14.1	50.6	0.65	20.5	73.7	0.81	25.5	91.7
250	0.36	17.4	62.8	0.52	25.5	91.7	0.75	37.0	133.3	0.94	46.0	165.6
300	0.40	28.4	102.1	0.58	41.3	148.8	0.85	60.0	216.0	1.05	74.5	268.2
350	0.44	42.8	153.9	0.65	62.2	223.9	0.94	90.2	324.7	1.16	111.9	402.9
400	0.49	61.0	219.4	0.71	88.6	318.9	1.02	128.3	462.0	1.27	159.1	572.8
450	0.52	83.3	299.9	0.76	121.0	435.5	1.10	175.1	630.2	1.36	217.0	781.0
500	0.56	110.1	396.5	0.81	159.8	575.2	1.18	231.1	831.8	1.46	286.2	1030.4
600	0.63	178.4	642.2	0.91	258.5	930.5	1.32	373.3	1343.8	1.63	462.1	1663.5
700	0.70	268.0	964.9	1.01	387.9	1396.6	1.45	559.6	2014.7	1.80	692.4	2492.5
800	0.76	381.2	1372.3	1.10	551.2	1984.3	1.58	794.4	2860.0	1.95	982.4	3536.6
900	0.82	519.9	1871.5	1.18	751.1	2704.1	1.70	1081.8	3894.4	2.10	1337.1	4813.7
1000	0.87	686.0	2469.6	1.26	990.5	3565.7	1.82	1425.5	5131.9	2.24	1761.4	6340.9
1100	0.93	881.4	3173.0	1.34	1271.8	4578.5	1.92	1829.3	6585.7	2.38	2259.6	8134.6
1200	0.98	1107.8	3988.1	1.41	1597.6	5751.4	2.03	2296.8	8268.4	2.51	2836.2	10210.2
1300	1.03	1366.9	4920.9	1.48	1970.3	7093.0	2.13	2831.2	10192.3	2.63	3495.2	12582.8
1400	1.08	1660.4	5977.3	1.55	2392.1	8611.7	2.23	3435.9	12369.3	2.75	4240.8	15266.7
1500	1.13	1989.7	7162.9	1.62	2865.4	10315.4	2.33	4114.1	14810.7	2.87	5076.7	18276.1
1600	1.17	2356.5	8483.2	1.69	3392.3	12212.1	2.42	4868.8	17527.5	2.99	6006.8	21624.5

D	S=	0.004		S=	0.005		S=	0.006		S=	0.007	
(mm)	v(m/s)	Q(l/s)	Q(m3/h)	v(m/s)	Q(l/s)	Q(m3/h)	v(m/s)	Q(l/s)	Q(m3/h)	v(m/s)	Q(l/s)	Q(m3/h)
50	0.37	0.7	2.6	0.42	0.8	3.0	0.47	0.9	3.3	0.51	1.0	3.6
80	0.51	2.6	9.3	0.58	2.9	10.5	0.64	3.2	11.6	0.70	3.5	12.6
100	0.60	4.7	16.9	0.68	5.3	19.1	0.75	5.9	21.1	0.81	6.4	22.9
125	0.69	8.5	30.7	0.78	9.6	34.6	0.86	10.6	38.2	0.94	11.5	41.5
150	0.78	13.8	49.8	0.88	15.6	56.2	0.97	17.2	62.0	1.06	18.7	67.3
200	0.95	29.7	107.0	1.07	33.5	120.5	1.17	36.9	132.8	1.27	40.0	144.1
250	1.09	53.6	193.1	1.23	60.4	217.3	1.35	66.5	239.4	1.47	72.1	259.7
300	1.23	86.8	312.4	1.38	97.7	351.6	1.52	107.5	387.1	1.65	116.6	419.8
350	1.35	130.3	469.1	1.52	146.6	527.7	1.68	161.3	580.8	1.82	174.9	629.7
400	1.47	185.2	666.7	1.66	208.3	749.8	1.82	229.2	825.0	1.98	248.4	894.4
450	1.59	252.5	908.8	1.78	283.8	1021.8	1.96	312.2	1124.1	2.13	338.4	1218.3
500	1.70	333.0	1198.7	1.91	374.3	1347.3	2.10	411.7	1482.0	2.27	446.1	1606.0
600	1.90	537.3	1934.1	2.14	603.7	2173.2	2.35	663.8	2389.7	2.54	719.2	2589.1
700	2.09	804.7	2896.8	2.35	903.9	3253.9	2.58	993.7	3577.2	2.80	1076.4	3875.0
800	2.27	1141.4	4108.9	2.55	1281.7	4614.2	2.80	1408.8	5071.8	3.04	1525.8	5493.0
900	2.44	1553.1	5591.1	2.74	1743.7	6277.5	3.01	1916.3	6898.8	3.26	2075.2	7470.8
1000	2.60	2045.4	7363.3	2.92	2296.0	8265.6	3.21	2522.9	9082.5	3.48	2731.8	9834.4
1100	2.76	2623.4	9444.1	3.10	2944.4	10599.8	3.40	3235.0	11645.8	3.69	3502.4	12608.7
1200	2.91	3292.1	11851.6	3.27	3694.4	13300.0	3.59	4058.6	14611.0	3.88	4393.8	15817.5
1300	3.06	4056.4	14603.0	3.43	4551.6	16385.7	3.77	4999.7	17999.1	4.08	5412.2	19483.8
1400	3.20	4920.9	17715.2	3.59	5521.0	19875.6	3.94	6064.1	21830.6	4.26	6563.8	23629.8
1500	3.33	5890.1	21204.3	3.74	6607.7	23787.7	4.11	7257.1	26125.4	4.44	7854.6	28276.7
1600	3.47	6968.3	25085.9	3.89	7816.5	28139.5	4.27	8584.1	30902.7	4.62	9290.4	33445.4

k(mm) = 0.05   T(°C) = 20

D	S=	0.008		S=	0.009		S=	0.010		S=	0.012	
(mm)	v(m/s)	Q(l/s)	Q(m3/h)	v(m/s)	Q(l/s)	Q(m3/h)	v(m/s)	Q(l/s)	Q(m3/h)	v(m/s)	Q(l/s)	Q(m3/h)
50	0.54	1.1	3.9	0.58	1.1	4.1	0.62	1.2	4.4	0.68	1.3	4.8
80	0.75	3.8	13.6	0.80	4.0	14.5	0.85	4.3	15.3	0.93	4.7	16.9
100	0.87	6.8	24.6	0.93	7.3	26.2	0.98	7.7	27.7	1.08	8.5	30.6
125	1.01	12.4	44.5	1.07	13.2	47.4	1.14	13.9	50.2	1.25	15.4	55.3
150	1.14	20.1	72.2	1.21	21.4	76.9	1.28	22.6	81.3	1.41	24.9	89.6
200	1.37	43.0	154.6	1.45	45.7	164.6	1.54	48.3	174.0	1.69	53.2	191.5
250	1.58	77.4	278.6	1.68	82.3	296.4	1.77	87.0	313.3	1.95	95.7	344.7
300	1.77	125.1	450.3	1.88	133.0	479.0	1.99	140.6	506.1	2.19	154.6	556.7
350	1.95	187.6	675.3	2.07	199.5	718.2	2.19	210.8	758.8	2.41	231.8	834.4
400	2.12	266.4	959.0	2.25	283.3	1019.7	2.38	299.2	1077.3	2.62	329.0	1184.4
450	2.28	362.8	1306.2	2.43	385.8	1388.8	2.56	407.5	1466.9	2.82	447.9	1612.5
500	2.44	478.2	1721.6	2.59	508.4	1830.3	2.73	537.0	1933.1	3.01	590.2	2124.6
600	2.73	770.8	2774.8	2.90	819.3	2949.5	3.06	865.2	3114.7	3.36	950.7	3422.4
700	3.00	1153.4	4152.3	3.19	1225.8	4413.0	3.36	1294.4	4659.8	3.69	1422.0	5119.0
800	3.25	1634.8	5885.4	3.46	1737.3	6254.2	3.65	1834.2	6603.3	4.01	2014.7	7252.9
900	3.49	2223.2	8003.6	3.71	2362.3	8504.3	3.92	2493.9	8978.1	4.31	2738.9	9860.0
1000	3.73	2926.3	10534.7	3.96	3109.1	11192.9	4.18	3282.1	11815.7	4.59	3604.1	12974.7
1100	3.95	3751.5	13505.4	4.19	3985.6	14348.2	4.43	4207.1	15145.6	4.86	4619.3	16629.5
1200	4.16	4705.9	16941.2	4.42	4999.2	17997.2	4.67	5276.8	18996.4	5.12	5793.2	20855.7
1300	4.37	5796.3	20866.7	4.64	6157.2	22166.1	4.90	6498.8	23395.5	5.37	7134.3	25683.3
1400	4.57	7029.3	25305.3	4.85	7466.6	26879.8	5.12	7880.4	28369.4	5.62	8650.4	31141.3
1500	4.76	8411.2	30280.2	5.06	8934.1	32162.6	5.34	9428.8	33943.7	5.86	10349.4	37257.8
1600	4.95	9948.1	35813.3	5.26	10566.2	38038.2	5.55	11150.9	40143.2	6.09	12238.8	44059.8

D	S=	0.014		S=	0.016		S=	0.018		S=	0.020	
(mm)	v(m/s)	Q(l/s)	Q(m3/h)	v(m/s)	Q(l/s)	Q(m3/h)	v(m/s)	Q(l/s)	Q(m3/h)	v(m/s)	Q(l/s)	Q(m3/h)
50	0.74	1.5	5.2	0.79	1.6	5.6	0.85	1.7	6.0	0.90	1.8	6.3
80	1.01	5.1	18.3	1.09	5.5	19.7	1.16	5.8	21.0	1.23	6.2	22.2
100	1.17	9.2	33.2	1.26	9.9	35.6	1.34	10.5	37.9	1.42	11.1	40.1
125	1.36	16.7	60.0	1.46	17.9	64.4	1.55	19.0	68.5	1.64	20.1	72.4
150	1.53	27.0	97.2	1.64	29.0	104.3	1.74	30.8	110.9	1.84	32.6	117.2
200	1.84	57.7	207.6	1.97	61.9	222.7	2.09	65.8	236.8	2.21	69.5	250.2
250	2.11	103.8	373.6	2.27	111.3	400.6	2.41	118.3	426.0	2.55	125.0	450.0
300	2.37	167.6	603.3	2.54	179.6	646.7	2.70	191.0	687.5	2.85	201.7	726.1
350	2.61	251.1	904.1	2.80	269.2	969.0	2.97	286.1	1029.9	3.14	302.1	1087.7
400	2.84	356.4	1283.0	3.04	381.9	1374.8	3.23	405.9	1461.2	3.41	428.6	1542.9
450	3.05	485.1	1746.5	3.27	519.8	1871.3	3.47	552.4	1988.7	3.67	583.2	2099.7
500	3.26	639.1	2300.9	3.49	684.7	2465.1	3.71	727.6	2619.4	3.91	768.2	2765.4
600	3.64	1029.3	3705.6	3.90	1102.6	3969.3	4.14	1171.4	4217.2	4.37	1236.6	4451.7
700	4.00	1539.4	5541.7	4.28	1648.7	5935.3	4.55	1751.4	6305.2	4.80	1848.7	6655.2
800	4.34	2180.7	7850.6	4.65	2335.4	8407.4	4.94	2480.7	8930.5	5.21	2618.2	9425.4
900	4.66	2964.3	10671.5	4.99	3174.2	11427.1	5.30	3371.4	12137.2	5.59	3558.0	12808.9
1000	4.97	3900.3	14041.1	5.32	4176.2	15034.2	5.65	4435.3	15967.2	5.96	4680.6	16850.0
1100	5.26	4998.6	17994.8	5.63	5351.7	19266.2	5.98	5683.5	20460.7	6.31	5997.4	21590.7
1200	5.54	6268.4	22566.4	5.93	6710.9	24159.2	6.30	7126.6	25655.7	6.65	7519.9	27071.5
1300	5.82	7718.9	27788.1	6.23	8263.3	29747.9	6.61	8774.8	31589.2	6.98	9258.6	33331.0
1400	6.08	9358.7	33691.4	6.51	10018.3	36065.8	6.91	10637.9	38296.5	7.29	11224.1	40406.8
1500	6.34	11196.3	40306.6	6.78	11984.8	43145.3	7.20	12725.6	45812.1	7.60	13426.4	48335.0
1600	6.58	13239.7	47663.0	7.05	14171.6	51017.7	7.48	15047.0	54169.3	7.90	15875.2	57150.7

k(mm) =   0.1   T(°C) =   20

D	S = 0.0005			S = 0.001			S = 0.002			S = 0.003		
(mm)	v(m/s)	Q(l/s)	Q(m3/h)	v(m/s)	Q(l/s)	Q(m3/h)	v(m/s)	Q(l/s)	Q(m3/h)	v(m/s)	Q(l/s)	Q(m3/h)
50	0.11	0.2	0.8	0.17	0.3	1.2	0.25	0.5	1.7	0.31	0.6	2.2
80	0.16	0.8	2.9	0.23	1.2	4.2	0.34	1.7	6.2	0.43	2.1	7.7
100	0.19	1.5	5.3	0.27	2.1	7.7	0.40	3.1	11.2	0.49	3.9	14.0
125	0.22	2.7	9.6	0.32	3.9	14.0	0.46	5.7	20.4	0.57	7.0	25.4
150	0.25	4.4	15.7	0.36	6.3	22.8	0.52	9.2	33.2	0.65	11.4	41.2
200	0.30	9.4	33.8	0.44	13.7	49.2	0.63	19.8	71.3	0.78	24.5	88.4
250	0.35	17.0	61.4	0.50	24.7	89.1	0.73	35.8	128.8	0.90	44.3	159.4
300	0.39	27.7	99.7	0.57	40.1	144.5	0.82	57.9	208.5	1.01	71.7	258.0
350	0.43	41.7	150.1	0.63	60.4	217.3	0.90	87.0	313.2	1.12	107.6	387.3
400	0.47	59.4	213.9	0.68	85.9	309.2	0.98	123.7	445.4	1.22	152.9	550.4
450	0.51	81.2	292.2	0.74	117.2	422.0	1.06	168.7	607.3	1.31	208.4	750.1
500	0.55	107.3	386.1	0.79	154.8	557.2	1.13	222.6	801.2	1.40	274.8	989.3
600	0.61	173.6	624.9	0.88	250.2	900.6	1.27	359.3	1293.4	1.57	443.3	1596.0
700	0.68	260.6	938.2	0.97	375.2	1350.7	1.40	538.3	1937.9	1.73	663.9	2390.2
800	0.74	370.4	1333.5	1.06	532.8	1918.0	1.52	763.8	2749.7	1.87	941.7	3390.0
900	0.79	504.9	1817.7	1.14	725.7	2612.3	1.63	1039.6	3742.7	2.01	1281.3	4612.6
1000	0.85	665.9	2397.3	1.22	956.4	3443.2	1.74	1369.5	4930.1	2.15	1687.3	6074.3
1100	0.90	855.2	3078.9	1.29	1227.6	4419.4	1.85	1756.9	6324.8	2.28	2164.1	7790.6
1200	0.95	1074.5	3868.2	1.36	1541.6	5549.6	1.95	2205.2	7938.7	2.40	2715.7	9776.4
1300	1.00	1325.4	4771.3	1.43	1900.6	6842.1	2.05	2717.7	9783.6	2.52	3346.1	12045.9
1400	1.05	1609.3	5793.6	1.50	2306.8	8304.6	2.14	3297.4	11870.7	2.64	4059.1	14612.9
1500	1.09	1928.0	6940.7	1.56	2762.5	9945.1	2.23	3947.5	14210.8	2.75	4858.6	17490.8
1600	1.14	2282.7	8217.8	1.63	3269.7	11770.8	2.32	4670.8	16814.7	2.86	5747.9	20692.6

D	S= 0.004			S= 0.005			S= 0.006			S= 0.007		
(mm)	v(m/s)	Q(l/s)	Q(m3/h)	v(m/s)	Q(l/s)	Q(m3/h)	v(m/s)	Q(l/s)	Q(m3/h)	v(m/s)	Q(l/s)	Q(m3/h)
50	0.36	0.7	2.5	0.41	0.8	2.9	0.45	0.9	3.2	0.49	1.0	3.4
80	0.50	2.5	9.0	0.56	2.8	10.1	0.62	3.1	11.2	0.67	3.4	12.1
100	0.58	4.5	16.3	0.65	5.1	18.4	0.72	5.6	20.2	0.78	6.1	22.0
125	0.67	8.2	29.6	0.75	9.2	33.3	0.83	10.2	36.6	0.90	11.0	39.7
150	0.75	13.3	48.0	0.85	15.0	54.0	0.93	16.5	59.4	1.01	17.9	64.4
200	0.91	28.6	102.8	1.02	32.1	115.6	1.12	35.3	127.1	1.22	38.3	137.8
250	1.05	51.5	185.4	1.18	57.9	208.3	1.30	63.6	229.1	1.40	68.9	248.2
300	1.18	83.3	299.8	1.32	93.5	336.8	1.45	102.8	370.2	1.58	111.4	401.0
350	1.30	125.0	449.9	1.46	140.3	505.2	1.60	154.2	555.2	1.74	167.0	601.3
400	1.41	177.5	639.2	1.59	199.3	717.5	1.74	219.0	788.4	1.89	237.1	853.7
450	1.52	241.9	870.9	1.71	271.5	977.4	1.88	298.3	1073.8	2.03	322.9	1162.5
500	1.62	319.0	1148.2	1.82	357.9	1288.5	2.00	393.1	1415.3	2.17	425.6	1532.1
600	1.82	514.4	1851.7	2.04	577.0	2077.2	2.24	633.7	2281.2	2.43	685.8	2469.0
700	2.00	770.0	2772.2	2.24	863.6	3109.0	2.46	948.3	3413.8	2.67	1026.2	3694.2
800	2.17	1091.9	3930.7	2.44	1224.3	4407.5	2.67	1344.1	4838.8	2.89	1454.3	5235.6
900	2.33	1485.3	5347.1	2.62	1665.2	5994.7	2.87	1827.9	6580.6	3.11	1977.6	7119.5
1000	2.49	1955.6	7040.2	2.79	2192.2	7891.8	3.06	2406.1	8662.1	3.31	2603.0	9370.8
1100	2.64	2507.8	9027.9	2.96	2810.8	10118.7	3.25	3084.8	11105.4	3.51	3337.0	12013.0
1200	2.78	3146.5	11327.4	3.12	3526.3	12694.7	3.42	3869.8	13931.4	3.70	4185.8	15069.1
1300	2.92	3876.4	13955.1	3.27	4343.9	15638.2	3.59	4766.8	17160.4	3.88	5155.7	18560.7
1400	3.05	4701.9	16927.0	3.42	5268.6	18966.9	3.76	5781.1	20811.9	4.06	6252.5	22509.0
1500	3.18	5627.4	20258.5	3.57	6305.1	22698.3	3.91	6918.0	24904.7	4.23	7481.8	26934.3
1600	3.31	6656.8	23964.6	3.71	7458.0	26848.9	4.07	8182.6	29457.3	4.40	8849.0	31856.6

k(mm) =　　　　0.1　　　T(°C) =　　　20

D	S=	0.008		S=	0.009		S=	0.010		S=	0.012	
(mm)	v(m/s)	Q(l/s)	Q(m3/h)	v(m/s)	Q(l/s)	Q(m3/h)	v(m/s)	Q(l/s)	Q(m3/h)	v(m/s)	Q(l/s)	Q(m3/h)
50	0.52	1.0	3.7	0.56	1.1	3.9	0.59	1.2	4.2	0.65	1.3	4.6
80	0.72	3.6	13.0	0.77	3.8	13.9	0.81	4.1	14.6	0.89	4.5	16.1
100	0.83	6.5	23.6	0.89	7.0	25.1	0.94	7.4	26.5	1.03	8.1	29.2
125	0.96	11.8	42.6	1.03	12.6	45.3	1.08	13.3	47.9	1.19	14.6	52.7
150	1.09	19.2	69.1	1.15	20.4	73.5	1.22	21.6	77.6	1.34	23.7	85.4
200	1.31	41.0	147.7	1.39	43.6	157.1	1.47	46.1	165.9	1.61	50.6	182.3
250	1.51	73.9	266.0	1.60	78.5	282.7	1.69	82.9	298.6	1.86	91.1	328.0
300	1.69	119.4	429.7	1.79	126.8	456.6	1.89	133.9	482.1	2.08	147.1	529.6
350	1.86	178.9	644.2	1.98	190.1	684.5	2.09	200.7	722.6	2.29	220.4	793.6
400	2.02	254.0	914.4	2.15	269.9	971.5	2.27	284.9	1025.6	2.49	312.8	1126.1
450	2.17	345.9	1245.2	2.31	367.4	1322.8	2.44	387.9	1396.3	2.68	425.8	1532.9
500	2.32	455.8	1640.8	2.47	484.2	1743.0	2.60	511.0	1839.7	2.86	561.0	2019.5
600	2.60	734.4	2643.8	2.76	780.0	2808.0	2.91	823.2	2963.4	3.20	903.5	3252.5
700	2.85	1098.7	3955.3	3.03	1166.8	4200.6	3.20	1231.3	4432.6	3.51	1351.2	4864.3
800	3.10	1557.0	5605.1	3.29	1653.4	5952.2	3.47	1744.6	6280.6	3.81	1914.3	6891.5
900	3.33	2117.0	7621.3	3.53	2248.0	8092.7	3.73	2371.9	8538.7	4.09	2602.3	9368.3
1000	3.55	2786.3	10030.5	3.77	2958.4	10650.3	3.97	3121.3	11236.7	4.36	3424.3	12327.4
1100	3.76	3571.7	12858.1	3.99	3792.2	13651.9	4.21	4000.8	14402.8	4.62	4388.8	15799.8
1200	3.96	4480.1	16128.2	4.21	4756.4	17123.1	4.44	5017.9	18064.3	4.87	5504.2	19815.2
1300	4.16	5517.9	19864.3	4.41	5858.0	21088.9	4.66	6179.8	22247.3	5.11	6778.4	24402.2
1400	4.35	6691.4	24088.9	4.61	7103.6	25573.0	4.87	7493.6	26976.0	5.34	8219.0	29588.5
1500	4.53	8006.6	28823.8	4.81	8499.6	30598.7	5.07	8966.0	32277.6	5.56	9833.6	35400.8
1600	4.71	9469.5	34090.2	5.00	10052.3	36188.3	5.27	10603.6	38173.0	5.78	11629.1	41864.9

D	S=	0.014		S=	0.016		S=	0.018		S=	0.020	
(mm)	v(m/s)	Q(l/s)	Q(m3/h)	v(m/s)	Q(l/s)	Q(m3/h)	v(m/s)	Q(l/s)	Q(m3/h)	v(m/s)	Q(l/s)	Q(m3/h)
50	0.71	1.4	5.0	0.76	1.5	5.4	0.81	1.6	5.7	0.85	1.7	6.0
80	0.97	4.9	17.5	1.04	5.2	18.8	1.10	5.5	20.0	1.17	5.9	21.1
100	1.12	8.8	31.6	1.20	9.4	33.9	1.28	10.0	36.1	1.35	10.6	38.1
125	1.29	15.9	57.1	1.39	17.0	61.2	1.47	18.1	65.1	1.56	19.1	68.7
150	1.45	25.7	92.5	1.56	27.5	99.1	1.66	29.3	105.3	1.75	30.9	111.2
200	1.75	54.8	197.5	1.87	58.8	211.5	1.99	62.4	224.8	2.10	65.9	237.3
250	2.01	98.7	355.1	2.15	105.7	380.4	2.29	112.3	404.1	2.41	118.5	426.6
300	2.25	159.2	573.3	2.41	170.5	613.9	2.56	181.1	652.1	2.70	191.2	688.3
350	2.48	238.6	858.9	2.66	255.5	919.7	2.82	271.3	976.8	2.98	286.4	1030.9
400	2.69	338.5	1218.6	2.88	362.4	1304.8	3.06	384.9	1385.7	3.23	406.2	1462.2
450	2.90	460.7	1658.7	3.10	493.3	1775.8	3.29	523.8	1885.8	3.48	552.7	1989.8
500	3.09	606.9	2185.0	3.31	649.7	2339.0	3.51	689.9	2483.8	3.71	728.0	2620.7
600	3.46	977.4	3518.5	3.70	1046.1	3766.1	3.93	1110.7	3998.7	4.14	1171.9	4218.7
700	3.80	1461.5	5261.5	4.06	1564.2	5631.2	4.32	1660.7	5978.5	4.55	1752.0	6307.1
800	4.12	2070.4	7453.4	4.41	2215.7	7976.6	4.68	2352.2	8468.0	4.94	2481.3	8932.9
900	4.42	2814.3	10131.4	4.73	3011.6	10841.8	5.03	3197.0	11509.2	5.30	3372.3	12140.4
1000	4.71	3703.0	13330.7	5.05	3962.4	14264.6	5.36	4206.1	15141.9	5.65	4436.6	15971.8
1100	4.99	4745.7	17084.6	5.34	5078.0	18280.8	5.67	5390.1	19404.3	5.98	5685.3	20467.0
1200	5.26	5951.5	21425.5	5.63	6367.9	22924.6	5.98	6759.1	24332.7	6.30	7129.1	25664.6
1300	5.52	7328.9	26384.2	5.91	7841.4	28229.2	6.27	8322.8	29962.2	6.61	8778.2	31601.5
1400	5.77	8886.2	31990.5	6.18	9507.3	34226.4	6.56	10090.7	36326.6	6.91	10642.6	38313.2
1500	6.02	10631.5	38273.2	6.44	11374.2	40947.1	6.83	12071.9	43458.7	7.20	12731.8	45834.4
1600	6.25	12572.3	45260.4	6.69	13450.3	48421.2	7.10	14275.0	51390.1	7.49	15055.1	54198.3

$$k(mm) = \quad 0.5 \quad T(°C) = \quad 20$$

D	S =	0.0005		S =	0.001		S =	0.002		S =	0.003	
(mm)	v(m/s)	Q(l/s)	Q(m3/h)	v(m/s)	Q(l/s)	Q(m3/h)	v(m/s)	Q(l/s)	Q(m3/h)	v(m/s)	Q(l/s)	Q(m3/h)
50	0.10	0.2	0.7	0.15	0.3	1.0	0.21	0.4	1.5	0.26	0.5	1.9
80	0.14	0.7	2.6	0.21	1.0	3.7	0.30	1.5	5.4	0.37	1.8	6.6
100	0.17	1.3	4.7	0.24	1.9	6.8	0.35	2.7	9.8	0.43	3.3	12.0
125	0.19	2.4	8.6	0.28	3.4	12.4	0.40	4.9	17.7	0.49	6.1	21.8
150	0.22	3.9	14.0	0.32	5.6	20.1	0.45	8.0	28.8	0.56	9.9	35.5
200	0.27	8.4	30.2	0.38	12.0	43.3	0.55	17.2	61.9	0.67	21.1	76.1
250	0.31	15.2	54.8	0.44	21.8	78.4	0.63	31.0	111.8	0.78	38.2	137.4
300	0.35	24.7	88.9	0.50	35.3	127.0	0.71	50.3	181.0	0.87	61.8	222.5
350	0.39	37.1	133.7	0.55	53.0	190.9	0.79	75.5	272.0	0.96	92.8	334.2
400	0.42	52.9	190.4	0.60	75.5	271.7	0.85	107.4	386.8	1.05	132.0	475.1
450	0.45	72.2	259.9	0.65	103.0	370.6	0.92	146.5	527.5	1.13	180.0	647.9
500	0.49	95.3	343.2	0.69	135.9	489.3	0.98	193.3	696.0	1.21	237.4	854.7
600	0.55	154.2	555.0	0.78	219.6	790.6	1.10	312.2	1124.1	1.36	383.3	1380.0
700	0.60	231.3	832.8	0.86	329.3	1185.6	1.22	468.0	1684.9	1.49	574.5	2068.2
800	0.65	328.6	1183.1	0.93	467.6	1683.5	1.32	664.4	2391.7	1.62	815.3	2935.2
900	0.70	447.8	1612.1	1.00	637.0	2293.1	1.42	904.6	3256.7	1.74	1110.1	3996.3
1000	0.75	590.5	2125.6	1.07	839.6	3022.5	1.52	1192.1	4291.6	1.86	1462.7	5265.6
1100	0.80	758.1	2729.3	1.13	1077.7	3879.8	1.61	1529.9	5507.6	1.98	1876.9	6757.0
1200	0.84	952.3	3428.4	1.20	1353.4	4872.4	1.70	1921.0	6915.4	2.08	2356.5	8483.4
1300	0.88	1174.5	4228.1	1.26	1668.8	6007.7	1.78	2368.2	8525.4	2.19	2904.9	10457.7
1400	0.93	1426.0	5133.4	1.32	2025.7	7292.6	1.87	2874.3	10347.4	2.29	3525.5	12691.9
1500	0.97	1708.1	6149.2	1.37	2426.2	8734.2	1.95	3442.0	12391.2	2.39	4221.6	15197.8
1600	1.01	2022.2	7280.1	1.43	2871.9	10338.8	2.03	4073.9	14666.1	2.48	4996.4	17987.0

D	S=	0.004		S=	0.005		S=	0.006		S=	0.007	
(mm)	v(m/s)	Q(l/s)	Q(m3/h)	v(m/s)	Q(l/s)	Q(m3/h)	v(m/s)	Q(l/s)	Q(m3/h)	v(m/s)	Q(l/s)	Q(m3/h)
50	0.31	0.6	2.2	0.35	0.7	2.4	0.38	0.7	2.7	0.41	0.8	2.9
80	0.42	2.1	7.7	0.48	2.4	8.6	0.52	2.6	9.5	0.57	2.8	10.3
100	0.49	3.9	14.0	0.55	4.3	15.7	0.61	4.8	17.2	0.66	5.2	18.6
125	0.57	7.0	25.3	0.64	7.9	28.4	0.70	8.6	31.1	0.76	9.4	33.7
150	0.65	11.4	41.1	0.72	12.8	46.1	0.79	14.0	50.5	0.86	15.2	54.7
200	0.78	24.5	88.2	0.87	27.4	98.8	0.96	30.1	108.3	1.04	32.5	117.1
250	0.90	44.2	159.1	1.01	49.5	178.2	1.11	54.3	195.4	1.20	58.7	211.3
300	1.01	71.5	257.5	1.13	80.1	288.3	1.24	87.8	316.2	1.34	95.0	341.9
350	1.12	107.4	386.7	1.25	120.3	432.9	1.37	131.9	474.7	1.48	142.5	513.2
400	1.22	152.7	549.7	1.36	170.9	615.3	1.49	187.4	674.7	1.61	202.6	729.3
450	1.31	208.2	749.4	1.47	233.0	838.8	1.61	255.5	919.7	1.74	276.1	994.1
500	1.40	274.6	988.6	1.57	307.4	1106.5	1.72	337.0	1213.2	1.85	364.2	1311.2
600	1.57	443.3	1595.9	1.75	496.1	1786.1	1.92	543.9	1958.0	2.08	587.8	2116.1
700	1.73	664.3	2391.4	1.93	743.4	2676.1	2.12	814.9	2933.5	2.29	880.6	3170.3
800	1.88	942.6	3393.5	2.10	1054.8	3797.3	2.30	1156.2	4162.4	2.49	1249.5	4498.1
900	2.02	1283.3	4619.8	2.26	1435.9	5169.3	2.47	1573.9	5666.0	2.67	1700.8	6122.8
1000	2.15	1690.8	6086.8	2.41	1891.8	6810.4	2.64	2073.5	7464.6	2.85	2240.6	8066.2
1100	2.28	2169.5	7810.3	2.55	2427.3	8738.4	2.80	2660.4	9577.5	3.03	2874.8	10349.1
1200	2.41	2723.7	9805.5	2.69	3047.3	10970.2	2.95	3339.8	12023.3	3.19	3608.8	12991.7
1300	2.53	3357.4	12086.8	2.83	3756.1	13522.1	3.10	4116.6	14819.8	3.35	4448.1	16013.2
1400	2.65	4074.6	14668.4	2.96	4558.3	16409.9	3.25	4995.7	17984.4	3.51	5397.8	19432.3
1500	2.76	4878.9	17564.0	3.09	5458.0	19648.8	3.38	5981.6	21533.6	3.66	6463.0	23266.9
1600	2.87	5774.1	20786.8	3.21	6459.3	23253.6	3.52	7078.8	25483.8	3.80	7648.5	27534.7

k(mm) =   0.5   T(°C) =   20

D	S=	0.008		S=	0.009		S=	0.010		S=	0.012	
(mm)	v(m/s)	Q(l/s)	Q(m3/h)	v(m/s)	Q(l/s)	Q(m3/h)	v(m/s)	Q(l/s)	Q(m3/h)	v(m/s)	Q(l/s)	Q(m3/h)
50	0.44	0.9	3.1	0.47	0.9	3.3	0.49	1.0	3.5	0.54	1.1	3.8
80	0.61	3.0	11.0	0.64	3.2	11.7	0.68	3.4	12.3	0.75	3.7	13.5
100	0.70	5.5	19.9	0.75	5.9	21.1	0.79	6.2	22.3	0.87	6.8	24.5
125	0.82	10.0	36.0	0.87	10.6	38.3	0.91	11.2	40.4	1.00	12.3	44.3
150	0.92	16.2	58.5	0.98	17.3	62.1	1.03	18.2	65.5	1.13	20.0	71.9
200	1.11	34.8	125.4	1.18	37.0	133.1	1.24	39.0	140.3	1.36	42.8	153.9
250	1.28	62.8	226.1	1.36	66.7	239.9	1.43	70.3	253.1	1.57	77.1	277.5
300	1.44	101.6	365.7	1.53	107.8	388.1	1.61	113.7	409.3	1.76	124.7	448.8
350	1.58	152.5	549.0	1.68	161.8	582.6	1.77	170.7	614.4	1.94	187.1	673.5
400	1.72	216.7	780.1	1.83	230.0	827.9	1.93	242.5	873.0	2.12	265.8	957.0
450	1.86	295.4	1063.3	1.97	313.4	1128.4	2.08	330.5	1189.9	2.28	362.3	1304.3
500	1.98	389.6	1402.5	2.11	413.4	1488.2	2.22	435.9	1569.3	2.43	477.8	1720.1
600	2.22	628.7	2263.3	2.36	667.1	2401.5	2.49	703.4	2532.3	2.73	771.0	2775.5
700	2.45	941.8	3390.6	2.60	999.3	3597.6	2.74	1053.7	3793.4	3.00	1154.9	4157.5
800	2.66	1336.3	4810.6	2.82	1417.8	5104.2	2.97	1494.9	5381.8	3.26	1638.4	5898.1
900	2.86	1818.9	6548.0	3.03	1929.8	6947.4	3.20	2034.7	7325.1	3.51	2229.9	8027.7
1000	3.05	2396.1	8626.1	3.24	2542.2	9152.1	3.41	2680.4	9649.5	3.74	2937.4	10574.7
1100	3.23	3074.3	11067.3	3.43	3261.6	11741.9	3.62	3438.9	12380.0	3.97	3768.5	13566.7
1200	3.41	3859.2	13893.1	3.62	4094.4	14739.7	3.82	4316.8	15540.5	4.18	4730.5	17029.8
1300	3.58	4756.7	17124.0	3.80	5046.5	18167.3	4.01	5320.6	19154.0	4.39	5830.4	20989.3
1400	3.75	5772.2	20779.9	3.98	6123.8	22045.7	4.19	6456.4	23242.9	4.60	7074.9	25469.7
1500	3.91	6911.2	24880.3	4.15	7332.1	26395.5	4.37	7730.2	27828.8	4.79	8470.7	30494.4
1600	4.07	8178.8	29443.6	4.32	8676.8	31236.6	4.55	9147.9	32932.4	4.99	10024.0	36086.5

D	S=	0.014		S=	0.016		S=	0.018		S=	0.020	
(mm)	v(m/s)	Q(l/s)	Q(m3/h)	v(m/s)	Q(l/s)	Q(m3/h)	v(m/s)	Q(l/s)	Q(m3/h)	v(m/s)	Q(l/s)	Q(m3/h)
50	0.59	1.2	4.2	0.63	1.2	4.4	0.67	1.3	4.7	0.70	1.4	5.0
80	0.81	4.1	14.6	0.86	4.3	15.6	0.92	4.6	16.6	0.97	4.9	17.5
100	0.94	7.4	26.5	1.00	7.9	28.3	1.06	8.4	30.1	1.12	8.8	31.7
125	1.08	13.3	47.9	1.16	14.2	51.2	1.23	15.1	54.4	1.30	15.9	57.4
150	1.22	21.6	77.7	1.31	23.1	83.1	1.39	24.5	88.2	1.46	25.8	93.0
200	1.47	46.2	166.4	1.57	49.4	178.0	1.67	52.5	188.9	1.76	55.3	199.2
250	1.70	83.3	299.9	1.82	89.1	320.8	1.93	94.6	340.4	2.03	99.7	359.0
300	1.91	134.7	485.0	2.04	144.1	518.8	2.16	152.9	550.5	2.28	161.3	580.5
350	2.10	202.2	727.9	2.25	216.3	778.5	2.39	229.5	826.1	2.51	242.0	871.1
400	2.29	287.3	1034.3	2.45	307.3	1106.1	2.59	326.0	1173.7	2.74	343.8	1237.5
450	2.46	391.5	1409.5	2.63	418.7	1507.4	2.79	444.3	1599.4	2.95	468.5	1686.4
500	2.63	516.3	1858.8	2.81	552.2	1987.9	2.98	585.9	2109.2	3.15	617.7	2223.9
600	2.95	833.1	2999.1	3.15	890.9	3207.3	3.34	945.2	3402.8	3.52	996.6	3587.8
700	3.24	1247.9	4492.3	3.47	1334.4	4804.0	3.68	1415.8	5096.7	3.88	1492.7	5373.6
800	3.52	1770.3	6373.0	3.77	1893.0	6814.9	4.00	2008.4	7230.1	4.21	2117.4	7622.7
900	3.79	2409.4	8673.7	4.05	2576.4	9275.1	4.30	2733.3	9839.9	4.53	2881.7	10374.1
1000	4.04	3173.8	11425.6	4.32	3393.8	12217.5	4.58	3600.4	12961.3	4.83	3795.8	13664.8
1100	4.28	4071.7	14658.0	4.58	4353.8	15673.8	4.86	4618.8	16627.8	5.12	4869.5	17530.2
1200	4.52	5111.0	18399.4	4.83	5465.1	19674.3	5.13	5797.7	20871.6	5.40	6112.2	22004.1
1300	4.75	6299.2	22677.1	5.07	6735.6	24248.1	5.38	7145.4	25723.5	5.66	7533.1	27119.1
1400	4.97	7643.7	27517.4	5.31	8173.2	29423.4	5.63	8670.4	31213.5	5.94	9140.8	32906.7
1500	5.18	9151.6	32945.8	5.54	9785.4	35227.5	5.87	10380.7	37370.5	6.19	10943.7	39397.5
1600	5.39	10829.7	38987.1	5.76	11579.7	41686.9	6.11	12284.0	44222.6	6.44	12950.3	46620.9

k(mm) =   1   T(°C) =   20

D	S =	0.0005		S =	0.001		S =	0.002		S =	0.003	
(mm)	v(m/s)	Q(l/s)	Q(m3/h)	v(m/s)	Q(l/s)	Q(m3/h)	v(m/s)	Q(l/s)	Q(m3/h)	v(m/s)	Q(l/s)	Q(m3/h)
50	0.09	0.2	0.7	0.13	0.3	1.0	0.19	0.4	1.4	0.24	0.5	1.7
80	0.13	0.7	2.4	0.19	0.9	3.4	0.27	1.4	4.9	0.33	1.7	6.0
100	0.15	1.2	4.3	0.22	1.7	6.2	0.31	2.5	8.9	0.39	3.0	10.9
125	0.18	2.2	7.9	0.26	3.1	11.3	0.37	4.5	16.2	0.45	5.5	19.9
150	0.20	3.6	12.9	0.29	5.1	18.5	0.41	7.3	26.3	0.51	9.0	32.3
200	0.25	7.8	27.9	0.35	11.1	39.8	0.50	15.7	56.6	0.62	19.3	69.6
250	0.29	14.1	50.6	0.41	20.0	72.1	0.58	28.5	102.5	0.71	35.0	125.8
300	0.32	22.8	82.2	0.46	32.5	117.0	0.65	46.2	166.2	0.80	56.7	204.0
350	0.36	34.4	123.8	0.51	48.9	176.1	0.72	69.5	250.0	0.89	85.2	306.8
400	0.39	49.0	176.3	0.55	69.6	250.7	0.79	98.9	355.9	0.97	121.3	436.7
450	0.42	66.9	240.8	0.60	95.1	342.3	0.85	135.0	485.9	1.04	165.6	596.0
500	0.45	88.4	318.2	0.64	125.6	452.1	0.91	178.2	641.6	1.11	218.6	787.1
600	0.51	143.1	515.0	0.72	203.2	731.4	1.02	288.2	1037.6	1.25	353.5	1272.6
700	0.56	214.8	773.3	0.79	305.0	1097.9	1.12	432.5	1557.1	1.38	530.4	1909.4
800	0.61	305.4	1099.3	0.86	433.4	1560.3	1.22	614.5	2212.3	1.50	753.5	2712.7
900	0.65	416.3	1498.8	0.93	590.7	2126.7	1.32	837.5	3014.9	1.61	1026.8	3696.5
1000	0.70	549.2	1977.2	0.99	779.2	2805.0	1.41	1104.4	3975.9	1.72	1354.0	4874.4
1100	0.74	705.5	2539.8	1.05	1000.7	3602.6	1.49	1418.3	5105.7	1.83	1738.7	6259.2
1200	0.78	886.6	3191.8	1.11	1257.4	4526.6	1.58	1781.8	6414.6	1.93	2184.3	7863.4
1300	0.82	1093.8	3937.8	1.17	1551.1	5583.9	1.66	2197.8	7912.2	2.03	2694.1	9698.9
1400	0.86	1328.5	4782.7	1.22	1883.7	6781.2	1.73	2668.9	9607.9	2.13	3271.4	11777.1
1500	0.90	1591.9	5731.0	1.28	2256.9	8124.9	1.81	3197.5	11510.9	2.22	3919.2	14109.1
1600	0.94	1885.3	6787.1	1.33	2672.6	9621.3	1.88	3786.1	13629.9	2.31	4640.5	16705.9

D	S=	0.004		S=	0.005		S=	0.006		S=	0.007	
(mm)	v(m/s)	Q(l/s)	Q(m3/h)	v(m/s)	Q(l/s)	Q(m3/h)	v(m/s)	Q(l/s)	Q(m3/h)	v(m/s)	Q(l/s)	Q(m3/h)
50	0.28	0.5	2.0	0.31	0.6	2.2	0.34	0.7	2.4	0.37	0.7	2.6
80	0.38	1.9	7.0	0.43	2.2	7.8	0.47	2.4	8.5	0.51	2.6	9.2
100	0.45	3.5	12.7	0.50	3.9	14.2	0.55	4.3	15.5	0.59	4.7	16.8
125	0.52	6.4	23.0	0.58	7.2	25.7	0.64	7.8	28.2	0.69	8.5	30.5
150	0.59	10.4	37.4	0.66	11.6	41.9	0.72	12.8	45.9	0.78	13.8	49.6
200	0.71	22.3	80.5	0.80	25.0	90.1	0.87	27.4	98.7	0.94	29.6	106.7
250	0.82	40.4	145.5	0.92	45.2	162.8	1.01	49.6	178.5	1.09	53.6	192.9
300	0.93	65.5	235.9	1.04	73.3	264.0	1.14	80.4	289.4	1.23	86.9	312.7
350	1.02	98.5	354.7	1.15	110.2	396.9	1.26	120.8	435.0	1.36	130.6	470.1
400	1.12	140.2	504.8	1.25	156.9	564.8	1.37	172.0	619.1	1.48	185.8	668.9
450	1.20	191.4	688.9	1.35	214.1	770.8	1.48	234.7	844.8	1.59	253.6	912.8
500	1.29	252.7	909.7	1.44	282.7	1017.7	1.58	309.8	1115.3	1.70	334.8	1205.2
600	1.44	408.5	1470.7	1.62	457.0	1645.2	1.77	500.8	1803.0	1.91	541.1	1948.1
700	1.59	612.9	2206.5	1.78	685.6	2468.2	1.95	751.3	2704.8	2.11	811.8	2922.4
800	1.73	870.7	3134.5	1.94	973.9	3506.1	2.12	1067.3	3842.1	2.29	1153.1	4151.1
900	1.86	1186.4	4271.1	2.09	1327.0	4777.4	2.29	1454.2	5235.1	2.47	1571.1	5656.0
1000	1.99	1564.4	5631.9	2.23	1749.8	6299.3	2.44	1917.4	6902.6	2.64	2071.5	7457.5
1100	2.11	2008.8	7231.7	2.36	2246.8	8088.5	2.59	2462.0	8863.1	2.80	2659.8	9575.4
1200	2.23	2523.6	9084.9	2.50	2822.5	10161.0	2.73	3092.7	11133.9	2.95	3341.3	12028.6
1300	2.34	3112.5	11205.1	2.62	3481.2	12532.2	2.87	3814.4	13731.9	3.10	4120.9	14835.2
1400	2.46	3779.4	13605.8	2.75	4226.9	15216.9	3.01	4631.5	16673.5	3.25	5003.6	18012.9
1500	2.56	4527.7	16299.6	2.87	5063.7	18229.5	3.14	5548.4	19974.2	3.39	5994.1	21578.7
1600	2.67	5360.9	19299.2	2.98	5995.5	21584.0	3.27	6569.3	23649.5	3.53	7097.0	25549.0

k(mm) = 1   T(°C) = 20

D (mm)	S= 0.008 v(m/s)	Q(l/s)	Q(m3/h)	S= 0.009 v(m/s)	Q(l/s)	Q(m3/h)	S= 0.010 v(m/s)	Q(l/s)	Q(m3/h)	S= 0.012 v(m/s)	Q(l/s)	Q(m3/h)
50	0.39	0.8	2.8	0.42	0.8	3.0	0.44	0.9	3.1	0.48	1.0	3.4
80	0.55	2.7	9.9	0.58	2.9	10.5	0.61	3.1	11.1	0.67	3.4	12.1
100	0.64	5.0	18.0	0.67	5.3	19.1	0.71	5.6	20.1	0.78	6.1	22.1
125	0.74	9.1	32.7	0.78	9.6	34.7	0.83	10.2	36.5	0.91	11.1	40.1
150	0.83	14.7	53.1	0.89	15.7	56.3	0.93	16.5	59.4	1.02	18.1	65.1
200	1.01	31.7	114.1	1.07	33.6	121.1	1.13	35.5	127.7	1.24	38.9	140.0
250	1.17	57.3	206.3	1.24	60.8	218.9	1.31	64.1	230.9	1.43	70.3	253.0
300	1.31	92.9	334.4	1.39	98.6	354.8	1.47	103.9	374.1	1.61	113.9	410.0
350	1.45	139.7	502.7	1.54	148.2	533.4	1.62	156.2	562.4	1.78	171.2	616.3
400	1.58	198.7	715.4	1.68	210.8	759.0	1.77	222.3	800.2	1.94	243.6	877.0
450	1.70	271.2	976.2	1.81	287.7	1035.6	1.91	303.3	1091.9	2.09	332.4	1196.6
500	1.82	358.0	1288.7	1.93	379.8	1367.3	2.04	400.4	1441.5	2.23	438.8	1579.6
600	2.05	578.7	2083.2	2.17	613.9	2210.0	2.29	647.2	2330.0	2.51	709.2	2553.1
700	2.26	868.0	3125.0	2.39	920.9	3315.2	2.52	970.9	3495.1	2.76	1063.8	3829.7
800	2.45	1233.0	4438.7	2.60	1308.0	4708.9	2.74	1379.0	4964.4	3.01	1511.0	5439.6
900	2.64	1679.9	6047.7	2.80	1782.1	6415.7	2.95	1878.8	6763.7	3.24	2058.6	7411.0
1000	2.82	2215.0	7973.9	2.99	2349.7	8459.0	3.15	2477.2	8917.8	3.46	2714.2	9771.1
1100	2.99	2844.0	10238.4	3.17	3017.0	10861.1	3.35	3180.6	11450.1	3.67	3484.9	12545.6
1200	3.16	3572.6	12861.3	3.35	3789.0	13643.5	3.53	3995.3	14383.2	3.87	4377.5	15759.1
1300	3.32	4406.2	15862.2	3.52	4674.1	16826.7	3.71	4927.5	17738.9	4.07	5398.4	19435.7
1400	3.48	5349.9	19259.7	3.69	5675.2	20430.7	3.89	5982.8	21538.2	4.26	6555.0	23598.1
1500	3.63	6408.9	23072.1	3.85	6798.5	24474.7	4.06	7167.0	25801.4	4.44	7852.5	28268.9
1600	3.77	7588.1	27317.1	4.00	8049.3	28977.6	4.22	8485.6	30548.2	4.62	9297.1	33469.4

D (mm)	S= 0.014 v(m/s)	Q(l/s)	Q(m3/h)	S= 0.016 v(m/s)	Q(l/s)	Q(m3/h)	S= 0.018 v(m/s)	Q(l/s)	Q(m3/h)	S= 0.020 v(m/s)	Q(l/s)	Q(m3/h)
50	0.52	1.0	3.7	0.56	1.1	4.0	0.59	1.2	4.2	0.63	1.2	4.4
80	0.73	3.6	13.1	0.78	3.9	14.0	0.82	4.1	14.9	0.87	4.4	15.7
100	0.84	6.6	23.9	0.90	7.1	25.5	0.96	7.5	27.1	1.01	7.9	28.6
125	0.98	12.0	43.3	1.05	12.9	46.3	1.11	13.7	49.1	1.17	14.4	51.8
150	1.11	19.6	70.4	1.18	20.9	75.3	1.26	22.2	79.9	1.32	23.4	84.2
200	1.34	42.0	151.3	1.43	44.9	161.8	1.52	47.7	171.7	1.60	50.3	181.0
250	1.55	75.9	273.4	1.65	81.2	292.4	1.76	86.2	310.2	1.85	90.8	327.1
300	1.74	123.1	443.0	1.86	131.6	473.8	1.98	139.6	502.6	2.08	147.2	529.9
350	1.92	185.0	665.9	2.06	197.8	712.1	2.18	209.9	755.5	2.30	221.2	796.5
400	2.09	263.2	947.5	2.24	281.4	1013.2	2.38	298.6	1074.8	2.50	314.8	1133.2
450	2.26	359.1	1292.8	2.41	384.0	1382.4	2.56	407.4	1466.5	2.70	429.5	1546.1
500	2.41	474.1	1706.7	2.58	506.9	1824.9	2.74	537.8	1935.9	2.89	566.9	2040.9
600	2.71	766.2	2758.4	2.90	819.3	2949.4	3.07	869.1	3128.8	3.24	916.2	3298.5
700	2.99	1149.3	4137.5	3.19	1228.9	4423.9	3.39	1303.6	4692.9	3.57	1374.3	4947.4
800	3.25	1632.4	5876.6	3.47	1745.4	6283.3	3.68	1851.5	6665.3	3.88	1951.8	7026.6
900	3.50	2224.0	8006.3	3.74	2377.9	8560.3	3.96	2522.4	9080.7	4.18	2659.1	9572.9
1000	3.73	2932.2	10555.8	3.99	3135.0	11286.2	4.23	3325.6	11972.2	4.46	3505.8	12621.0
1100	3.96	3764.7	13552.9	4.24	4025.2	14490.6	4.49	4269.8	15371.3	4.74	4501.2	16204.2
1200	4.18	4729.0	17024.4	4.47	5056.1	18202.1	4.74	5363.4	19308.3	5.00	5654.0	20354.5
1300	4.39	5832.2	20996.0	4.70	6235.6	22448.3	4.98	6614.5	23812.3	5.25	6972.9	25102.5
1400	4.60	7081.2	25492.5	4.92	7571.0	27255.7	5.22	8031.0	28911.7	5.50	8466.1	30478.0
1500	4.80	8482.8	30537.9	5.13	9069.4	32650.0	5.44	9620.4	34633.6	5.74	10141.6	36509.8
1600	5.00	10043.3	36155.8	5.34	10737.8	38656.2	5.67	11390.2	41004.6	5.97	12007.2	43225.8

k(mm) =   5   T(°C) =   20

D	S =	0.0005		S =	0.001		S =	0.002		S =	0.003	
(mm)	v(m/s)	Q(l/s)	Q(m3/h)	v(m/s)	Q(l/s)	Q(m3/h)	v(m/s)	Q(l/s)	Q(m3/h)	v(m/s)	Q(l/s)	Q(m3/h)
50	0.07	0.1	0.5	0.10	0.2	0.7	0.14	0.3	1.0	0.17	0.3	1.2
80	0.10	0.5	1.8	0.14	0.7	2.5	0.20	1.0	3.6	0.24	1.2	4.4
100	0.12	0.9	3.3	0.16	1.3	4.6	0.23	1.8	6.6	0.29	2.2	8.1
125	0.14	1.7	6.0	0.19	2.4	8.5	0.27	3.4	12.1	0.34	4.1	14.8
150	0.16	2.7	9.9	0.22	3.9	14.0	0.31	5.5	19.9	0.38	6.8	24.4
200	0.19	6.0	21.6	0.27	8.5	30.6	0.38	12.0	43.3	0.47	14.8	53.1
250	0.22	11.0	39.4	0.32	15.5	55.9	0.45	22.0	79.1	0.55	26.9	97.0
300	0.25	17.9	64.4	0.36	25.3	91.3	0.51	35.9	129.2	0.62	44.0	158.4
350	0.28	27.1	97.5	0.40	38.3	138.0	0.56	54.3	195.5	0.69	66.5	239.5
400	0.31	38.7	139.4	0.44	54.8	197.4	0.62	77.6	279.5	0.76	95.1	342.5
450	0.33	53.1	191.1	0.47	75.2	270.6	0.67	106.4	383.1	0.82	130.4	469.4
500	0.36	70.4	253.3	0.51	99.6	358.6	0.72	141.0	507.6	0.88	172.8	622.0
600	0.40	114.5	412.0	0.57	162.0	583.4	0.81	229.4	825.7	0.99	281.0	1011.6
700	0.45	172.6	621.3	0.63	244.3	879.6	0.90	345.8	1244.8	1.10	423.6	1525.1
800	0.49	246.2	886.3	0.69	348.5	1254.6	0.98	493.2	1775.5	1.20	604.2	2175.2
900	0.53	336.7	1212.0	0.75	476.5	1715.6	1.06	674.4	2427.7	1.30	826.1	2974.1
1000	0.57	445.3	1603.1	0.80	630.3	2269.0	1.14	891.9	3210.8	1.39	1092.6	3933.4
1100	0.60	573.4	2064.1	0.85	811.5	2921.4	1.21	1148.3	4133.8	1.48	1406.7	5064.1
1200	0.64	722.1	2599.4	0.90	1021.9	3678.8	1.28	1445.9	5205.4	1.57	1771.3	6376.7
1300	0.67	892.5	3213.1	0.95	1263.1	4547.2	1.35	1787.2	6433.9	1.65	2189.3	7881.6
1400	0.71	1085.9	3909.2	1.00	1536.7	5532.2	1.41	2174.3	7827.4	1.73	2663.5	9588.6
1500	0.74	1303.3	4691.8	1.04	1844.3	6639.4	1.48	2609.4	9393.8	1.81	3196.5	11507.3
1600	0.77	1545.7	5564.5	1.09	2187.3	7874.2	1.54	3094.6	11140.6	1.89	3790.8	13647.0

D	S=	0.004		S=	0.005		S=	0.006		S=	0.007	
(mm)	v(m/s)	Q(l/s)	Q(m3/h)	v(m/s)	Q(l/s)	Q(m3/h)	v(m/s)	Q(l/s)	Q(m3/h)	v(m/s)	Q(l/s)	Q(m3/h)
50	0.20	0.4	1.4	0.22	0.4	1.5	0.24	0.5	1.7	0.26	0.5	1.8
80	0.28	1.4	5.1	0.31	1.6	5.7	0.34	1.7	6.2	0.37	1.9	6.7
100	0.33	2.6	9.3	0.37	2.9	10.4	0.40	3.2	11.4	0.44	3.4	12.3
125	0.39	4.8	17.1	0.43	5.3	19.2	0.48	5.8	21.0	0.51	6.3	22.7
150	0.44	7.8	28.1	0.49	8.7	31.5	0.54	9.6	34.5	0.59	10.4	37.3
200	0.54	17.0	61.4	0.61	19.1	68.6	0.66	20.9	75.2	0.72	22.6	81.2
250	0.63	31.1	112.0	0.71	34.8	125.3	0.78	38.1	137.2	0.84	41.2	148.3
300	0.72	50.8	182.9	0.80	56.8	204.5	0.88	62.3	224.1	0.95	67.3	242.1
350	0.80	76.8	276.6	0.89	85.9	309.4	0.98	94.1	338.9	1.06	101.7	366.1
400	0.87	109.9	395.6	0.98	122.9	442.4	1.07	134.6	484.7	1.16	145.4	523.6
450	0.95	150.6	542.1	1.06	168.4	606.2	1.16	184.5	664.2	1.25	199.3	717.5
500	1.02	199.6	718.4	1.14	223.1	803.3	1.25	244.5	880.1	1.34	264.1	950.7
600	1.15	324.5	1168.4	1.28	362.9	1306.5	1.41	397.6	1431.3	1.52	429.5	1546.1
700	1.27	489.3	1761.3	1.42	547.1	1969.5	1.56	599.4	2157.7	1.68	647.4	2330.8
800	1.39	697.8	2512.2	1.55	780.3	2809.0	1.70	854.8	3077.4	1.84	923.4	3324.2
900	1.50	954.1	3434.8	1.68	1066.9	3840.7	1.84	1168.8	4207.6	1.98	1262.5	4545.0
1000	1.61	1261.8	4542.6	1.80	1410.9	5079.3	1.97	1545.7	5564.6	2.13	1669.7	6010.8
1100	1.71	1624.5	5848.3	1.91	1816.5	6539.3	2.09	1990.0	7164.0	2.26	2149.6	7738.4
1200	1.81	2045.6	7364.2	2.02	2287.3	8234.2	2.22	2505.8	9020.8	2.39	2706.7	9744.1
1300	1.90	2528.4	9102.1	2.13	2827.0	10177.3	2.33	3097.1	11149.4	2.52	3345.4	12043.4
1400	2.00	3075.9	11073.3	2.23	3439.3	12381.4	2.45	3767.8	13564.0	2.64	4069.9	14651.5
1500	2.09	3691.4	13289.0	2.34	4127.4	14858.8	2.56	4521.7	16278.0	2.76	4884.2	17583.1
1600	2.18	4377.8	15760.0	2.43	4894.9	17621.6	2.67	5362.4	19304.6	2.88	5792.3	20852.3

$$k(mm) = \quad 5 \quad T(°C) = \quad 20$$

D	S=	0.008		S=	0.009		S=	0.010		S=	0.012	
(mm)	v(m/s)	Q(l/s)	Q(m3/h)	v(m/s)	Q(l/s)	Q(m3/h)	v(m/s)	Q(l/s)	Q(m3/h)	v(m/s)	Q(l/s)	Q(m3/h)
50	0.28	0.5	2.0	0.29	0.6	2.1	0.31	0.6	2.2	0.34	0.7	2.4
80	0.40	2.0	7.2	0.42	2.1	7.6	0.44	2.2	8.0	0.49	2.4	8.8
100	0.47	3.7	13.2	0.50	3.9	14.0	0.52	4.1	14.8	0.57	4.5	16.2
125	0.55	6.7	24.3	0.58	7.2	25.8	0.61	7.5	27.2	0.67	8.3	29.7
150	0.63	11.1	39.9	0.66	11.7	42.3	0.70	12.4	44.6	0.77	13.6	48.8
200	0.77	24.1	86.8	0.81	25.6	92.1	0.86	27.0	97.1	0.94	29.6	106.4
250	0.90	44.0	158.5	0.95	46.7	168.2	1.00	49.2	177.3	1.10	53.9	194.2
300	1.02	71.9	258.8	1.08	76.3	274.6	1.14	80.4	289.4	1.25	88.1	317.1
350	1.13	108.7	391.5	1.20	115.3	415.2	1.26	121.6	437.7	1.38	133.2	479.6
400	1.24	155.5	559.8	1.31	164.9	593.8	1.38	173.9	626.0	1.52	190.5	685.8
450	1.34	213.1	767.1	1.42	226.0	813.7	1.50	238.3	857.7	1.64	261.0	939.7
500	1.44	282.3	1016.4	1.53	299.5	1078.2	1.61	315.7	1136.5	1.76	345.9	1245.1
600	1.62	459.2	1653.0	1.72	487.0	1753.4	1.82	513.4	1848.3	1.99	562.5	2024.9
700	1.80	692.2	2491.8	1.91	734.2	2643.1	2.01	774.0	2786.2	2.20	847.9	3052.4
800	1.96	987.2	3553.9	2.08	1047.1	3769.7	2.20	1103.8	3973.8	2.41	1209.3	4353.3
900	2.12	1349.7	4859.1	2.25	1431.7	5154.1	2.37	1509.2	5433.1	2.60	1653.3	5952.0
1000	2.27	1785.0	6426.1	2.41	1893.4	6816.2	2.54	1995.9	7185.2	2.78	2186.5	7871.4
1100	2.42	2298.1	8273.1	2.56	2437.6	8775.3	2.70	2569.5	9250.3	2.96	2814.9	10133.7
1200	2.56	2893.7	10417.3	2.71	3069.3	11049.6	2.86	3235.5	11647.7	3.13	3544.5	12760.0
1300	2.69	3576.5	12875.5	2.86	3793.6	13657.0	3.01	3998.9	14396.1	3.30	4380.8	15770.9
1400	2.83	4351.0	15663.7	3.00	4615.1	16614.4	3.16	4864.9	17513.7	3.46	5329.5	19186.1
1500	2.95	5221.6	18797.8	3.13	5538.5	19938.7	3.30	5838.3	21017.8	3.62	6395.8	23024.8
1600	3.08	6192.5	22292.9	3.27	6568.3	23645.9	3.44	6923.8	24925.6	3.77	7584.9	27305.7

D	S=	0.014		S=	0.016		S=	0.018		S=	0.020	
(mm)	v(m/s)	Q(l/s)	Q(m3/h)	v(m/s)	Q(l/s)	Q(m3/h)	v(m/s)	Q(l/s)	Q(m3/h)	v(m/s)	Q(l/s)	Q(m3/h)
50	0.37	0.7	2.6	0.39	0.8	2.8	0.42	0.8	2.9	0.44	0.9	3.1
80	0.52	2.6	9.5	0.56	2.8	10.1	0.59	3.0	10.8	0.63	3.2	11.3
100	0.62	4.9	17.5	0.66	5.2	18.7	0.70	5.5	19.8	0.74	5.8	20.9
125	0.73	8.9	32.1	0.78	9.5	34.4	0.83	10.1	36.5	0.87	10.7	38.4
150	0.83	14.7	52.8	0.89	15.7	56.4	0.94	16.6	59.8	0.99	17.5	63.1
200	1.02	31.9	114.9	1.09	34.1	122.9	1.15	36.2	130.4	1.22	38.2	137.4
250	1.19	58.3	209.8	1.27	62.3	224.3	1.35	66.1	237.9	1.42	69.7	250.8
300	1.35	95.2	342.6	1.44	101.7	366.2	1.53	107.9	388.5	1.61	113.8	409.5
350	1.50	143.9	518.0	1.60	153.8	553.8	1.70	163.2	587.5	1.79	172.0	619.3
400	1.64	205.8	740.8	1.75	220.0	792.0	1.86	233.3	840.1	1.96	246.0	885.5
450	1.77	282.0	1015.1	1.90	301.4	1085.2	2.01	319.7	1151.1	2.12	337.1	1213.4
500	1.90	373.6	1345.0	2.03	399.4	1437.9	2.16	423.7	1525.2	2.27	446.6	1607.8
600	2.15	607.6	2187.2	2.30	649.5	2338.4	2.44	689.0	2480.3	2.57	726.3	2614.6
700	2.38	915.9	3297.1	2.54	979.1	3524.9	2.70	1038.6	3738.9	2.84	1094.8	3941.3
800	2.60	1306.2	4702.4	2.78	1396.5	5027.2	2.95	1481.2	5332.4	3.11	1561.4	5621.0
900	2.81	1785.9	6429.2	3.00	1909.3	6873.3	3.18	2025.1	7290.5	3.36	2134.7	7685.1
1000	3.01	2361.8	8502.5	3.21	2525.0	9089.8	3.41	2678.2	9641.5	3.59	2823.1	10163.3
1100	3.20	3040.6	10946.1	3.42	3250.6	11702.2	3.63	3447.9	12412.4	3.82	3634.5	13084.1
1200	3.39	3828.6	13782.9	3.62	4093.1	14735.0	3.84	4341.5	15629.3	4.05	4576.4	16475.0
1300	3.57	4732.0	17035.1	3.81	5058.9	18211.9	4.04	5365.9	19317.1	4.26	5656.2	20362.4
1400	3.74	5756.7	20724.1	4.00	6154.3	22155.6	4.24	6527.8	23500.1	4.47	6881.0	24771.8
1500	3.91	6908.5	24870.5	4.18	7385.7	26588.4	4.43	7833.8	28201.9	4.67	8257.8	29727.9
1600	4.07	8192.9	29494.5	4.36	8758.8	31531.7	4.62	9290.3	33445.2	4.87	9793.0	35254.9

Appendix 7

Spreadsheet hydraulic lessons – Overview

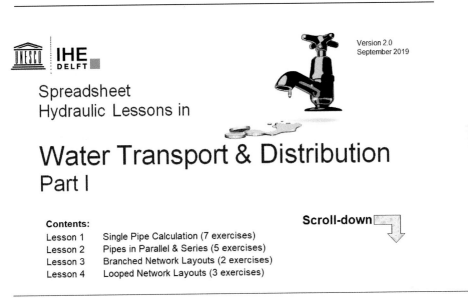

Version 2.0
September 2019

IHE
DELFT

Spreadsheet
Hydraulic Lessons in

Water Transport & Distribution
Part I

Contents:

Lesson 1 — Single Pipe Calculation (7 exercises)
Lesson 2 — Pipes in Parallel & Series (5 exercises)
Lesson 3 — Branched Network Layouts (2 exercises)
Lesson 4 — Looped Network Layouts (3 exercises)

Scroll-down

Disclaimer

This application has been developed with due care and attention, and is solely for educational and training purposes. In its default format, the worksheets include protection that can prevent unintentional deletion of cells with formulae. If this protection has been disabled, any inserting, deleting or cutting and pasting of the rows and columns can lead to damage or disappearance of the formulae, resulting in inaccurate calculations. Therefore, the author and IHE Delft are not responsible and assume no liability whatsoever for any results or any use made of the results obtained from this application in its original or modified format.

Introduction

The spreadsheet hydraulic lessons have been developed as an aid for steady state hydraulic calculations of simple water transport and distribution networks. These are to be carried out while solving the workshop problems that should normally be calculated manually; the spreadsheet serves here as a fast check of the results. Moreover, the spreadsheet lessons will help teachers to demonstrate a wider range of problems in a clear way, as well as to allow the students to continue analysing them at home. Ultimately, a real understanding of the hydraulic concepts will be reached through 'playing' with the data.

Over forty problems have been classified in eight groups/worksheets according to the contents of the book 'Introduction to Urban Water Distribution' by N. Trifunovic. This book covers a core curriculum of ' Water Transport and Distribution', which is a three-week module in the Water Supply Engineering specialisation at IHE Delft. Outside the regular MSc programme, this module is also offered as a stand-alone short course/online learning package.

To be able to use the spreadsheet, brief accompanying instructions for each exercise are given in the 'About' worksheet (see below).
Each layout covers approximately one full screen (30 rows) consisting of drawings, tables and graphs. In the tables:

The green colour indicates input cells. Except for the headers, these cells are unprotected and their contents are used for calculations.
The brown colour indicates output cells. These cells contain fixed formulae and are therefore protected.

Moreover, some intermediate calculations have been moved further to the right in the worksheet, being irrelevant for educational purposes. These are also protected by left unhidden for the sake of better understanding of the calculation process in each lesson.

Each lesson serves as a kind of chess problem, in which 'check-mate' should be reached within a few, correct moves. This suggests a study process where thinking takes more time than the execution, which was the main concept in the development of the exercises. Any simplifications that have been introduced (neglected minor losses, pump curve definition, etc.) were meant to facilitate this process. In addition, the worksheets have been designed without complicated routines or macros; only a superficial knowledge of spreadsheets is required to be able to use them effectively.

This is the second edition and any suggestions for improvement or extension are obviously welcome.

N. Trifunovic

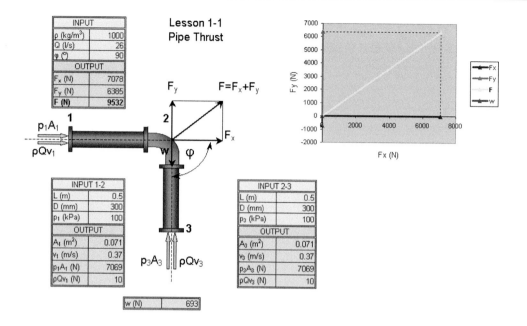

Lesson 1-1   Pipe Thrust

Contents:
Calculation of the pipe thrust in a pipe bend.

Goal:
Sensitivity analysis of the forces acting on the pipe bend, namely from the water pressure, flow conveyance and the water weight.

Abbreviations:

INPUT	
ρ (kg/m³)	Mass density of water
Q (l/s)	Flow rate
φ (°)	Angle of bend (degrees)
OUTPUT	
F_x (N)	Thrust force in horizontal direction
F_y (N)	Thrust force in vertical direction
F (N)	Combined force
w (N)	Weight of water in the isolated section

INPUT	
L (m)	Pipe length
D (mm)	Pipe diameter
p_1 (kPa)	Pressure in cross-section 1 (3)
OUTPUT	
A_1 (m²)	Cross-section area 1 (3)
v_1 (m/s)	Velocity in cross-section 1 (3)
p_1A_1 (N)	Force from the pressure acting on cross-section 1 (3)
ρQv_1 (N)	Dynamic force acting on cross-section 1 (3)

Notes:
The calculation ultimately yields the combined force F from the balance of all forces acting in the cross-sections 1 & 3, including the weight of the water.
The water weight is calculated from the pipe lengths & diameters in the sections 1-2-3. If this is to be neglected, some of these parameters should be set to 0.
The thrust forces (horizontal, vertical and combined) including the force from the water weight are plotted in the graph.

Lesson 1-2
Hydraulic Grade Line

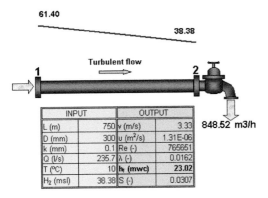

61.40

38.38

Turbulent flow

1   2

848.52 m3/h

INPUT		OUTPUT	
L (m)	750	v (m/s)	3.33
D (mm)	300	u (m²/s)	1.31E-06
k (mm)	0.1	Re (-)	765651
Q (l/s)	235.7	λ (-)	0.0162
T (°C)	10	h_f (mwc)	23.02
H₂ (msl)	38.38	S (-)	0.0307

Lesson 1-2   Hydraulic Grade Line

Contents:
Calculation of the friction losses in a single pipe (application of the Darcy-Weisbach formula).

Goal:
Sensitivity analysis of the basic hydraulic parameters, namely the pipe length, diameter, internal roughness and flow rate, and water temperature.

Abbreviations:

L (m)	Pipe length		v (m/s)	Flow velocity
D (mm)	Pipe diameter		u (m²/s)	Kinematic viscosity
k (mm)	Internal roughness		Re (-)	Reynolds number
Q (l/s)	Flow rate		λ (-)	Darcy-Weisbach friction factor
T (°C)	Water temperature (degrees Celsius)		h_f (mwc)	Friction loss (metres of water column)
H₂ (msl)	Downstream piezometric head (metres above sea level)		S (-)	Hydraulic gradient

Notes:
The calculation ultimately yields the upstream piezometric head required to maintain the specified downstream head for the specified values of L,D,k & Q.

Lesson 1-3
Friction Loss Formulas

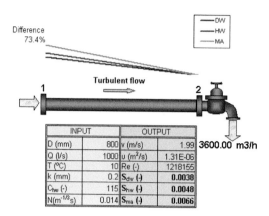

INPUT		OUTPUT	
D (mm)	800	v (m/s)	1.99
Q (l/s)	1000	u (m²/s)	1.31E-06
T (°C)	10	Re (-)	1218155
k (mm)	0.2	S_{dw} (-)	0.0038
C_{hw} (-)	115	S_{hw} (-)	0.0048
N(m$^{-1/3}$s)	0.014	S_{ma} (-)	0.0066

3600.00 m3/h

Lesson 1-3   Friction Loss Formulae

Contents:
Single pipe calculation of the hydraulic gradients using the Darcy-Weisbach, Hazen-Williams and Manning formulae.

Goal:
Comparison of the calculation accuracy and sensitivity of the Darcy-Weisbach, Hazen-Williams and Manning friction factors.

Abbreviations:

D (mm)	Pipe diameter		v (m/s)	Flow velocity
Q (l/s)	Flow rate		u (m²/s)	Kinematic viscosity
T (°C)	Water temperature		Re (-)	Reynolds number
k (mm)	Internal roughness		S_{dw} (-)	Hydraulic gradient determined by the Darcy-Weisbach formula
C_{hw} (-)	Hazen-Williams friction factor		S_{hw} (-)	Hydraulic gradient determined by the Hazen-Williams formula
N(m$^{-1/3}$s)	Manning friction factor		S_{ma} (-)	Hydraulic gradient determined by the Manning formula

Notes:
The percentage shows the difference between the lowest and the highest value of the three hydraulic gradients.

Lesson 1-4
Maximum Capacity

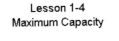

INPUT		OUTPUT	
L (m)	500	h_f (mwc)	5.00
D (mm)	200	u (m²/s)	1.24E-06
k (mm)	0.5	Re (-)	199147
S (-)	0.01	λ (-)	0.0258
T (°C)	12	v (m/s)	1.23
H_2 (msl)	25	Q (l/s)	38.74
Assumption			
v (m/s)	1.23		

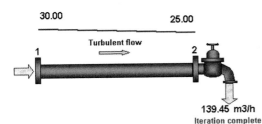

30.00   25.00

Turbulent flow

1   2

139.45 m3/h
Iteration complete

Lesson 1-4 Maximum Capacity

Contents:
Single pipe calculation using the Darcy-Weisbach formula.

Goal:
Determination of the maximum flow rate in a pipe of a specified diameter and hydraulic gradient.

Abbreviations:

L (m)	Pipe length		h_f (mwc)	Friction loss
D (mm)	Pipe diameter		u (m²/s)	Kinematic viscosity
k (mm)	Internal roughness		Re (-)	Reynolds number
S (-)	Hydraulic gradient		λ (-)	Darcy-Weisbach friction factor
T (°C)	Water temperature		v (m/s)	Calculated flow velocity
H_2 (msl)	Downstream piezometric head		Q (l/s)	Flow rate

v (m/s)	Assumed flow velocity

Notes:
The iterative procedure starts by assuming the flow velocity (commonly 1 m/s) required for determination of the Reynolds number i.e. the friction factor.
The velocity calculated afterwards by the Darcy-Weisbach formula serves as an input for the next iteration.
The iterative process is achieved by typing the value of the calculated velocity into the cell of the assumed velocity.
The message **Iteration complete** appears once the difference between the velocities in two iterations drops below 0.01 m/s.

Lesson 1-5
Optimal Diameter

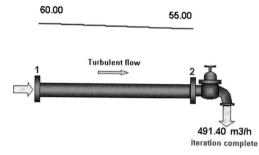

INPUT		OUTPUT	
L (m)	500	h$_f$ (mwc)	5.00
k (mm)	0.1	u (m^2/s)	1.31E-06
Q (l/s)	136.5	**D (mm)**	**303**
S (-)	0.01	Re (-)	438667
T (°C)	10	λ (-)	0.0168
H$_2$ (msl)	55	v (m/s)	1.88
Assumption			
v (m/s)	1.89		

491.40 m3/h
Iteration complete

Lesson 1-5   Optimal Diameter

Contents:
Single pipe calculation using the Darcy-Weisbach formula.

Goal:
Determination of the optimal pipe diameter for a specified flow rate and hydraulic gradient.

Abbreviations:

L (m)	Pipe length		h$_f$ (mwc)	Friction loss
k (mm)	Internal roughness		u (m^2/s)	Kinematic viscosity
Q (l/s)	Flow rate		D (mm)	Pipe diameter
S (-)	Hydraulic gradient		Re (-)	Reynolds number
T (°C)	Water temperature		λ (-)	Darcy-Weisbach friction factor
H$_2$ (msl)	Downstream piezometric head		v (m/s)	Calculated flow velocity

v (m/s)	Assumed flow velocity

Notes:
This is the same iterative procedure as in Lesson 1-3, except that the pipe diameter is determined from the assumed/calculated velocity (and specified flow rate). The message **Iteration complete** appears once the difference between the velocities in two iterations drops below 0.01 m/s.

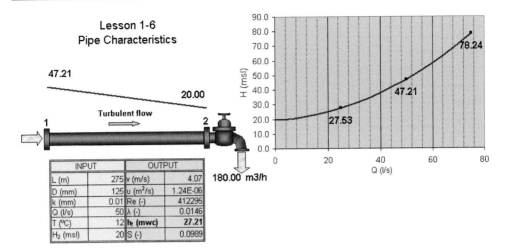

Lesson 1-6
Pipe Characteristics

47.21

20.00

Turbulent flow

78.24

47.21

27.53

INPUT		OUTPUT	
L (m)	275	v (m/s)	4.07
D (mm)	125	u (m²/s)	1.24E-06
k (mm)	0.01	Re (-)	412295
Q (l/s)	50	λ (-)	0.0146
T (°C)	12	h_f (mwc)	27.21
H₂ (msl)	20	S (-)	0.0989

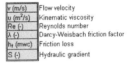

180.00 m3/h

Lesson 1-6 Pipe Characteristics

Contents:
Friction loss calculation in a single pipe of specified length, diameter and roughness.

Goal:
Determination of the pipe characteristics diagram.

Abbreviations:

L (m)	Pipe length		v (m/s)	Flow velocity
D (mm)	Pipe diameter		u (m²/s)	Kinematic viscosity
k (mm)	Internal roughness		Re (-)	Reynolds number
Q (l/s)	Flow rate		λ (-)	Darcy-Weisbach friction factor
T (°C)	Water temperature		h_f (mwc)	Friction loss
H₂ (msl)	Downstream piezometric head		S (-)	Hydraulic gradient

Notes:
The friction loss is calculated for the flow range 0-1.5Q (specified) in the same way as in Lesson 1-1, and the results are plotted in the graph.
The three points selected in the graph show the upstream heads required to maintain the specified downstream head for 0.5Q, Q and 1.5Q, respectively.
It is assumed the reference level at the pipe axis makes the downstream piezometric head equal to the pressure and the static head of the pipe characteristics.
The friction loss at the same curve represents its dynamic head.

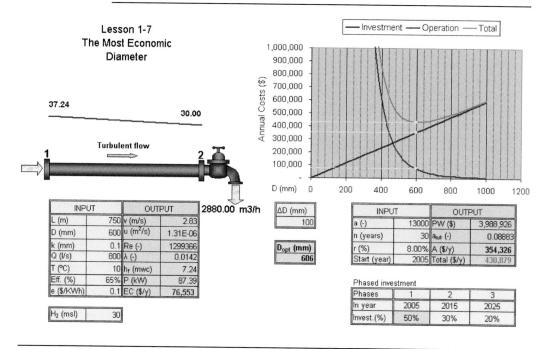

Lesson 1-7
The Most Economic
Diameter

INPUT		OUTPUT	
L (m)	750	v (m/s)	2.83
D (mm)	600	u (m²/s)	1.31E-06
k (mm)	0.1	Re (-)	1299366
Q (l/s)	800	λ (-)	0.0142
T (°C)	10	h_f (mwc)	7.24
Eff. (%)	65%	P (kW)	87.39
e ($/KWh)	0.1	EC ($/y)	76,553

H₂ (msl)	30

2880.00 m3/h

ΔD (mm)
100

D_opt (mm)
606

INPUT		OUTPUT	
a (-)	13000	PW ($)	3,988,926
n (years)	30	a_n/r (-)	0.08883
r (%)	8.00%	A ($/y)	354,326
Start (year)	2005	Total ($/y)	430,879

Phased investment

Phases	1	2	3
In year	2005	2015	2025
Invest.(%)	50%	30%	20%

Lesson 1-7   The Most Economic Diameter

Contents:
Calculation of annual investment and operational costs for a single pipe, based on the present worth method.

Goal:
To select the most economic pipe diameter for the given flow rate and specified conditions of the bank loan. The analysis includes a possibility of phased investment.

Abbreviations:

L (m)	Pipe length		v (m/s)	Flow velocity
D (mm)	Pipe diameter		u (m²/s)	Kinematic viscosity
k (mm)	Internal roughness		Re (-)	Reynolds number
Q (l/s)	Flow rate		λ (-)	Darcy-Weisbach friction factor
T (°C)	Water temperature (degrees Celsius)		h_f (mwc)	Friction loss (metres of water column)
Eff. (%)	Assumed pump efficiency at flow Q		P (kW)	Power consumption at flow Q
e ($/KWh)	Energy tariff		EC ($/y)	Annual energy (operational) costs
H₂ (msl)	Downstream piezometric head (metres above sea level)		D_opt (mm)	Optimal diameter (Equation 4.9, Paragraph 4.1.2 in the book)
ΔD (mm)	Diameter increment (X-axis on the diagram)		PW ($)	Present worth of the total investment (at the start of repayment)
a (-)	Pipe cost factor (Cost = aD in $ for D expressed in metres)		a_n/r (-)	Annuity
n (years)	Loan repayment period		A ($/y)	Annual instalment for the loan repayment (investment cost)
r (%)	Interest rate		Total ($/y)	Total annual investment and operational costs
Start (year)	Year in which the repayment starts (blank or 0 means 'immediately')			
In year	Year of the phased invetment (blank or 0 means 'no phased investment')			
Invest. (%)	Percentage of the total investment in particular phase (blank or 0 means 'no phased investment')			

Notes:
The simplified procedure is based on the assumption that all input parameters remain constant throughout the loan repayment period (average values).
In addition, the effect of inflation has been neglected. The loan repayment can start immediately or be delayed for a selected number of years, which increases the annual instalments. On the other hand, phased investments can reduce these instalments.
The optimum diameter is calculated based on the fixed friction factor; the result serves as an indication for selection of the input diameter.
The diagram with the investment and operational costs has been plotted in 20 points for the range of diameters between 0 and 10ΔD. The range of the Y-axis has been fixed between 0 and 1,000,000 $. In the case of larger pipes, this range may need to be widened and therefore the worksheet protection has to be temporarily removed ('Excel' menu command **Tools>>Protection>>Unprotect Sheet**).

Lesson 2-1a
Maximum Capacity

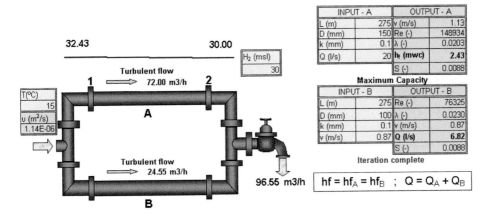

INPUT - A		OUTPUT - A	
L (m)	275	v (m/s)	1.13
D (mm)	150	Re (-)	148934
k (mm)	0.1	λ (-)	0.0203
Q (l/s)	20	h_f (mwc)	2.43
		S (-)	0.0088

Maximum Capacity

INPUT - B		OUTPUT - B	
L (m)	275	Re (-)	76325
D (mm)	100	λ (-)	0.0230
k (mm)	0.1	v (m/s)	0.87
v (m/s)	0.87	Q (l/s)	6.82
		S (-)	0.0088

Iteration complete

$$hf = hf_A = hf_B \quad ; \quad Q = Q_A + Q_B$$

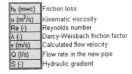

Lesson 2-1a Pipes in Parallel - Maximum Capacity

Contents:
Hydraulic calculation of two pipes connected in parallel.

Goal:
Resulting from the demand growth, a new pipe (B) of a specified diameter is to be laid in parallel, next to the existing one (A).
The task is to find the maximum flow rate in this pipe by maintaining the same hydraulic gradient as in the existing pipe.

Abbreviations:

L (m)	Pipe length	h_f (mwc)	Friction loss
D (mm)	Pipe diameter	u (m²/s)	Kinematic viscosity
k (mm)	Internal roughness	Re (-)	Reynolds number
Q (l/s)	Flow rate in the existing pipe	λ (-)	Darcy-Weisbach friction factor
T (°C)	Water temperature	v (m/s)	Calculated flow velocity
H₂ (msl)	Downstream piezometric head	Q (l/s)	Flow rate in the new pipe
v (m/s)	Assumed flow velocity in the new pipe	S (-)	Hydraulic gradient

Notes:
The friction loss in the existing pipe is calculated as in Lesson 1-1. Its hydraulic gradient is used as an input for calculation of the maximum capacity in the new pipe.
The same iterative procedure as in Lesson 1-3 applies for the new pipe.
The message **Iteration complete** appears once the difference between the velocities in two iterations drops below 0.01 m/s.

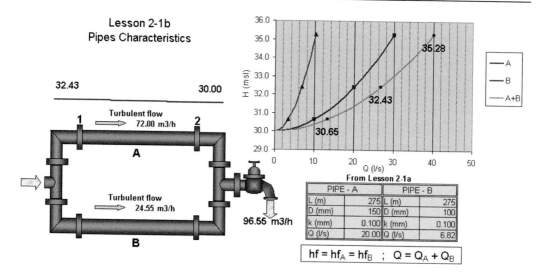

Lesson 2-1b
Pipes Characteristics

32.43   30.00

Turbulent flow
72.00 m3/h

A

Turbulent flow
24.55 m3/h

B

96.55 m3/h

35.28

32.43

30.65

From Lesson 2-1a

PIPE - A		PIPE - B	
L (m)	275	L (m)	275
D (mm)	150	D (mm)	100
k (mm)	0.100	k (mm)	0.100
Q (l/s)	20.00	Q (l/s)	6.82

$$hf = hf_A = hf_B \quad ; \quad Q = Q_A + Q_B$$

Lesson 2-1b Pipes in Parallel - Pipe Characteristics

Contents:
Hydraulic calculation of two pipes connected in parallel.

Goal:
Determination of the pipe characteristics diagrams for the system from Lesson 2-1b.

Abbreviations:

L (m)	Pipe length
D (mm)	Pipe diameter
k (mm)	Internal roughness
Q (l/s)	Flow rate in the existing pipe

Notes:
The pipe characteristics diagram is presented for each of the pipes and both of them operating in parallel, in the range 0-1.5Q (=$Q_A + Q_B$).
The three points selected in the graph show the upstream head required to maintain the specified downstream head for 0.5Q, Q and 1.5Q, respectively.
The flow rate in each of the pipes can be determined from this diagram.

Lesson 2-2
Optimal Diameter

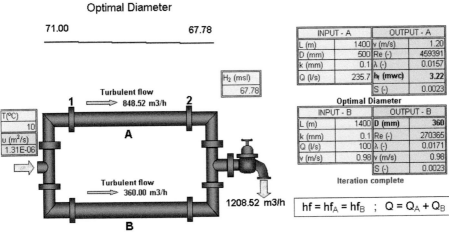

71.00   67.78

INPUT - A		OUTPUT - A	
L (m)	1400	v (m/s)	1.20
D (mm)	500	Re (-)	459391
k (mm)	0.1	λ (-)	0.0157
Q (l/s)	235.7	hf (mwc)	3.22
		S (-)	0.0023

Optimal Diameter

INPUT - B		OUTPUT - B	
L (m)	1400	D (mm)	360
k (mm)	0.1	Re (-)	270365
Q (l/s)	100	λ (-)	0.0171
v (m/s)	0.98	v (m/s)	0.98
		S (-)	0.0023

Iteration complete

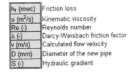

$$hf = hf_A = hf_B \quad ; \quad Q = Q_A + Q_B$$

Lesson 2-2   Pipes in Parallel - Optimal Diameter

Contents:
Hydraulic calculation of two pipes connected in parallel.

Goal:
Resulting from the demand growth, a new pipe (B) is to be laid in parallel, next to the existing one (A).
The task is to determine the optimal diameter of this pipe for a given flow rate, by maintaining the same hydraulic gradient as in the existing pipe.

Abbreviations:

L (m)	Pipe length
D (mm)	Diameter of the existing pipe
k (mm)	Internal roughness
Q (l/s)	Flow rate
T (°C)	Water temperature
H₂ (msl)	Downstream piezometric head
v (m/s)	Assumed flow velocity in the new pipe

hf (mwc)	Friction loss
υ (m²/s)	Kinematic viscosity
Re (-)	Reynolds number
λ (-)	Darcy-Weisbach friction factor
v (m/s)	Calculated flow velocity
D (mm)	Diameter of the new pipe
S (-)	Hydraulic gradient

Notes:
The friction loss in the existing pipe is calculated as in Lesson 1-1. Its hydraulic gradient is used as an input for calculation of the maximum capacity in the new pipe.
The same iterative procedure as in Lesson 1-4 applies for the new pipe.
The message **Iteration complete** appears once the difference between the velocities in two iterations drops below 0.01 m/s.

Lesson 2-3
Equivalent Diameter

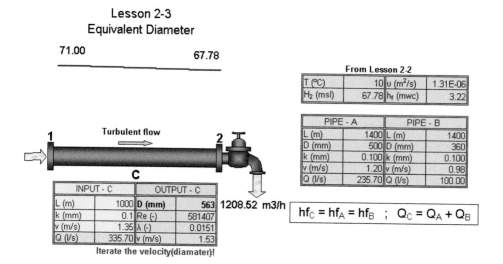

71.00   67.78

From Lesson 2-2

T (°C)	10	u (m²/s)	1.31E-06
H₂ (msl)	67.78	h_f (mwc)	3.22

PIPE - A		PIPE - B	
L (m)	1400	L (m)	1400
D (mm)	500	D (mm)	360
k (mm)	0.100	k (mm)	0.100
v (m/s)	1.20	v (m/s)	0.98
Q (l/s)	235.70	Q (l/s)	100.00

Turbulent flow

C

INPUT - C		OUTPUT - C	
L (m)	1000	D (mm)	563
k (mm)	0.1	Re (-)	581407
v (m/s)	1.35	λ (-)	0.0151
Q (l/s)	335.70	v (m/s)	1.53

Iterate the velocity(diamater)!

1208.52 m3/h

$$hf_C = hf_A = hf_B \quad ; \quad Q_C = Q_A + Q_B$$

Lesson 2-3   Pipes in Parallel - Equivalent Diameter

Contents:
Calculation of a hydraulically equivalent pipe.

Goal:
As an alternative to the system in Lesson 2-2, one larger pipe can be laid instead of two parallel pipes.
The task is to determine the optimal diameter of this pipe for a given flow rate, by maintaining the same hydraulic gradient as in the existing pipe.

Abbreviations:
The same as in Lesson 2-2.

Notes:
The total flow rate and hydraulic gradient from Lesson 2-2 are used as an input for calculation of the optimal pipe diameter.
The same iterative procedure as in Lesson 1-4 applies.
The message **Iteration complete** appears once the difference between the velocities in two iterations drops below 0.01 m/s.

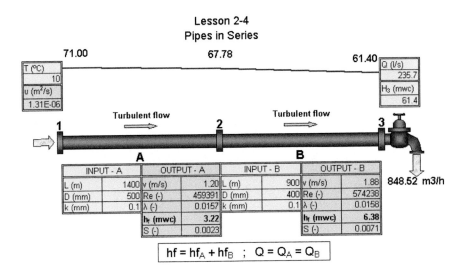

Lesson 2-4
Pipes in Series

71.00   67.78   61.40

T (°C)
10

u (m²/s)
1.31E-06

Q (l/s)
235.7

H₃ (mwc)
61.4

Turbulent flow   **Turbulent flow**

1   2   3

A   B

848.52 m3/h

INPUT - A		OUTPUT - A		INPUT - B		OUTPUT - B	
L (m)	1400	v (m/s)	1.20	L (m)	900	v (m/s)	1.88
D (mm)	500	Re (-)	459391	D (mm)	400	Re (-)	574238
k (mm)	0.1	λ (-)	0.0157	k (mm)	0.1	λ (-)	0.0158
		h_f (mwc)	**3.22**			**h_f (mwc)**	**6.38**
		S (-)	0.0023			S (-)	0.0071

$$hf = hf_A + hf_B \quad ; \quad Q = Q_A = Q_B$$

Lesson 2-4   Pipes in Series - Hydraulic Grade Line

Contents:
Calculation of the friction losses in two pipes connected in series.

Goal:
Resulting from the system expansion, a new pipe (B) of a specified diameter is to be laid in series, after the existing one (A).
The task is to determine the piezometric head at the upstream side (H₁) required to maintain the minimum head at the downstream side (H₃).

Abbreviations:

L (m)	Pipe length		v (m/s)	Flow velocity
D (mm)	Pipe diameter		Re (-)	Reynolds number
k (mm)	Internal roughness		λ (-)	Darcy-Weisbach friction factor
Q (l/s)	Flow rate		h_f (mwc)	Friction loss
H₃ (msl)	Downstream piezometric head		S (-)	Hydraulic gradient
T (°C)	Water temperature		u (m²/s)	Kinematic viscosity

Notes:
By maintaining the same flow rate in pipes A & B, the same calculation procedure applies as in Lesson 1-1.

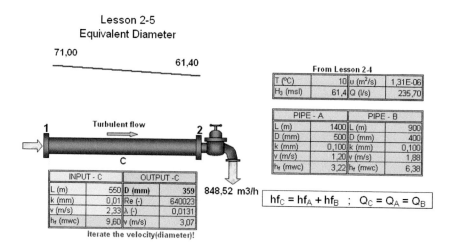

Lesson 2-5
Equivalent Diameter

71,00

61,40

Turbulent flow

1

2

C

848,52 m3/h

From Lesson 2-4

T (°C)	10	u (m²/s)	1,31E-06
H₃ (msl)	61,4	Q (l/s)	235,70

PIPE - A		PIPE - B	
L (m)	1400	L (m)	900
D (mm)	500	D (mm)	400
k (mm)	0,100	k (mm)	0,100
v (m/s)	1,20	v (m/s)	1,88
hf (mwc)	3,22	hf (mwc)	6,38

INPUT - C		OUTPUT -C	
L (m)	550	D (mm)	359
k (mm)	0,01	Re (-)	640023
v (m/s)	2,33	λ (-)	0,0131
hf (mwc)	9,60	v (m/s)	3,07

Iterate the velocity(diameter)!

$$hf_C = hf_A + hf_B \quad ; \quad Q_C = Q_A = Q_B$$

Lesson 2-5   Pipes in Series - Equivalent Diameter

Contents:
Calculation of a hydraulically equivalent pipe.

Goal:
As an alternative to the system in Lesson 2-4, one longer pipe is to be laid instead of the two serial pipes.
The task is to determine the optimal diameter of this pipe for a given flow rate, by maintaining the existing head difference between the points 1 & 3.

Abbreviations:
The same as in Lesson 2-4.

Notes:
The flow rate and piezometric head difference (H_1-H_3 i.e. hf_A+hf_B) from Lesson 2-4 are used as an input for calculation of the optimal pipe diameter.
The same iterative procedure applies as in Lesson 1-4.
The message **Iteration complete** appears once the difference between the velocities in two iterations drops below 0.01 m/s.

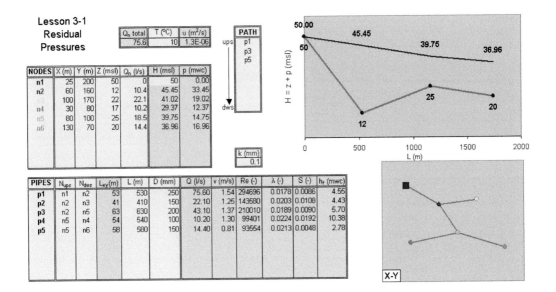

Lesson 3-1
Residual
Pressures

Q_h total	T (°C)	u (m²/s)
75.6	10	1.3E-06

PATH

ups p1
     p3
     p5
dws

NODES	X (m)	Y (m)	Z (msl)	Q_h (l/s)	H (msl)	p (mwc)
n1	25	200	50	0	50	0.00
n2	60	160	12	10.4	45.45	33.45
	100	170	22	22.1	41.02	19.02
n4	30	80	17	10.2	29.37	12.37
n5	80	100	25	18.5	39.75	14.75
n6	130	70	20	14.4	36.96	16.96

k (mm)
0.1

PIPES	N_{ups}	N_{dws}	L_{xy}(m)	L (m)	D (mm)	Q (l/s)	v (m/s)	Re (-)	λ (-)	S (-)	h_f (mwc)
p1	n1	n2	53	530	250	75.60	1.54	294696	0.0178	0.0086	4.55
p2	n2	n3	41	410	150	22.10	1.25	143580	0.0203	0.0108	4.43
p3	n2	n5	63	630	200	43.10	1.37	210010	0.0189	0.0090	5.70
p4	n5	n4	54	540	100	10.20	1.30	99401	0.0224	0.0192	10.38
p5	n5	n6	58	580	150	14.40	0.81	93554	0.0213	0.0048	2.78

X-Y

Lesson 3-1   Branched Network Layouts - Residual Pressures

Contents:
Calculation of the friction losses in a branched network configuration.

Goal:
Pressures in the system should be determined for a specified network configuration, distribution of nodal demands and piezometric head fixed in a number of nodes.

Abbreviations:

NODES	Node data
X (m)	Horizontal co-ordinate
Y (m)	Vertical co-ordinate
Z (msl)	Altitude
Q_h (l/s)	Nodal demand
H (msl)	Piezometric head
p (mwc)	Nodal pressure

Q_h total	Total demand of the system
T (°C)	Water temperature
u (m²/s)	Kinematic viscosity

PIPES	Pipe data
N_{ups}	Upstream node
N_{dws}	Downstream node
L_{xy}(m)	Length calculated from the X/Y co-ordinates
L (m)	Length adopted for the hydraulic calculation
D (mm)	Diameter
Q (l/s)	Flow rate
v (m/s)	Flow velocity
Re (-)	Reynolds number
λ (-)	Darcy-Weisbach friction factor
S (-)	Hydraulic gradient
h_f (mwc)	Friction loss

PATH	Pipes selected to be plotted with their piezometric heads.
	(starting from the upstream- to the downstream pipes)
k (mm)	Internal roughness (uniform for all pipes)

Notes:
The nodes are plotted based on the X/Y input (origin of the graph is at the lower left corner). Any node name can be used.
The first node in the list of nodes simulates the source and therefore has a fixed piezometric head.
The pipes are plotted based on the N_{ups}/N_{dws} input. This input determines connectivity between the nodes and hence the flow rates/directions.
From the determined pipe flows, the friction losses and consequently the nodal heads/pressures will be calculated.
Each node may appear only once as a downstream node (N_{dws}). Doing otherwise suggests a system consisting of more than one source, or from loops.

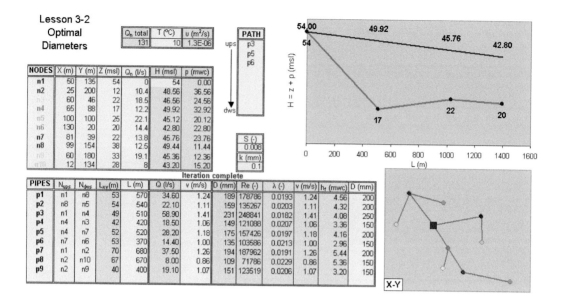

Lesson 3-2   Branched Network Layouts - Optimal Diameters

Contents:
Hydraulic calculation of a branched network configuration.

Goal:
The pipe diameters in the system should be determined for a specified network configuration, distribution of nodal demands and uniform (= design) hydraulic gradient.

Abbreviations:

NODES	Node data
X (m)	Horizontal co-ordinate
Y (m)	Vertical co-ordinate
Z (msl)	Altitude
Q_n (l/s)	Nodal demand
H (msl)	Piezometric head
p (mwc)	Nodal pressure

Q_n total	Total demand of the system
T (°C)	Water temperature
υ (m²/s)	Kinematic viscosity

PATH	Pipes selected to be plotted with their piezometric heads.
	(starting from the upstream to the downstream pipes)
S (-)	Design hydraulic gradient (uniform for all pipes)
k (mm)	Internal roughness (uniform for all pipes)

PIPES	Pipe data
N_{ups}	Upstream node
N_{dws}	Downstream node
L_{xy}(m)	Length calculated from the X/Y co-ordinates
L (m)	Length adopted for hydraulic calculation
Q (l/s)	Flow rate
v (m/s)	Flow velocity of iteration '1'
D (mm)	Calculated diameter
Re (-)	Reynolds number
λ (-)	Darcy-Weisbach friction factor
v (m/s)	Flow velocity of iteration 'i+1'
h_f (mwc)	Friction loss
D (mm)	Adopted diameter (manufactured size)

Notes:
The procedure of the network building is the same as in Lesson 3-1. <u>The order of the nodes from upstream to downstream has to be respected in the list of pipes.</u>
The first node in the list of nodes simulates the source and therefore has a fixed piezometric head.
The hydraulic calculation follows the principles of the single pipe calculation from Lesson 1-4; the iteration procedure has to be conducted for all pipes.
This can be done at once, by copying the entire column of the 'i+1' velocities, and pasting it subsequently to the column of '1' velocities.
<u>Only the 'Excel' command **Edit>>Paste Special>>Values** should be used in this case</u>; the ordinary 'Paste' command also copies the cell formulae, which is wrong!
The message **Iteration complete** appears once the total difference between the velocities in two iterations drops below 0.01 m/s.

k (mm) | 0.1

Lesson 4-1
Method of
Balancing Heads

Qn total	T (°C)	u (m²/s)
75.6	10	1.3E-06

NODES	X (m)	Y (m)	Z (msl)	Qn (l/s)	H (msl)	p (mwc)
n1	25	200	50	0	50	0.00
n2	60	160	12	10.4	47.88	35.88
n3	100	170	22	22.1	44.22	22.22
n4	30	80	17	10.2	67.70	50.70
n5	80	100	25	18.5	46.53	21.53
n6	130	70	20	14.4	45.28	25.28

X-Y

PIPES	N1cw	N2cw	Lxy(m)	L (m)	D (mm)	Q (l/s)	v (m/s)	hf (mwc)	Q (l/s)
p1	n1	n2	53	530	250	50.60	1.03	2.12	61.56
p2	n2	n5	63	630	200	20.20	0.64	1.36	29.95
p3	n5	n4	54	540	100	-14.80	-1.88	-21.17	-3.84
p4	n4	n1	120	1200	100	-25.00	-3.18	-129.89	-14.04
						0.00			0.00
LOOP 1				δQ (l/s)=	10.96		Sum=	-147.59	

Iterate Q!

PIPES	N1cw	N2cw	Lxy(m)	L (m)	D (mm)	Q (l/s)	v (m/s)	hf (mwc)	Q (l/s)
p2	n5	n2	63	630	200	-20.20	-0.64	-1.36	-29.95
p5	n2	n3	41	410	150	20.00	1.13	3.66	21.21
p6	n3	n6	104	1040	100	-2.10	-0.27	-1.06	-0.89
p7	n6	n5	58	580	150	-16.50	-0.93	-3.60	-15.29
						0.00			0.00
LOOP 2				δQ (l/s)=	1.21		Sum=	-2.35	

Iterate Q!

PIPES	N1cw	N2cw	Lxy(m)	L (m)	D (mm)	Q (l/s)	v (m/s)	hf (mwc)	Q (l/s)
									0.00
									0.00
									0.00
									0.00
									0.00
LOOP 3				δQ (l/s)=	0.00		Sum=	0.00	

Iteration complete

Lesson 4-1   Looped Network Layouts - Method of Balancing Heads

Contents:
Hydraulic calculation of a looped network configuration by the Hardy-Cross Method of Balancing Heads (Loop Oriented Method).

Goal:
The flows and pressures in the system should be determined for a specified network configuration, nodal demands and piezometric head fixed in a source node.

Abbreviations:

NODES	Node data
X (m)	Horizontal co-ordinate
Y (m)	Vertical co-ordinate
Z (msl)	Altitude
Qn (l/s)	Nodal demand
H (msl)	Piezometric head
p (mwc)	Nodal pressure

Qn total	Total demand of the system (l/s)
T (°C)	Water temperature
u (m²/s)	Kinematic viscosity

k (mm)	Internal roughness (uniform)

PIPES	Pipe data per loop
N1cw	Pipe node nr.1 (clockwise direction)
N2cw	Pipe node nr.2 (clockwise direction)
Lxy(m)	Length calculated from the X/Y co-ordinates
L (m)	Length adopted for hydraulic calculation
D (mm)	Diameter
Q (l/s)	Flow rate of iteration 'i'
v (m/s)	Flow velocity
hf (mwc)	Friction loss
Q (l/s)	Flow rate of iteration 'i+1'

δQ (l/s)	Flow rate correction. $Q_{i+1} = Q_i + \delta Q$
Sum	Sum of friction losses in the loop (clockwise direction)

Notes:
The table with the nodal data is prepared in the same way as in Lessons 3-1 and 3-2.
The pipes are plotted based on the N1cw/N2cw input. As a convention, this input has to be made in a clockwise direction for each loop.
The pipes shared by neighbouring loops should appear in both tables (with opposite flow directions).
The first node in the list of nodes and pipes (in loop 1) simulates the source and therefore has a fixed piezometric head.
To provide a correct spreadsheet calculation of nodal piezometric heads, the tables of loops 2 & 3 should start with a previously filled (shared) pipe.
The iterative process starts by distributing the pipe flows 'i' arbitrarily, but satisfying the Continuity Eequation in each node.
Negative flows, velocities and friction losses indicate anti-clockwise flow direction.
The flow correction (δQ) is calculated from the friction losses/piezometric heads, and flows for iteration 'i+1' are determined for all loops simultaneously.
Both δQ corrections of the neighbouring loops are applied in case of the shared pipes (with opposite sign).
The iteration proceeds by copying the entire column of the 'i+1' flows, and pasting it subsequently to the column of 'i' flows.
Only the 'Excel' command **Edit>>Paste Special>>Values** should be used in this case; the ordinary 'Paste' command also copies the cell formulae, which is wrong!
The message **Iteration complete** appears once the sum of friction losses in the loop drops below 0.01 mwc.

Lesson 4-2
Method of
Balancing Flows

Q$_n$ total	T (°C)	u (m²/s)
75.6	10	1.3E-06

δH total
14.19

Iterate H!

NODES	X (m)	Y (m)	Z (msl)	Q$_n$ (l/s)	H (msl)	p (mwc)	δQ (l/s)	δH (msl)	H (msl)
n1	25	200	50	0	50	0.00	-55.92	0.00	50.00
n2	60	160	12	10.4	48.00	36.00	-19.78	-1.04	46.96
n3	100	170	22	22.1	46.00	24.00	-13.85	-3.43	42.57
n4	30	80	17	10.2	42.00	25.00	-8.68	-5.85	36.15
n5	80	100	25	18.5	40.00	15.00	18.67	2.70	42.70
n6	130	70	20	14.4	38.00	18.00	3.95	1.16	39.16

X-Y

Iterate v!

PIPES	N1	N2	L$_{xy}$(m)	L (m)	D (mm)	h$_f$ (mwc)	S (-)	v (m/s)	Re (-)	λ (-)	v (m/s)	Q (l/s)	Q/h$_f$
p1	n1	n2	53	530	250	2.00	0.0038	1.44	275127	0.0179	1.02	49.96	24.98
p2	n2	n5	63	630	200	8.00	0.0127	0.21	31496	0.0245	1.42	44.76	5.60
p3	n2	n3	41	410	150	2.00	0.0049	0.88	100560	0.0211	0.83	14.58	7.29
p4	n3	n6	104	1040	100	8.00	0.0077	0.88	67443	0.0233	0.81	6.32	0.79
p5	n4	n5	54	540	100	2.00	0.0037	1.10	84241	0.0227	0.57	4.44	2.22
p6	n5	n6	58	580	150	2.00	0.0034	0.64	73812	0.0219	0.68	12.03	6.01
p7	n1	n4	120	1200	100	8.00	0.0067	1.11	85335	0.0227	0.76	5.96	0.75
								0.00			0.00		
								0.00			0.00		
								0.00			0.00		

k (mm)
0.1

Lesson 4-2   Looped Network Layouts - Method of Balancing Flows

Contents:
Hydraulic calculation of a looped network configuration by the Method of Balancing Flows (Node Oriented Method).

Goal:
The flows and pressures in the system should be determined for a specified network configuration, nodal demands and piezometric head fixed in a source node.

Abbreviations:

NODES	Node data
X (m)	Horizontal co-ordinate
Y (m)	Vertical co-ordinate
Z (msl)	Altitude
Q$_n$ (l/s)	Nodal demand
H (msl)	Piezometric head of iteration 'i'
p (mwc)	Nodal pressure
δQ (l/s)	Balance of the flow continuity equation
δH (msl)	Piezometric head correction. H$_{i+1}$ = H$_i$ + δH
H (msl)	Piezometric head of iteration 'i+1'

PIPES	Pipe data
N1	Pipe node nr.1
N2	Pipe node nr.2
L$_{xy}$(m)	Length calculated from the X/Y co-ordinates
L (m)	Length adopted for hydraulic calculation
D (mm)	Diameter
h$_f$ (mwc)	Friction loss
S (-)	Hydraulic gradient
v (m/s)	Flow velocity of iteration 'i'
Re (-)	Reynolds number
λ (-)	Darcy-Weisbach friction factor
v (m/s)	Flow velocity of iteration 'i+1'
Q (l/s)	Flow rate
Q/h$_f$	Ratio used for calculation of δH-corrections

Q$_n$ total	Total demand of the system
T (°C)	Water temperature
u (m²/s)	Kinematic viscosity

δH total	Sum of all δH-corrections

k (mm)	Internal roughness (uniform for all pipes)

Notes:
The table with the nodal data is prepared in the same way as in Lesson 4-1

The pipes are plotted based on the N1/N2 input. Unlike in Lesson 4-1, the order of nodes/pipes is not crucial in this case.

The first node in the list of nodes simulates the source and has therefore fixed piezometric head.

The heads in other nodes are distributed arbitrarily in the first iteration, except that no nodes should be allocated the same value.

The calculation starts by iterating the velocities in order to determine the pipe flows for given piezometric heads. This is done in the same way as in Lesson 3-2.

The message **Iteration complete** appears once the total difference between the velocities in two iterations drops below 0.01 m/s.

After the pipe flows have been determined, the correction (δH) is calculated and the iteration of piezometric heads proceeds.

A consecutive iteration is done node by node, by typing the current 'H$_{i+1}$' value into 'H$_i$' cell. Copying the entire column does not lead to a convergence.

The new values of the nodal piezometric heads should result in a gradual reduction of the 'δH total' value; the velocities (flows) have to be re-iterated.

The message **Iteration complete** appears once the sum of δH-corrections for all nodes drops below 0.01 mwc.

Lesson 4-3
Linear Theory

Qn total	T (°C)	υ (m²/s)		δH total
72.9	10	1.3E-06		0.06

Iteration complete

NODES	X (m)	Y (m)	Z (msl)	Qn (l/s)	H (msl)	p (mwc)	δQ (l/s)	H (msl)
n1	40	70	54	0	60	6.00	-72.59	60.00
n2	80	20	12	10.4	55.88	55.87	-0.25	55.87
n3	40	20	17	12.2	58.07	41.07	-0.11	58.06
n4	80	70	25	22.1	56.06	31.06	-0.12	56.06
n5	120	70	20	14.4	52.64	32.64	0.04	52.65
n6	120	20	22	13.8	50.04	28.04	0.13	50.08

ω (-)
1

k (mm)	X-Y
0.1	

Iteration complete

PIPES	N1	N2	Lxy(m)	L (m)	D (mm)	Q(l/s)	v (m/s)	Re (-)	λ (-)	1/U	H1/U	H2/U	hf (mwc)	S (-)	v (m/s)	Q (l/s)
p1	n1	n4	40	400	200	45.05	1.43	219531	0.0188	0.0114	0.69	0.64	3.94	0.0098	1.43	45.06
p2	n4	n2	50	500	150	3.54	0.20	23017	0.0265	0.0196	1.10	1.09	0.18	0.0004	0.20	3.53
p3	n2	n3	40	400	150	-15.44	-0.87	100333	0.0211	0.0071	0.39	0.41	-2.19	0.0055	0.87	-15.44
p4	n3	n1	50	500	200	-27.54	-0.88	134170	0.0197	0.0143	0.83	0.86	-1.93	0.0039	0.88	-27.54
p5	n4	n5	40	400	150	19.54	1.11	126952	0.0206	0.0057	0.32	0.30	3.42	0.0086	1.11	19.54
p6	n5	n6	50	500	100	5.11	0.65	49790	0.0242	0.0020	0.10	0.10	2.60	0.0052	0.65	5.10
p7	n6	n2	40	400	100	-8.83	-1.12	86071	0.0227	0.0015	0.08	0.08	-5.84	0.0146	1.12	-8.83
																0.00
																0.00
																0.00

Lesson 4-3   Looped Network Layouts - Linear Theory

Contents:
Hydraulic calculation of a looped network configuration based on the linear theory (solution by the Newton-Raphson/successive over-relaxation method).

Goal:
For specified network configuration, nodal demands and piezometric head fixed in a source node, the flows and pressures in the system should be determin

Abbreviations:

NODES	Node data
X (m)	Horizontal co-ordinate
Y (m)	Vertical co-ordinate
Z (msl)	Altitude
Qn (l/s)	Nodal demand
H (msl)	Piezometric head of iteration "i"
p (mwc)	Nodal pressure
δQ (l/s)	Balance of the flow continuity equation
H (msl)	Piezometric head of iteration "i+1"

Qn total	Total demand of the system
T (°C)	Water temperature
υ (m²/s)	Kinematic viscosity

δH total	Total error between two iterations (δH = ABS(Hᵢ₊₁ - Hᵢ))

Total error between two iterations ($\delta H = ABS(H_{i+1} - H_i)$)

ω (-)	Successive over-relaxation factor (value range 1.0-2.0)

PIPES	Pipe data
N1	Pipe node nr.1 name
N2	Pipe node nr.2 name
Lxy(m)	Length calculated from the X/Y co-ordinates
L (m)	Length adopted for hydraulic calculation
D (mm)	Diameter
Q(l/s)	Flow rate of iteration "i"
v (m/s)	Flow velocity of iteration "i"
Re (-)	Reynolds number
λ (-)	Darcy-Weisbach friction factor
1/U	Linearisation coefficient
H1/U	Ratio used for calculation of "Hᵢ₊₁" (from N1)
H2/U	Ratio used for calculation of "Hᵢ₊₁" (from N2)
hf (mwc)	Friction loss
S (-)	Hydraulic gradient
v (m/s)	Flow velocity of iteration "i+1"
Q (l/s)	Flow rate of iteration "i+1"

k (mm)	Internal roughness (uniform)

Remarks:
The table with the nodal and pipe data is prepared in the same way as in Lesson 4-2.
The first node in the list of nodes simulates the source and has therefore fixed piezometric head.
The heads in other nodes are distributed arbitrarily in the 1ˢᵗ iteration, except that no nodes should be allocated the same value.
The pipe flows in the 1ˢᵗ iteration are also distributed arbitrarily (commonly to fit the velocities around 1 m/s).
The calculation starts by iterating piezometric heads in the nodes, in order to determine the pipe flows in the next iteration.
A consecutive iteration is done node by node, by typing the current "Hᵢ₊₁" value into "Hᵢ" cell.
Alternative approach, by copying the entire column ("Excel" command "Edit>>Paste Special (Values)"), is likely to yield slower convergence.
The new values of nodal piezometric heads should result in gradual reduction of the "δH total" value; the velocities (flows) have to be re-iterated.
That is done by copying the entire "Qᵢ₊₁" column into "Qᵢ" cells ("Excel" command "Edit>>Paste Special (Values)").
Messages **Iteration complete** appear once the total difference between the heads (flows) in two iterations drops below 0.1 mwc (l/s).

UNESCO IHE DELFT

Version 2.0
September 2019

Spreadsheet
Hydraulic Lessons in

Water Transport & Distribution
Part II

Contents:

Scroll-down

Disclaimer

This application has been developed with due care and attention, and is solely for educational and training purposes. In its default format, the worksheets include protection that can prevent unintentional deletion of cells with formulae. If this protection has been disabled, any inserting, deleting or cutting and pasting of the rows and columns can lead to damage or disappearance of the formulae, resulting in inaccurate calculations. Therefore, the author and IHE Delft are not responsible and assume no liability whatsoever for any results or any use made of the results obtained from this application in its original or modified format.

Introduction

The spreadsheet hydraulic lessons have been developed as an aid for steady state hydraulic calculations of simple water transport and distribution networks. These are to be carried out while solving the workshop problems that should normally be calculated manually; the spreadsheet serves here as a fast check of the results. Moreover, the spreadsheet lessons will help teachers to demonstrate a wider range of problems in a clear way, as well as to allow the students to continue analysing them at home. Ultimately, a real understanding of the hydraulic concepts will be reached through 'playing' with the data.

Over forty problems have been classified in eight groups/worksheets according to the contents of the book 'Introduction to Urban Water Distribution' by N. Trifunovic. This book covers a core curriculum of ' Water Transport and Distribution', which is a three-week module in the Water Supply Engineering specialisation at IHE Delft. Outside the regular MSc programme, this module is also offered as a stand-alone short course/online learning package.

To be able to use the spreadsheet, brief accompanying instructions for each exercise are given in the 'About' worksheet (see below). Each layout covers approximately one full screen (30 rows) consisting of drawings, tables and graphs. In the tables:

The green colour indicates input cells. Except for the headers, these cells are unprotected and their contents are used for calculations. The brown colour indicates output cells. These cells contain fixed formulae and are therefore protected.

Moreover, some intermediate calculations have been moved further to the right in the worksheet, being irrelevant for educational purposes. These are also protected by left unhidden for the sake of better understanding of the calculation process in each lesson.

Each lesson serves as a kind of chess problem, in which 'check-mate' should be reached within a few, correct moves. This suggests a study process where thinking takes more time than the execution, which was the main concept in the development of the exercises. Any simplifications that have been introduced (neglected minor losses, pump curve definition, etc.) were meant to facilitate this process. In addition, the worksheets have been designed without complicated routines or macros; only a superficial knowledge of spreadsheets is required to be able to use them effectively.

This is the second edition and any suggestions for improvement or extension are obviously welcome.

N. Trifunovic

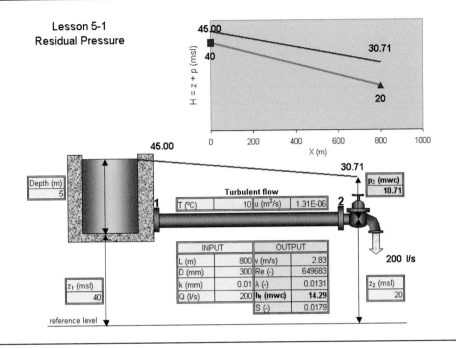

Lesson 5-1   Gravity Supply - Residual Pressure

Contents:
Calculation of the friction losses in a gravity-fed system (application of the Darcy-Weisbach formula).

Goal:
Pressure analysis for various demands, locations of the reservoir and consumers' points, and change in pipe parameters.

Abbreviations:

L (m)	Pipe length		v (m/s)	Flow velocity
D (mm)	Pipe diameter		Re (-)	Reynolds number
k (mm)	Internal roughness		λ (-)	Darcy-Weisbach friction factor
Q (l/s)	Flow rate		h_f (mwc)	Friction loss (metres of water column)
T (°C)	Water temperature (degrees Celsius)		S (-)	Hydraulic gradient
Depth (m)	Water depth in the reservoir		u (m²/s)	Kinematic viscosity
z_1 (msl)	Elevation of the reservoir bottom (metres above sea level)		p_2 (mwc)	Pressure at the discharge point
z_2 (msl)	Elevation of the discharge point			

Notes:
The calculation ultimately yields the pressure at the discharge point. The graph shows the actual pipe route with its hydraulic grade line.

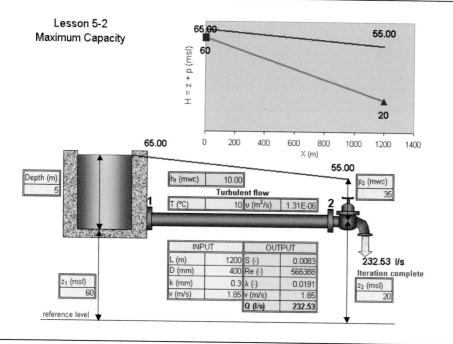

Lesson 5-2   Gravity Supply - Maximum Capacity

Contents:
Hydraulic calculation of a gravity-fed system.

Goal:
Determination of the maximum discharge at the required minimum pressure and various positions of the reservoir.

Abbreviations:

L (m)	Pipe length		S (-)	Hydraulic gradient
D (mm)	Pipe diameter		Re (-)	Reynolds number
k (mm)	Internal roughness		λ (-)	Darcy-Weisbach friction factor
Q (l/s)	Flow rate		v (m/s)	Calculated flow velocity
T (°C)	Water temperature		Q (l/s)	Flow rate (discharge)
Depth (m)	Water depth in the reservoir		υ (m²/s)	Kinematic viscosity
z_1 (msl)	Elevation of the reservoir bottom		h_f (mwc)	Friction loss
z_2 (msl)	Elevation of the discharge point			
v (m/s)	Assumed flow velocity			
p_2 (mwc)	Pressure required at the discharge point			

Notes:
The iterative procedure starts by assuming the flow velocity (commonly at 1 m/s) required for determination of the Reynolds number i.e. the friction factor.
The velocity calculated afterwards by the Darcy-Weisbach formula serves as an input for the next iteration.
The iterative process is achieved by typing the value of the calculated velocity into the cell of the assumed velocity.
The message **Iteration complete** appears once the difference between the velocities in two iterations drops below 0.01 m/s.

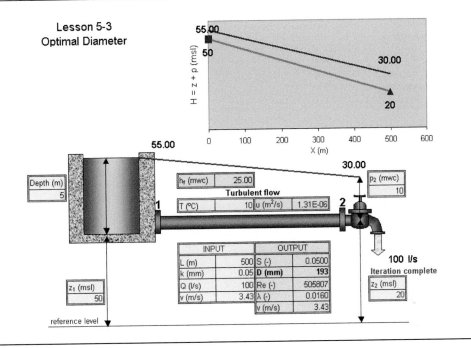

Lesson 5-3   Gravity Supply - Optimal Diameter

Contents:
Hydraulic calculation of a gravity-fed system.

Goal:
Determination of the optimal pipe diameter at the required minimum pressure and various positions of the reservoir.

Abbreviations:
The same as in Lesson 5-2.

Notes:
The same iterative procedure is used A311 as in Lesson 5-2, except that the pipe diameter is determined from the assumed/calculated velocity (and specified flow rate) A353
The message **Iteration complete** appears once the difference between the velocities in two iterations drops below 0.01 m/s.

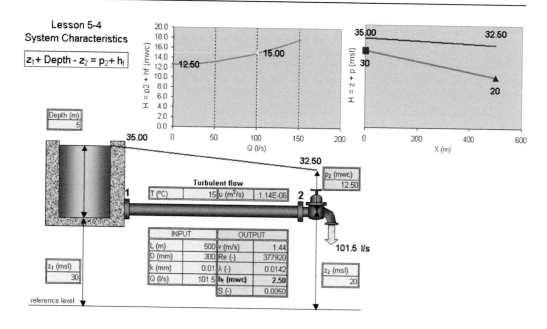

Lesson 5-4   Gravity Supply - System Characteristics

Contents:
Friction loss calculation of a gravity-fed system.

Goal:
Determination of the system characteristics diagram.

Abbreviations:

L (m)	Pipe length	v (m/s)	Flow velocity
D (mm)	Pipe diameter	Re (-)	Reynolds number
k (mm)	Internal roughness	λ (-)	Darcy-Weisbach friction factor
Q (l/s)	Flow rate	h_f (mwc)	Friction loss
T (°C)	Water temperature	S (-)	Hydraulic gradient
Depth (m)	Water depth in the reservoir	p_2 (mwc)	Pressure at the discharge point
z_1 (msl)	Elevation of the reservoir bottom	u (m²/s)	Kinematic viscosity
z_2 (msl)	Elevation of the discharge point		

Notes:
The friction loss is calculated for the flow range 0 to1.5Q (specified) and the results are plotted on the graph.
The point on the graph shows the upstream head required to maintain the downstream pressure for flow Q.
That head equals the elevation difference between the water surface in the reservoir and the discharge point.
The static head equals the downstream pressure, which fluctuates when the system parameters are modified.

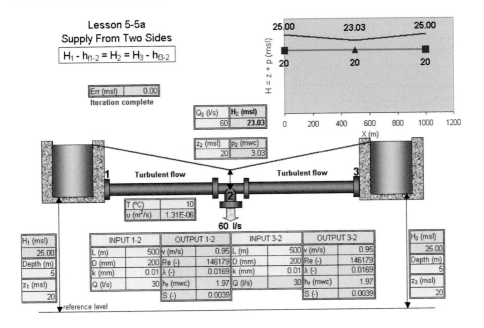

Lesson 5-5a Gravity Supply - Supply from Two Sides

Contents:
Friction loss calculation of a gravity system fed from two sides.

Goal:
To calculate the contribution from each source, based on various locations of the supply and discharge points, and changes in pipe parameters.

Abbreviations:

L (m)	Pipe length		v (m/s)	Flow velocity
D (mm)	Pipe diameter		Re (-)	Reynolds number
k (mm)	Internal roughness		λ (-)	Darcy-Weisbach friction factor
Q_2 (l/s)	Discharge		h_f (mwc)	Friction loss
Q (l/s)	Flow rate in pipe 1-2		S (-)	Hydraulic gradient
T (°C)	Water temperature		Q (l/s)	Flow rate in pipe 3-2
$z_{1,3}$ (msl)	Elevation of the reservoir bottoms		u (m²/s)	Kinematic viscosity
z_2 (msl)	Elevation of the discharge point		H_{1-3} (msl)	Piezometric heads
Depth (m)	Water depth in the reservoirs		Err (msl)	The difference in H_2 calculated from both sides

Notes:
The process of trial and error consists of altering the flow in pipe 1-2, which has implications for the head-losses in both pipes.
The process ends when the difference in H_2 calculated from the left and right side drops below 0.01 msl and the message **Iteration complete** appears.
Negative values for pipe flows indicate a change of direction i.e. the water that is flowing to the reservoir.

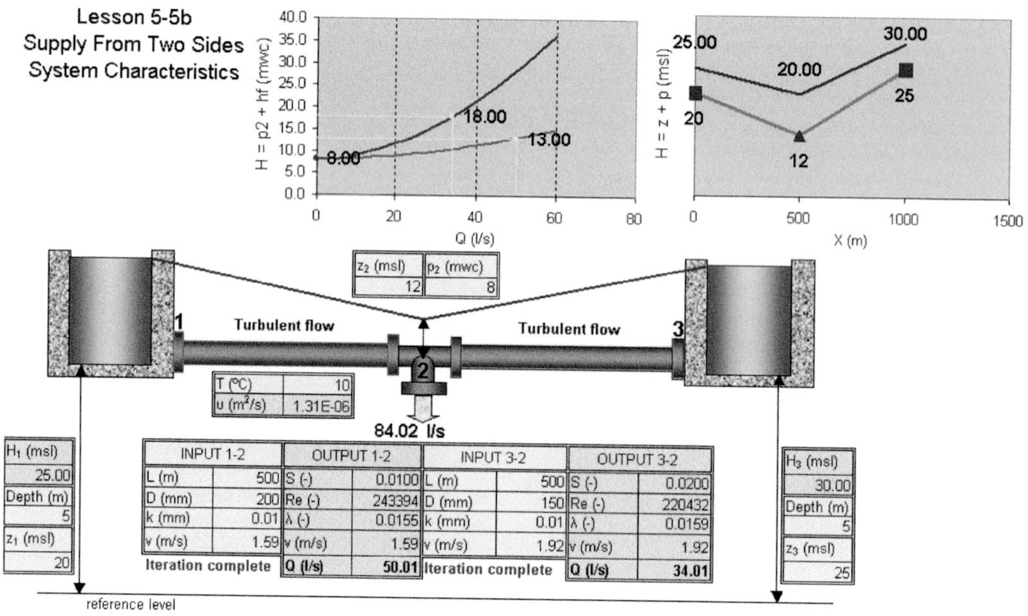

Lesson 5-5b
Supply From Two Sides
System Characteristics
H = p2 + hf (mwc)
Q (l/s)
18.00
13.00
8.00
H = z + p (msl)
X (m)
25.00
30.00
20.00
20
25
12
z2 (msl) 12
p2 (mwc) 8
Turbulent flow
Turbulent flow
1
2
3
T (°C) 10
υ (m²/s) 1.31E-06
84.02 l/s
H1 (msl) 25.00
Depth (m) 5
z1 (msl) 20
INPUT 1-2
L (m) 500
D (mm) 200
k (mm) 0.01
v (m/s) 1.59
OUTPUT 1-2
S (-) 0.0100
Re (-) 243394
λ (-) 0.0155
v (m/s) 1.59
Q (l/s) 50.01
Iteration complete
INPUT 3-2
L (m) 500
D (mm) 150
k (mm) 0.01
v (m/s) 1.92
OUTPUT 3-2
S (-) 0.0200
Re (-) 220432
λ (-) 0.0159
v (m/s) 1.92
Q (l/s) 34.01
Iteration complete
H3 (msl) 30.00
Depth (m) 5
z3 (msl) 25
reference level

Lesson 5-5b Gravity Supply - Supply from Two Sides, System Characteristics

Contents:

Friction loss calculation of a gravity system fed from two sides.

Goal:

To calculate the contribution from each source, based on various locations of the supply and discharge points, and changes in pipe parameters.

Abbreviations:

The same as in Lesson 5-5a.

Notes:

The contribution from each source to the discharge is analysed by constructing pipe characteristics for both pipes.
The same procedure as in Lesson 5-4 is applied here.

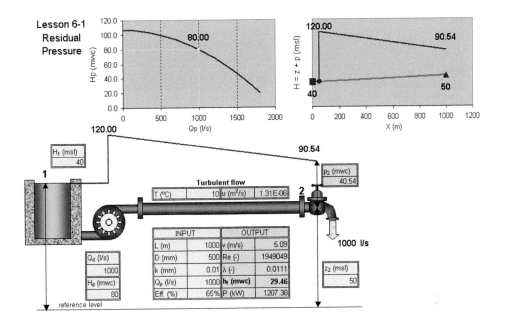

Lesson 6-1   Pumped Supply - Residual Pressure

Contents:
Calculation of the friction losses in a pumped system.

Goal:
Calculation of pumping heads and friction losses in a pumped system.

Abbreviations:

L (m)	Pipe length
D (mm)	Pipe diameter
k (mm)	Internal roughness
Q_p (l/s)	Pumped discharge
Eff. (%)	Pump efficiency at Qp
Q_d (l/s)	Pump duty flow
H_d (mwc)	Pump duty head
H_1 (msl)	Piezometric head at the suction side of the pump
z_2 (msl)	Elevation of the discharge point
T (°C)	Water temperature

v (m/s)	Flow velocity
Re (-)	Reynolds number
λ (-)	Darcy-Weisbach friction factor
h_f (mwc)	Friction loss
P (kW)	Pump power consumption at Qp
p_2 (mwc)	Pressure at the discharge point
υ (m²/s)	Kinematic viscosity

Notes:
The pump curve is approximated by the formula $H_p=c-aQ_p^b$, where $c=4H_d/3$, b=2 and $a=H_d/3/Q_d^2$.

The pump graph is plotted for the Q_p range between 0 and $1.8Q_d$. The duty head and duty flow indicate the working point of the maximum pump efficiency.

The pump raises head H_1 to head H_1+H_p from where the pipe friction loss is going to be deducted.

The calculation ultimately yields the pressure at the discharge point. The graph shows the actual pipe route with its hydraulic grade line.

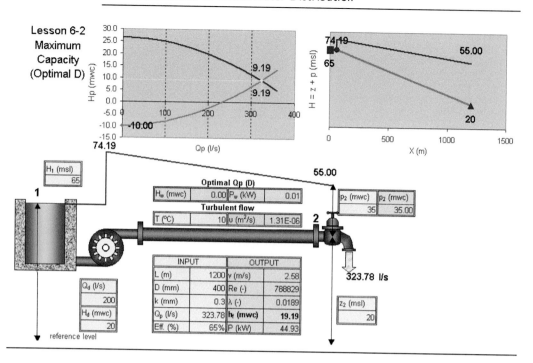

Lesson 6-2   Pumped Supply - Maximum Capacity/Optimal Diameter

Contents:
Calculation of pumping heads and friction losses in a pumped system.

Goal:
Relation between the pump and system characteristics.

Abbreviations:

L (m)	Pipe length
D (mm)	Pipe diameter
k (mm)	Internal roughness
Q_p (l/s)	Pumped discharge
Eff. (%)	Pump efficiency at Q_p
Q_d (l/s)	Pump duty flow
H_d (mwc)	Pump duty head
H_1 (msl)	Piezometric head at the suction side of the pump
z_2 (msl)	Elevation of the discharge point
p_2 (mwc)	Minimum pressure required at the discharge point
T (°C)	Water temperature

v (m/s)	Flow velocity
Re (-)	Reynolds number
λ (-)	Darcy-Weisbach friction factor
h_f (mwc)	Friction loss
P (kW)	Pump power consumption at Q_p
p_2 (mwc)	Actual pressure at the discharge point
H_w (mwc)	Excessive pumping head
P_w (kW)	Excessive pumping power
u (m²/s)	Kinematic viscosity

Notes:
The pump curve is plotted in the same way as in Lesson 6-1. The system characteristics curve is plotted for the same range of flows.
The diagram shows the difference between the pumping head and the head required to deliver the minimum pressure at the discharge point.
The optimal working point is in the intersection between the pump- and system characteristics curves (H_w=0).

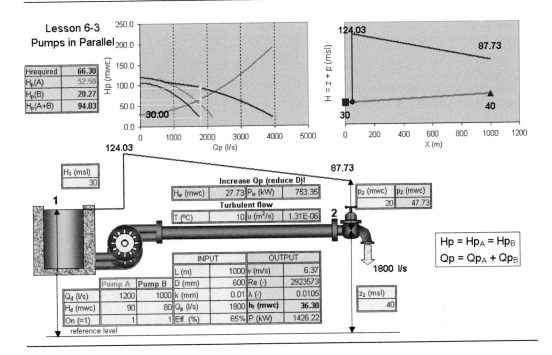

Lesson 6-3   Pumped Supply - Pumps in Parallel

Contents:
Calculation of pumping heads and friction losses for parallel arrangement of the pumps.

Goal:
Relation between the pump- and system characteristics for various pump sizes and operational modes.

Abbreviations:

L (m)	Pipe length	v (m/s)	Flow velocity
D (mm)	Pipe diameter	Re (-)	Reynolds number
k (mm)	Internal roughness	λ (-)	Darcy-Weisbach friction factor
Q_p (l/s)	Pumped discharge	h_f (mwc)	Friction loss
Eff. (%)	Efficiency at Q_p	P (kW)	Pump power consumption at Q_p
Q_d (l/s)	Pump duty flow (A,B)	p_2 (mwc)	Actual pressure at the discharge point
H_d (mwc)	Pump duty head (A,B)	H_w (mwc)	Excessive pumping head
On (=1)	The pump is 'off' unless the cell input =1	P_w (kW)	Excessive pumping power
H_1 (msl)	Piezometric head at the suction side of the pump	u (m²/s)	Kinematic viscosity
z_2 (msl)	Elevation of the discharge point	Hrequired	Minimum required head at Q_p
p_2 (mwc)	Minimum pressure required at the discharge point	H_p(A)	Pumping head A at Q_p
T (°C)	Water temperature	H_p(B)	Pumping head B at Q_p
		H_p(A+B)	Head of pumps A & B both in operation at Q_p

Notes:
The pump and system characteristics curves are plotted in the same way as in Lesson 6-2.
The diagram shows the difference between the pumping head and the head required to deliver the minimum pressure at the discharge point in all three cases: single operation of pumps A & B and their joint operation.
The hydraulic grade line on the graph is plotted for actual operation of the pumps.

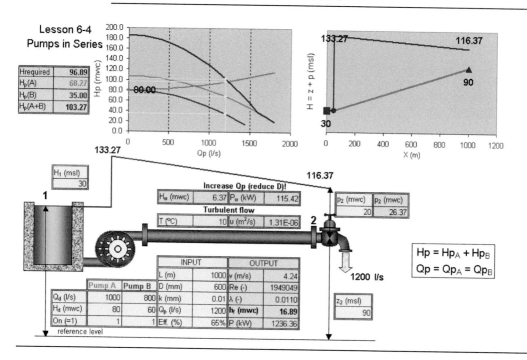

Lesson 6-4   Pumped Supply - Pumps in Series

Contents:
Calculation of pumping heads and friction losses for serial arrangement of pumps.

Goal:
Relation between the pump- and system characteristics curves for various pump sizes and operational modes.

Abbreviations:
The same as in Lesson 6-3.

Notes:
The same as in Lesson 6-3.

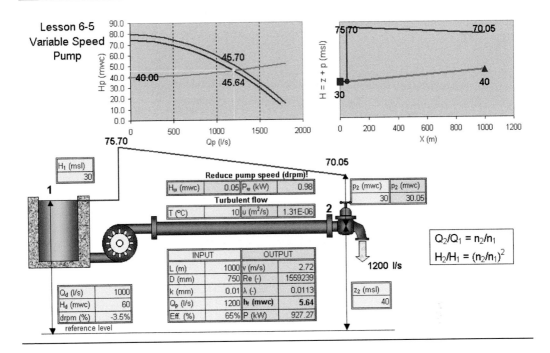

Lesson 6-5   Pumped Supply - Variable Speed Pump

Contents:
Calculation of pumping heads and friction losses in a system with variable speed pump.

Goal:
Analysis of effects caused by modification of the pump speed.

Abbreviations:

L (m)	Pipe length
D (mm)	Pipe diameter
k (mm)	Internal roughness
Q_p (l/s)	Pumped discharge
Eff. (%)	Pump efficiency at Q_p
Q_d (l/s)	Pump duty flow
H_d (mwc)	Pump duty head
drpm (%)	Pump speed increase/decrease
H_1 (msl)	Piezometric head at the suction side of the pump
z_2 (msl)	Elevation of the discharge point
p_2 (mwc)	Minimum pressure required at the discharge point
T (°C)	Water temperature

v (m/s)	Flow velocity
Re (-)	Reynolds number
λ (-)	Darcy-Weisbach friction factor
h_f (mwc)	Friction loss
P (kW)	Pump power consumption at Q_p
p_2 (mwc)	Actual pressure at the discharge point
H_w (mwc)	Excessive pumping head
P_w (kW)	Excessive pumping power
υ (m²/s)	Kinematic viscosity

Notes:
The diagram shows the difference in pumping heads between the pump curve at the original and modified speeds.

**Lesson 6-6
Pump Station
Layout**

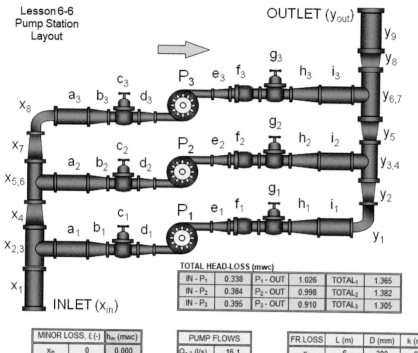

TOTAL HEAD-LOSS (mwc)					
IN - P$_1$	0.338	P$_1$ - OUT	1.026	TOTAL$_1$	1.365
IN - P$_2$	0.384	P$_2$ - OUT	0.998	TOTAL$_2$	1.382
IN - P$_3$	0.395	P$_3$ - OUT	0.910	TOTAL$_3$	1.305

MINOR LOSS, ξ (-)		h$_m$ (mwc)
x$_{in}$	0	0.000
x$_3$	0.89	0.038
x$_4$	0.02	0.000
x$_6$	0.92	0.039
x$_7$	0.1	0.004
x$_8$	0.15	0.006
b$_1$	0.02	0.002
c$_1$	0.3	0.026
d$_1$	0.2	0.240
e$_1$	0.1	0.052
f$_1$	0.8	0.418
g$_1$	0.35	0.183
h$_1$	0.03	0.006
b$_2$	0.02	0.002
c$_2$	0.3	0.026
d$_2$	0.2	0.240
e$_2$	0.1	0.052
f$_2$	0.8	0.418
g$_2$	0.35	0.183
h$_2$	0.03	0.006
b$_3$	0.02	0.002
c$_3$	0.3	0.026
d$_3$	0.2	0.240
e$_3$	0.1	0.052
f$_3$	0.8	0.418
g$_3$	0.35	0.183
h$_3$	0.03	0.006
y$_1$	0.12	0.026
y$_2$	0.03	0.003
y$_4$	0.28	0.025
y$_5$	0.07	0.006
y$_7$	0.02	0.002
y$_8$	0.01	0.001
y$_{out}$	0	0.000

PUMP FLOWS	
Q$_{p,3}$ (l/s)	16.1
Q$_{p,2}$ (l/s)	16.1
Q$_{p,1}$ (l/s)	16.1

T (°C)	10
υ (m²/s)	1.31E-06

Recommended velocities:
Feeder main (x): 0.6-0.8 m/s
Suction pipe (a): 0.8-1.2 m/s
Pressure pipe (i): 1.5-2.0 m/s
Disch. Header (y): 1.2-1.7 m/s

FR.LOSS	L (m)	D (mm)	k (mm)	v (m/s)	h$_f$ (mwc)
x$_1$	6	300	0.1	0.68	0.009
x$_2$	4	300	0.1	0.68	0.006
x$_5$	4	250	0.1	0.66	0.007
a$_1$	3	150	0.1	0.91	0.018
i$_1$	3	100	0.1	2.05	0.138
a$_2$	3	150	0.1	0.91	0.018
i$_2$	3	100	0.1	2.05	0.138
a$_3$	3	150	0.1	0.91	0.018
i$_3$	3	100	0.1	2.05	0.138
y$_3$	4	125	0.1	1.31	0.059
y$_6$	4	175	0.1	1.34	0.041
y$_9$	6	200	0.1	1.54	0.068

DIAMETERS REDUCERS - SUCTION SIDE (mm)				v$_{b-ds}$ (m/s)	v$_{d-ds}$ (m/s)
150	b$_1$	125	d$_1$	65	1.31	4.85
150	b$_2$	125	d$_2$	65	1.31	4.85
150	b$_3$	125	d$_3$	65	1.31	4.85

DIAMETERS ENLARGERS - PRESSURE SIDE (mm)				v$_{e-ds}$ (m/s)	v$_{h-ds}$ (m/s)
50	e$_1$	80	h$_1$	100	3.20	2.05
50	e$_2$	80	h$_2$	100	3.20	2.05
50	e$_3$	80	h$_3$	100	3.20	2.05

Lesson 6-6   Pumped Supply - Pump Station Layout

Contents:
Calculation of friction and minor losses for the given layout of three pump units in parallel arrangement.

Goal:
Analysis of effects on head-losses caused by modification of the pump flow.

Abbreviations:

L (m)	Pipe length	x_{in}	The inlet	$b_{1,2,3}$	Reducer	v (m/s)	Flow velocity in the pipe	
D (mm)	Pipe diameter	x_3	T-branch	$c_{1,2,3}$	Valve	$v_{b,d\text{-}ds}$ (m/s)	Flow velocity downstream of reducer b,d	
k (mm)	Internal roughness	x_4	Reducer	$d_{1,2,3}$	Reducer	$v_{e,h\text{-}ds}$ (m/s)	Flow velocity downstream of enalrger e,h	
$Q_{p,1,2,3}$ (l/s)	Pumped discharge (per pump)	x_6	T-branch	$e_{1,2,3}$	Enlarger	h_f (mwc)	Friction loss	
x_1	Feeder main	x_7	Reducer	$f_{1,2,3}$	Non-return valve	h_m (mwc)	Minor loss	
x_2	Feeder main	x_8	Pipe bend	$g_{1,2,3}$	Valve	IN - $P_{1,2,3}$	Total loss from the inlet to pump $P_{1,2,3}$	
x_5	Feeder main	y_1	Pipe bend	$h_{1,2,3}$	Enlarger	$P_{1,2,3}$ - OUT	Total loss from $P_{1,2,3}$ to the outlet	
$a_{1,2,3}$	Suction pipe (per pump)	y_2	Enlarger			$TOTAL_{1,2,3}$	Total loss from the inlet to the outlet	
$l_{1,2,3}$	Pressure pipe (per pump)	y_4	T-branch			υ (m²/s)	Kinematic viscosity	
y_3	Discharge header	y_5	Enlarger					
y_6	Discharge header	y_7	T-branch					
y_9	Discharge header	y_8	Enlarger					
T (°C)	Water temperature	y_{out}	The outlet					

Notes:
The selection of minor loss coefficients can be done from the references given in Appendix 5 of the book 'Introduction to Urban Water Distribution'.
Minor variation in the total head-loss values calculated per pump route mostly result from inconsistent selection of the minor loss coefficients.

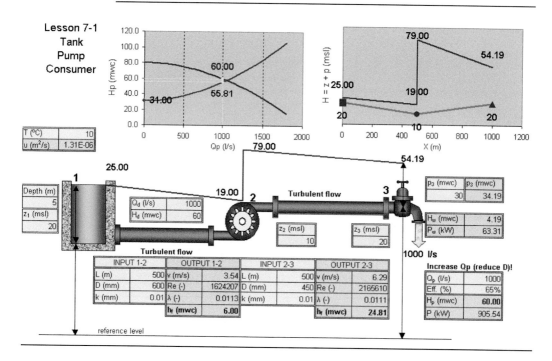

Lesson 7-1　Combined Supply - Tank, Pump, Consumer

Contents:
Calculation of pumping heads and friction losses in a system combining the gravity and pumped supply.

Goal:
Determination of residual pressure/maximum capacity/optimal diameters for various scenarios.

Abbreviations:
The same as in Lessons 5 & 6.

Remarks:
The system characteristics in the diagram is plotted for the pipe on the pressure side of the pump.

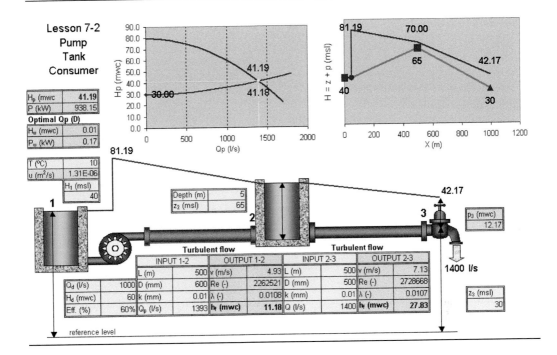

Lesson 7-2 Combined Supply - Pump, Tank, Consumer

Contents:
Calculation of pumping heads and friction losses in a system combining the gravity and pumped supply.

Goal:
Determination of residual pressure/maximum capacity/optimal diameters for various scenarios.

Abbreviations:
The same as in Lessons 5 & 6.

Remarks:
The system characteristics in the diagram is plotted for the pipe on the upstream side of the reservoir.
The static head of that pipe is equal to the difference between the water levels in the two reservoirs.
The system is hydraulically disconnected and the maximum discharge capacity is dependant from the gravity part only.

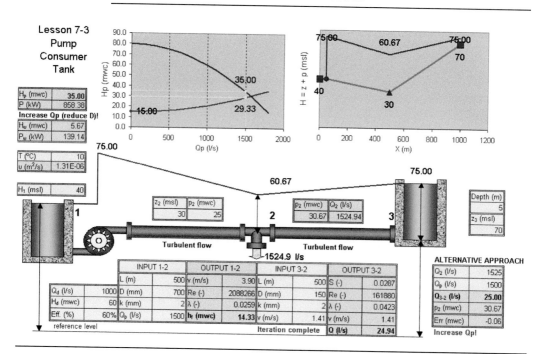

Lesson 7-3　Combined Supply - Pump, Consumer, Tank

Contents:
Calculation of pumping heads and friction losses in a system combining gravity and pumped supply.

Goal:
Determination of residual pressure/maximum capacity/optimal diameters for various scenarios.

Abbreviations:
The same as in Lessons 5 & 6.

Notes:
For a specified pumping flow, the contribution from the reservoir will be calculated through an iteration of velocities in pipe 3-2.

A negative value of the hydraulic gradient (S) in pipe 3-2 indicates a reversed flow direction i.e. filling of the reservoir (night regime).

By the alternative approach (see the table 'ALTERNATIVE APPROACH'), the contribution from both supplying points will be determined for specified demand. This is reached through a process of trial and error, with 'Err' indicating the difference in head H_2 calculated from both sides.

The pump is switched off by leaving the cells for H_d or/and Q_d empty.

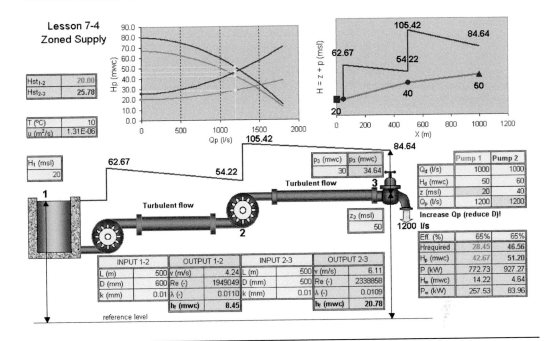

Lesson 7-4   Combined Supply - Zoned Supply

Contents:
Calculation of pumping heads and friction losses in a system combining the gravity and pumped supply.

Goal:
Determination of residual pressure/maximum capacity/optimal diameters for booster pumping.

Abbreviations:
The same as in Lessons 5 & 6. In addition, "Hst$_{1-2}$" and "Hst$_{3-2}$" indicate the static heads of pipes 1-2 & 3-2, respectively.

Remarks:
The diagram shows the pump characteristics of each pump with its corresponding pipe characteristics.
The optimal working point for specified discharge and minimum downstream pressure is in the intersection of the red coloured curves.
The green coloured curves are monitored for avoiding under-pressure in the part of the system between the two pumps.
The pumps are switched off by leaving cells for H$_d$ or/and Q$_d$ empty.

Lesson 8-1
Flow Conversion

VOLUME	m³	litre	ft³	gal	galUS
	1	2	3	4	5
m³	1	1000	35.315	219.97	264.19
litre	0.0010	1	0.0353	0.2200	0.2642
ft³	0.0283	28.317	1	6.2288	7.4811
gal	0.0045	4.5461	0.1605	1	1.2011
galUS	0.0038	3.7851	0.1337	0.8326	1

Flow Rate	Units
100	l/s

converted into:

Units	Flow Rate
m3/h	360

TIME	sec	min	hour	day
	1	2	3	4
sec	1	0.0167	2.8E-04	1.2E-05
min	60	1	0.0167	6.9E-04
hour	3600	60	1	0.0417
day	86400	1440	24	1

FLOW	m³/s	l/s	m³/h	Ml/d	ft³/s	gpm	mgd	gpmUS	mgdUS
	1	2	3	4	5	6	7	8	9
m³/s	1	1000	3600	86.4	35.315	13198	19.005	15852	22.826
l/s	0.001	1	3.6	0.0864	0.0353	13.198	0.019	15.852	0.0228
m³/h	2.8E-04	0.2778	1	0.0240	0.0098	3.6661	0.005	4.4032	0.0063
Ml/d	0.0116	11.574	41.667	1	0.4087	152.76	0.220	183.47	0.2642
ft³/s	0.0283	28.317	101.94	2.4466	1	373.73	0.538	448.87	0.6464
gpm	7.6E-05	0.0758	0.2728	0.0065	0.0027	1	0.001	1.2011	0.0017
mgd	0.0526	52.617	189.42	4.5461	1.8581	694.44	1	834.06	1.2011
gpmUS	6.3E-05	0.0631	0.2271	0.0055	0.0022	0.8326	0.0012	1	0.0014
mgdUS	0.0438	43.809	157.71	3.7851	1.5471	578.20	0.8326	694.44	1

Hour	Demand	
	l/s	m3/h
1	989.6	3562.56
2	945.9	3405.24
3	902.2	3247.92
4	727.6	2619.36
5	844	3038.4
6	1164.2	4191.12
7	1571.7	5658.12
8	1600.8	5762.88
9	1775.4	6391.44
10	1964.6	7072.56
11	2066.4	7439.04
12	2110.1	7596.36
13	1600.8	5762.88
14	1309.7	4714.92
15	1091.4	3929.04
16	945.9	3405.24
17	1062.3	3824.28
18	1455.2	5238.72
19	1746.3	6286.68
20	2139.2	7701.12
21	2110.1	7596.36
22	2037.3	7334.28
23	1746.3	6286.68
24	1018.7	3667.32
Total	34925.7	125733
Average	1455.24	5238.86

Lesson 8-1 Flow Conversion

Contents:
Comparison of the flow units.

Goal:
Conversion of the common flow units.

Abbreviations:

m³/s	cubic metre per second
l/s	litre per second
m3/h	cubic metre per hour
Ml/d	mega-litre per day
ft³/s	cubic feet per second
gpm	gallon per minute (Imperial)
mgd	mega-gallon per day (Imperial)
gpmUS	gallon per minute (US)
mgdUS	mega-gallon per day (US)

Notes:
Both individual values and the series of 24 hourly flows are converted from/to the units specified in the table (by typing).

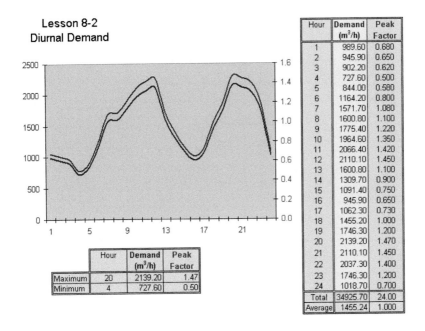

Lesson 8-2
Diurnal Demand

	Hour	Demand (m³/h)	Peak Factor
Maximum	20	2139.20	1.47
Minimum	4	727.60	0.50

Hour	Demand (m³/h)	Peak Factor
1	989.60	0.680
2	945.90	0.650
3	902.20	0.620
4	727.60	0.500
5	844.00	0.580
6	1164.20	0.800
7	1571.70	1.080
8	1600.80	1.100
9	1775.40	1.220
10	1964.60	1.350
11	2066.40	1.420
12	2110.10	1.450
13	1600.80	1.100
14	1309.70	0.900
15	1091.40	0.750
16	945.90	0.650
17	1062.30	0.730
18	1455.20	1.000
19	1746.30	1.200
20	2139.20	1.470
21	2110.10	1.450
22	2037.30	1.400
23	1746.30	1.200
24	1018.70	0.700
Total	34925.70	24.00
Average	1455.24	1.000

Lesson 8-2   Diurnal Demand

Contents:
Diurnal demand diagram.

Goal:
Determination of hourly peak factors.

Notes:
The peak factors are calculated as ratios between the actual and average hourly demand.

Lesson 8-3
Demand Categories

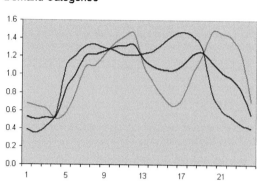

Hour	Cat.1 (m³/h)	Peak Fact.1	Cat.2 (m³/h)	Peak Fact.2	Demand (m³/h)	Cum. Peak F.
1	989.60	0.680	579.00	0.389	1568.60	0.533
2	945.90	0.650	523.00	0.351	1468.90	0.499
3	902.20	0.620	644.00	0.432	1546.20	0.525
4	727.60	0.500	835.00	0.561	1562.60	0.531
5	844.00	0.580	1650.00	1.108	2494.00	0.847
6	1164.20	0.800	1812.00	1.217	2976.20	1.011
7	1571.70	1.080	1960.00	1.316	3531.70	1.200
8	1600.80	1.100	1992.00	1.338	3592.80	1.220
9	1775.40	1.220	1936.00	1.300	3711.40	1.261
10	1964.60	1.350	1887.00	1.267	3851.60	1.308
11	2066.40	1.420	1821.00	1.223	3887.40	1.320
12	2110.10	1.450	1811.00	1.216	3921.10	1.332
13	1600.80	1.100	1837.00	1.234	3437.80	1.168
14	1309.70	0.900	1884.00	1.265	3193.70	1.085
15	1091.40	0.750	2011.00	1.351	3102.40	1.054
16	945.90	0.650	2144.00	1.440	3089.90	1.049
17	1062.30	0.730	2187.00	1.469	3249.30	1.104
18	1455.20	1.000	2132.00	1.432	3587.20	1.218
19	1746.30	1.200	1932.00	1.297	3678.30	1.249
20	2139.20	1.470	1218.00	0.818	3357.20	1.140
21	2110.10	1.450	898.00	0.603	3008.10	1.022
22	2037.30	1.400	786.00	0.528	2823.30	0.959
23	1746.30	1.200	657.00	0.441	2403.30	0.816
24	1018.70	0.700	601.00	0.404	1619.70	0.550
Total	34925.70	24.00	35737.00	24.00	70662.70	24.00
Average	1455.24	1.000	1489.04	1.000	2944.28	1.000

Peak Factors 1		Peak Factors 2		Cum. Peak Factors	
Maximum	1.470	Maximum	1.469	Maximum	1.332
at hour	20	at hour	17	at hour	12
Minimum	0.500	Minimum	0.351	Minimum	0.499
at hour	4	at hour	2	at hour	2

Lesson 8-3   Demand Categories

Contents:
Diurnal demand diagram.

Goal:
Determination of hourly peak factors for various demand categories and the overall demand pattern.

Notes:
The peak factors of the cumulative demand diagram have no direct relation to the peak factors of the categories composing this diagram.

Lesson 8-4
Seasonal Variations

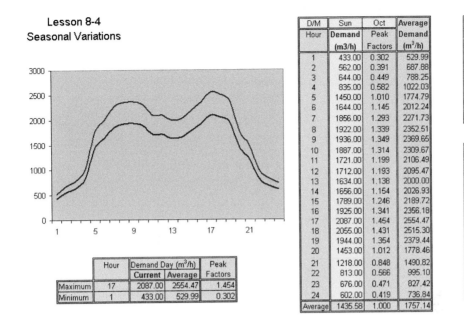

D/M	Sun	Oct	Average
Hour	Demand	Peak	Demand
	(m3/h)	Factors	(m³/h)
1	433.00	0.302	529.99
2	562.00	0.391	687.88
3	644.00	0.449	788.25
4	835.00	0.582	1022.03
5	1450.00	1.010	1774.79
6	1644.00	1.145	2012.24
7	1856.00	1.293	2271.73
8	1922.00	1.339	2352.51
9	1936.00	1.349	2369.65
10	1887.00	1.314	2309.67
11	1721.00	1.199	2106.49
12	1712.00	1.193	2095.47
13	1634.00	1.138	2000.00
14	1656.00	1.154	2026.93
15	1789.00	1.246	2189.72
16	1925.00	1.341	2356.18
17	2087.00	1.454	2554.47
18	2055.00	1.431	2515.30
19	1944.00	1.354	2379.44
20	1453.00	1.012	1778.46
21	1218.00	0.848	1490.82
22	813.00	0.566	995.10
23	676.00	0.471	827.42
24	602.00	0.419	736.84
Average	1435.58	1.000	1757.14

Week	Peak
	Factors
Mon	1.140
Tue	1.010
Wed	0.950
Thu	0.980
Fri	1.040
Sat	1.020
Sun	0.860
Average	1.000

Year	Peak
	Factors
Jan	0.960
Feb	0.970
Mar	1.010
Apr	1.030
May	1.040
Jun	1.050
Jul	1.040
Aug	1.030
Sep	0.990
Oct	0.950
Nov	0.960
Dec	0.970
Average	1.000

	Hour	Demand Day (m³/h)		Peak
		Current	Average	Factors
Maximum	17	2087.00	2554.47	1.454
Minimum	1	433.00	529.99	0.302

Lesson 8-4   Seasonal Variations

Contents:
Diurnal demand diagram.

Goal:
Comparison of the diurnal demand patterns for various periods of the week/year.

Abbreviations:

D/M	Day/Month

Notes:
Recalculation of the hourly demand and corresponding peak factors takes place by typing the day and month from the tables for weekly and monthly peak factors.
The last peak factor value specified in the weekly and annual table is calculated so that the value can fill up to 7 and 12 respectively,
which generates the average equal to 1.0.

Lesson 8-5
Demand Calculation

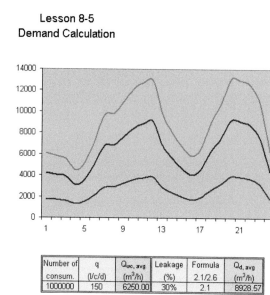

Hour	Peak Factors	Water Quantities (m³/h)		
		Q_{wc}	Q_{wl}	Q_d
1	0.680	4250.00	1821.43	6071.43
2	0.650	4062.50	1741.07	5803.57
3	0.620	3875.00	1660.71	5535.71
4	0.500	3125.00	1339.29	4464.29
5	0.580	3625.00	1553.57	5178.57
6	0.800	5000.00	2142.86	7142.86
7	1.080	6750.00	2892.86	9642.86
8	1.100	6875.00	2946.43	9821.43
9	1.220	7625.00	3267.86	10892.86
10	1.350	8437.50	3616.07	12053.57
11	1.420	8875.00	3803.57	12678.57
12	1.450	9062.50	3883.93	12946.43
13	1.100	6875.00	2946.43	9821.43
14	0.900	5625.00	2410.71	8035.71
15	0.750	4687.50	2008.93	6696.43
16	0.650	4062.50	1741.07	5803.57
17	0.730	4562.50	1955.36	6517.86
18	1.000	6250.00	2678.57	8928.57
19	1.200	7500.00	3214.29	10714.29
20	1.470	9187.50	3937.50	13125.00
21	1.450	9062.50	3883.93	12946.43
22	1.400	8750.00	3750.00	12500.00
23	1.200	7500.00	3214.29	10714.29
24	0.700	4375.00	1875.00	6250.00
Average	1.000	6250.00	2678.57	8928.57

Number of consum.	q (l/c/d)	$Q_{wc, avg}$ (m³/h)	Leakage (%)	Formula 2.1/2.6	$Q_{d, avg}$ (m³/h)
1000000	150	6250.00	30%	2.1	8928.57

Lesson 8-5   Demand Calculation

Contents:
Diurnal demand diagram.

Goal:
Calculation of the leakage component for specified diurnal pattern and leakage percentage.

Abbreviations:

Number of consum.	Number of consumers in the area

q (l/c/d)	Specific consumption (unit per capita consumption)

$Q_{wc, avg}$ (m³/h)	Average consumption

Leakage (%)	Leakage expressed as a percentage of water production

Formula 2.1/2.6	Formula used for the demand calculation, as discussed in Chapter 2

$Q_{d, avg}$ (m³/h)	Average demand

Q_{wc}	Q_{wl}	Q_d	Hourly water consumption, water leakage and water demand, respectively

Notes:
The demand is calculated in two ways: a) with the leakage proportional to the peak factor values (Formula 2.1 in Chapter 2 of the book), and b) with constant leakage, independent from the peak factors (Formula 2.6). Both approaches yield the same average demand but the range of the hourly flows will be wider in the first approach. This adds a safety factor to the result although it reflects the reality to a lesser extent. The last hourly peak factor value specified in the table is calculated so that the value can fill up to 24, which generates the average equal to 1.0.

Lesson 8-6
Design Demand

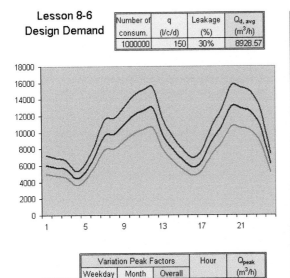

Number of consum.	q (l/c/d)	Leakage (%)	$Q_{d, avg}$ (m³/h)
1000000	150	30%	8928.57

Hour	Peak Factors	Demand Day (m³/h)		
		Avg.	Max.	Min.
1	0.680	6071.43	7267.50	4960.36
2	0.650	5803.57	6946.88	4741.52
3	0.620	5535.71	6626.25	4522.68
4	0.500	4464.29	5343.75	3647.32
5	0.580	5178.57	6198.75	4230.89
6	0.800	7142.86	8550.00	5835.71
7	1.080	9642.86	11542.50	7878.21
8	1.100	9821.43	11756.25	8024.11
9	1.220	10892.86	13038.75	8899.46
10	1.350	12053.57	14428.13	9847.77
11	1.420	12678.57	15176.25	10358.39
12	1.450	12946.43	15496.88	10577.23
13	1.100	9821.43	11756.25	8024.11
14	0.900	8035.71	9618.75	6565.18
15	0.750	6696.43	8015.63	5470.98
16	0.650	5803.57	6946.88	4741.52
17	0.730	6517.86	7801.88	5325.09
18	1.000	8928.57	10687.50	7294.64
19	1.200	10714.29	12825.00	8753.57
20	1.470	13125.00	15710.63	10723.13
21	1.450	12946.43	15496.88	10577.23
22	1.400	12500.00	14962.50	10212.50
23	1.200	10714.29	12825.00	8753.57
24	0.700	6250.00	7481.25	5106.25
Average	1.000	8928.57	10687.50	7294.64

	Variation Peak Factors			Hour	Q_{peak} (m³/h)
	Weekday	Month	Overall		
Maximum	1.140	1.050	1.197	20	15710.63
Minimum	0.860	0.950	0.817	4	3647.32

Lesson 8-6   Design Demand

Contents:
Diurnal demand diagram.

Goal:
Establishing the range of demands/peak factors that can occur during the year.

Abbreviations:

Q_{peak} (m³/h)	Absolute maximum and minimum peak demand

Notes:
Weekly and monthly factors compose the average consumption during the maximum/minimum consumption day. Maximum/minimum hourly peak factors in combination with the seasonal factors give the range of demands in the system between the maximum consumption hour of the maximum consumption day and the minimum consumption hour of the minimum consumption day.
The leakage component is calculated applying the first approach from the previous lesson.
The last hourly peak factor value specified in the table is calculated so that the value can fill up to 24, which generates the average equal to 1.0.

Lesson 8-7
Combined Demand

CATEGORIES	q (m³/d/ha)
A Residential area, appartments	90
B Individual houses	55
C Shopping area	125
D Offices	80
E Schools, Universities	100
F Hospitals	160
G Hotels	150
H Public green areas	15

	Q (m³/h)	Inhab.	Q(l/c/d)
District 1	666.77	86251	186
District 2	651.97	74261	211
District 3	216.76	18542	281
District 4	288.90	42149	165
District 5	161.05	22156	174
District 6	99.74	9958	240
District 7	58.03	8517	164
District 8	67.95	12560	130
TOTAL	2211.16	274394	193

CONSUMPTION CATEGORIES

		A	B	C	D	E	F	G	H
District 1	p (%)	37%	23%	10%	0%	4%	0%	0%	26%
Area 1 (ha)	c (%)	100%	100%	100%	0%	100%	0%	0%	40%
250	Ac(ha)	92.5	57.5	25	0	10	0	0	26
District 2	p (%)	20%	5%	28%	11%	12%	0%	5%	19%
Area 2 (ha)	c (%)	100%	100%	95%	100%	100%	0%	100%	80%
185	Ac(ha)	37	9.25	49.21	20.35	22.2	0	9.25	28.12
District 3	p (%)	10%	18%	3%	0%	0%	42%	0%	27%
Area 3 (ha)	c (%)	100%	100%	100%	0%	0%	100%	0%	35%
57	Ac(ha)	5.7	10.26	1.71	0	0	23.94	0	5.3865
District 4	p (%)	25%	28%	20%	2%	15%	0%	0%	10%
Area 4 (ha)	c (%)	100%	100%	95%	100%	100%	0%	0%	36%
88	Ac(ha)	22	24.64	16.72	1.76	13.2	0	0	3.168
District 5	p (%)	50%	0%	11%	0%	10%	0%	0%	29%
Area 5 (ha)	c (%)	100%	0%	100%	0%	100%	0%	0%	65%
54	Ac(ha)	27	0	5.94	0	5.4	0	0	10.179
District 6	p (%)	24%	11%	13%	15%	13%	8%	0%	16%
Area 6 (ha)	c (%)	100%	100%	100%	100%	100%	100%	0%	35%
29	Ac(ha)	6.96	3.19	3.77	4.35	3.77	2.32	0	1.624
District 7	p (%)	22%	28%	8%	19%	6%	0%	10%	7%
Area 7 (ha)	c (%)	100%	100%	100%	100%	100%	0%	100%	50%
17	Ac(ha)	3.74	4.76	1.36	3.23	1.02	0	1.7	0.595
District 8	p (%)	0%	0%	0%	0%	55%	20%	15%	10%
Area 8 (ha)	c (%)	0%	0%	0%	0%	85%	100%	100%	45%
16	Ac(ha)	0	0	0	0	7.48	3.2	2.4	0.72

Lesson 8-7   Combined Demand

Contents:
Calculation of the demand from the mixture of consumption categories in one district.

Goal:
To find out the total demand in the area based on the contribution from each consumption category and the network coverage in the districts.

Abbreviations:

District 1	Name of the district
Area 1 (ha)	Total area of the district (in hectars)
p (%)	Percentage of the district area (surface) covered by a particular consumption category
c (%)	Percentage of the district area (surface) covered/supplied by the distribution network
Ac(ha)	Area of the district covered by the distribution network (in hectares)
Q (m³/h)	Total demand in the district
Inhab.	Number of inhabitants in the district
Q(l/c/d)	Specific demand in the district

Notes:
An arbitrary list of categories and unit consumptions has been used.
A name can be used for each district; it copies itself into the table with the calculations.
Leakage can either be assumed as being included in the figure per category, or should be specified as a separate category.
The consumption figures indicated per category are assumed to be constant in all districts.
The contribution of the last category in each district is calculated so that the value can fill up to 100%, which represents the total district area.

Lesson 8-8
Demand Growth

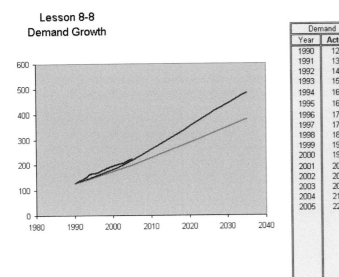

Demand	
Year	Actual
1990	126.00
1991	135.00
1992	142.00
1993	151.00
1994	163.00
1995	166.00
1996	170.00
1997	177.00
1998	185.00
1999	192.00
2000	196.00
2001	201.00
2002	206.00
2003	208.00
2004	218.00
2005	225.00

Up to	Growth	Demand Forecast	
year	(%)	Linear	Expon.
1990	-	126.00	126.00
2005	3.80%	197.82	220.46
2025	3.00%	316.51	398.18
2035	2.00%	379.81	485.37

Lesson 8-8   Demand Growth

Contents:
Demand forecast from the given historical data.

Goal:
Fitting of the actual growth pattern with the linear and exponential model.

Notes:
The first value and the corresponding year of the actual pattern are taken as the starting point in both models.
Time intervals in the tables can differ but the empty 'Year'-cells are allowed only at the end of the series.
The shorter time intervals make less difference between the results of the two models.

Lesson 8-9
Demand Frequency

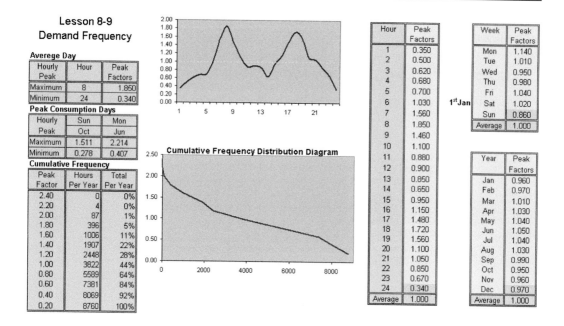

Averege Day

Hourly Peak	Hour	Peak Factors
Maximum	8	1.850
Minimum	24	0.340

Peak Consumption Days

Hourly Peak	Sun Oct	Mon Jun
Maximum	1.511	2.214
Minimum	0.278	0.407

Cumulative Frequency

Peak Factor	Hours Per Year	Total Per Year
2.40	0	0%
2.20	4	0%
2.00	87	1%
1.80	396	5%
1.60	1006	11%
1.40	1907	22%
1.20	2448	28%
1.00	3822	44%
0.80	5589	64%
0.60	7381	84%
0.40	8069	92%
0.20	8760	100%

Hour	Peak Factors
1	0.350
2	0.500
3	0.620
4	0.680
5	0.700
6	1.030
7	1.560
8	1.850
9	1.460
10	1.100
11	0.880
12	0.900
13	0.850
14	0.650
15	0.950
16	1.150
17	1.480
18	1.720
19	1.560
20	1.100
21	1.050
22	0.850
23	0.670
24	0.340
Average	1.000

1st Jan

Week	Peak Factors
Mon	1.140
Tue	1.010
Wed	0.950
Thu	0.980
Fri	1.040
Sat	1.020
Sun	0.860
Average	1.000

Year	Peak Factors
Jan	0.960
Feb	0.970
Mar	1.010
Apr	1.030
May	1.040
Jun	1.050
Jul	1.040
Aug	1.030
Sep	0.990
Oct	0.950
Nov	0.960
Dec	0.970
Average	1.000

Lesson 8-9   Demand Frequency

Contents:
Calculation of the demand frequency distribution for given diurnal-, weekly and annual demand pattrens.

Goal:
Determination of the cumulative frequency diagram that should help to adopt the design peak factor.

Notes:
Maximum/minimum hourly peak factors in combination with the seasonal factors give the range of absolute values for the hourly peak factors in the system between the maximum consumption hour of the maximum consumption day and the minimum consumption hour of the minimum consumption day.
The last peak factor value specified in each input table is calculated so that the value can fill up to 24, 7 and 12 respectively, which generates the average equal to 1.0.
After the calculation of the absolute values, all peak factor values are compared and counted. The figures in the Cumulative Frequency table indicate the number of hours/percentage of the total time, when the absolute peak factor was above the corresponding value in the table.
The cell filled with 1st Jan should be cut and pasted next to the week day which was assumed on January 1.

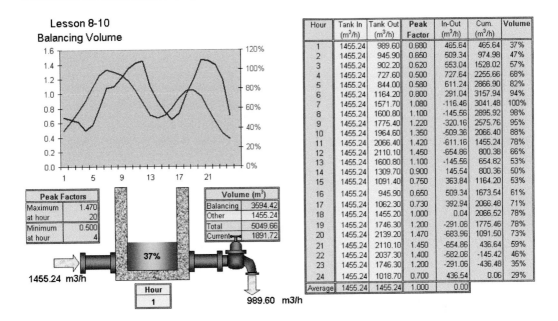

Lesson 8-10 Balancing Volume

Contents:
The diurnal demand diagram and volume variation of the balancing reservoir.

Goal:
The determination of the balancing volume in the reservoir and its variation during the day.

Notes:
The reservoir for which the volume is calculated is the only one existing in the system, and it is assumed that it balances the demand of the entire area.
In the case of favourable topography, the total balancing volume can be shared between a few smaller reservoirs.

Lesson 8-11

Nodal Demands

	Q1tot (l/s)	Q2tot (l/s)	Q3tot (l/s)	Qtot (l/s)
	120	180	65	365

NODES	X (m)	Y (m)	QL$_1$ (l/s)	QL$_2$ (l/s)	QL$_3$ (l/s)	Q$_n$ (l/s)
n1	25	135	35.38	0.00	0.00	35.38
n2	25	200	30.57	0.00	0.00	30.57
n3	60	200	24.62	54.81	0.00	79.43
n4	65	88	29.43	40.03	0.00	69.46
n5	100	100	0.00	35.19	16.98	52.17
n6	150	180	0.00	49.97	22.04	72.01
n7	170	80	0.00	0.00	15.52	15.52
n8	140	45	0.00	0.00	10.46	10.46
n9			0.00	0.00	0.00	0.00
n10			0.00	0.00	0.00	0.00

PIPES	Node 1	Node 2	L$_{xy}$(m)	L (m)	Q (l/s)
p1	n1	n2	65	680	38.49
p2	n2	n3	35	400	22.64
p3	n3	n4	112	470	26.60
p4	n4	n1	62	570	32.26
					0.00
LOOP 1			Sum=	2120	120.00

PIPES	Node 1	Node 2	L$_{xy}$(m)	L (m)	Q (l/s)
p3	n4	n3	112	1120	60.18
p5	n3	n6	92	920	49.43
p6	n6	n5	94	940	50.51
p7	n5	n4	37	370	19.88
					0.00
LOOP 2			Sum=	3350	180.00

PIPES	Node 1	Node 2	L$_{xy}$(m)	L (m)	Q (l/s)
p6	n5	n6	94	940	21.14
p8	n6	n7	102	1020	22.94
p9	n7	n8	46	360	8.10
p10	n8	n5	68	570	12.82
					0.00
LOOP 3			Sum=	2890	65.00

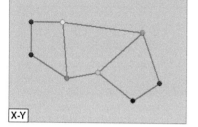

X-Y

Lesson 8-11 Nodal Demand

Contents:
Spatial demand distribution.

Goal:
Calculation of nodal demands based on simplified procedure for concentrating the demand into the junctions between the pipes.

Abbreviations:

NODES	Node data
X (m)	Horizontal co-ordinate
Y (m)	Vertical co-ordinate
QL$_1$ (l/s)	Contribution to the nodal demand from loop 1
QL$_2$ (l/s)	Contribution to the nodal demand from loop 2
QL$_3$ (l/s)	Contribution to the nodal demand from loop 3
Q$_n$ (l/s)	Average (baseline) nodal demand

PIPES	Pipe data per loop
Node 1	Pipe node nr.1
Node 2	Pipe node nr.2
L$_{xy}$(m)	Length calculated from the X/Y co-ordinates
L (m)	Length adopted for hydraulic calculation
Q (l/s)	Flow supplied from the pipe to the loop

Notes:
The nodes are plotted based on the X/Y input (the origin of the graph is at the lower left corner). Any node name can be used.
The pipes are plotted based on the Node 1/Node 2 input. This input determines the connection between the nodes and hence the flow rates/directions.
Half of the pipe flow (Q) is allocated to each of the corresponding nodes.

Lesson 8-12
Rapid NRW Assessment

INPUT DATA
Water Utility   **Name?**

i1	System input volume (SIV), m³/d	50,000
i2	Billed metered consumption (BMC), m³/d	15,000
i3	Billed unmetered consumption (BUC), m³/d	25,000
i4	Unbilled metered consumption (UMC), m³/d	1,000
i5	Unbilled unmetered consumption (UUC), %SIV	1.0%
i6	Water theft, %BC	5.0%
i7	Customer meters under-reading, %MC	5.0%
i8	Total length of mains (Lm), km	400
i9	Total number of service connections (Nc)	12,000
i10	Average length of single service connection, m	-
i11	Average pressure (p), mwc	15
i12	Average supply time per day, hrs	12
i13	Average tariff, EUR/m³	0.25
i14	Variable production cost, EUR/m³	0.10
i15	Unsatisfied demand*, m³/d	3,000

*-water which could be sold if physical losses were reduced

OUTPUT RESULTS
Water Utility   **Name?**

Volumes

			%SIV
r1	Metered consumption (MC), m³/d	16,000	32.0%
r2	Billed consumption (BC), m³/d	40,000	80.0%
r3	Unbilled consumption (UC), m³/d	1,500	3.0%
r4	Authorised consumption (AC), m³/d	41,500	83.0%
r5	Commercial/apparent losses, m³/d	2,800	5.6%
r6	Physical/real losses, m³/d	5,700	11.4%
r7	NRW, m³/d	10,000	20.0%

Water Loss Performance Indicators

r8	NRW (w.s.p), l/d/conn	1,667	
r9	Physical/real losses (w.s.p), l/d/conn	950	
r10	Commercial/apparent losses, % of authorised con	6.7%	
r11	UARL (MAAPL) (w.s.p) m³/d	126	
r12	Infrastructure leakage index (ILI)	45.2	

Financial Aspects

r13	Annual value of commercial losses, EUR/y	255,500	40.7%
r14	Annual value of physical losses, EUR/y*	372,300	59.3%
r15	Total annual value of NRW, EUR/y	627,800	100.0%

*-taking additional water sales into account

WATER BALANCE TABLE (m³/d-%SIV)
Water Utility   **Name?**

		Billed metered consumption		
	Billed authorised consumption	15,000 30.0%		Revenue water
	40,000 80.0%	Billed unmetered consumption		40,000 80.0%
Authorised consumption		25,000 50.0%		
41,500 83.0%		Unbilled metered consumption		
	Unbilled authorised consumption	1,000 2.0%		
	1,500 3.0%	Unbilled unmetered consumption		
		500 1.0%		
		Unauthorised consumption		
	Apparent (commercial) losses	2,000 4.0%		
	2,800 5.6%	Metering inaccuracies		
Water losses (UFW)		800 1.6%		Non-revenue water
8,500 17.0%		Leakage in mains		10,000 20.0%
		1,000 2.0%		
		Leakage/oveflow at storage tanks		
	Real (physical) losses	500 1.0%		
	5,700 11.4%	Leakage on service connections		
		4,200 8.4%		

Lesson 8-12 Rapid NRW Assessment

Contents:
Calculation of water balance and ILI index, accoriding to the IWA guidelines.

Goal:
To assess the performance indicators related to NRW and leakage, and get impressions about the most useful immediate interventions.

Abbreviations:
All input and output parameters are given with full names in the table, except for:

(w.s.p)   Means: "when the system is pressurised". Applies in case of intermittent supply situations where the parameter with this note
            is recalculated assuming continuous supply.

Notes:

Unsatisfied demand*, m³/d   The shortage of the demand that could be (partly) supplied in case of reduction of the physical (real) losses.

Annual value of physical losses, EUR/y*   The figure assumes the loss of revenue for the sales of unsatisfied demand,
            and the loss resulting from the production costs of the difference between the real loss and that demand.

EPANET – Version 2

(based on the EPANET 2 Users Manual by L.A. Rossman)[1]

8.1 Installation

EPANET 2 is a computer programme that performs hydraulic and water quality simulations of drinking water distribution systems. For a basic set of input data related to the geometry of the network and the water demand levels in it, the programme is able to determine the flow of water in each pipe, the pressure at each pipe junction, the flows and heads at each pump, and the water depth in each storage tank. Additional information on water quality parameters serves to calculate the concentration of a substance throughout a distribution system. In addition to substance concentrations, water age and source tracing can also be simulated. The user is able to edit EPANET 2 files and, after running a simulation, display the results on a colour-coded map of the distribution system and generate additional tabular and graphical views of these results.

The calculations made by EPANET can help in solving all sorts of practical problems, such as:

- the design of new extensions and pumping stations,
- the analysis of pumps' energy and cost,
- the optimal operation which can guarantee delivery of sufficient water quality and pressure,
- the assessment of network reliability,
- the development of effective flushing programmes,
- the use of satellite treatment, such as re-chlorination at storage tanks,
- the diagnosis of water quality problems, etc.

EPANET 2 software (release 12, dated January 2018) and its manual are in the public domain and can be downloaded from the website of the US Environmental Protection Agency. This site is easily accessible through any search engine (e.g. 'Google'), by using the keyword 'epanet'.

The setup file contains a self-extracting setup programme. The installation starts after double-clicking on this file. The default programme directory is *c:\ Program Files\EPANET 2.0.12*, which can be changed during the process that will be fully completed in just a few seconds. After the programme has been installed,

---

1 Rossman, L.A., EPANET 2 Users Manual, EPA/600/R-00/057, Water Supply and Water Resources Division, U.S. Environmental Protection Agency, Cincinnati, OH, September 2000.

the Start Menu will have a new item named EPANET 2.0.12. To start the programme, choose **Start>>Programs>>Epanet 2.0.12>>Epanet 2.0**.

8.2 Using the programme

After running the programme, a screen with the menu bar, toolbars, blank network map window and the browser window will appear.

Menu Bar

The *Menu Bar* located across the top of the EPANET workspace contains a collection of menus used to control the program. These include:

File Menu

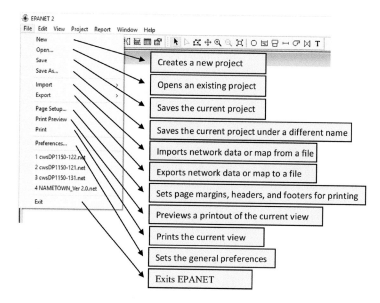

Figure 8.1 File menu options

Edit Menu

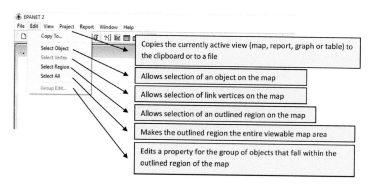

Figure 8.2 Edit menu options

View Menu

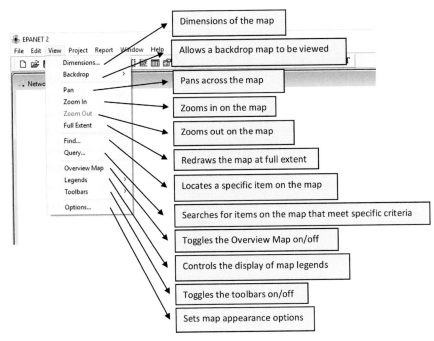

Figure 8.3 View menu options

Project Menu

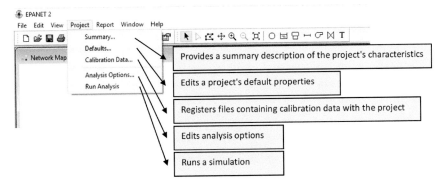

Figure 8.4 Project menu options

Report Menu

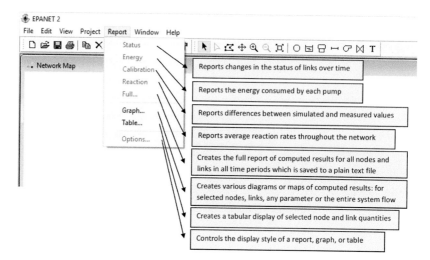

Figure 8.5 Report menu options

Window Menu

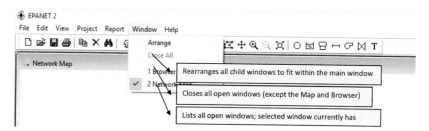

Figure 8.6 Window menu options

Help Menu

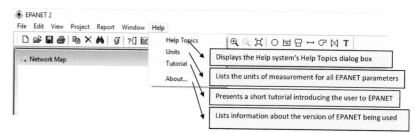

Figure 8.7 Help menu options

The *Toolbars* located below the Menu Bar provide shortcuts to commonly used operations. The toolbars can be made visible or invisible by selecting **View>>Toolbars**. *Toolbars*

There are two such toolbars:

Standard Toolbar: contains speed buttons for commonly used commands.   *Standard Toolbar*

   ◻ Opens a new project (**File>>New**)

   📂 Opens an existing project (**File>>Open**)

   💾 Saves the current project (**File>>Save**)

   🖨 Prints the currently active window (**File>>Print**)

   📑 Copies selection to the clipboard or to a file (**Edit>>Copy To**)

   ✗ Deletes the currently selected item

   🔍 Finds a specific item on the map (**View>>Find**)

   ⚡ Runs a simulation (**Project>>Run Analysis**)

   ?{} Runs a visual query on the map (**View>>Query**)

   📈 Creates a new graph view of results (**Report>>Graph**)

   ⊞ Creates a new table view of results (**Report>>Table**)

   🗐 Modifies options for the currently active view (**View>>Options** or **Report>>Options**)

Map Toolbar: contains buttons for working with the Network Map.   *Map Toolbar*

   ▲ Selects an object on the map (**Edit>>Select>>Object**)

   ▷ Selects link vertex points (**Edit>>Select Vertex**)

   ⌕ Selects a region on the map (**Edit>>Select Region**)

   ✛ Pans across the map (**View>>Pan**)

   🔍 Zooms in on the map (**View>>Zoom In**)

   🔍 Zooms out on the map (**View>>Zoom Out**)

   ⛶ Draws map at full extent (**View>>Full Extent**)

   ○ Adds a junction to the map

   ⊟ Adds a reservoir to the map (fixed surface level, unknown dimensions)

   ⊟ Adds a tank to the map (variable surface level, known dimensions)

   ⊢⊣ Adds a pipe to the map

   ⊂ Adds a pump to the map

   ⋈ Adds a valve to the map

   T Adds a label to the map

Network Map

The *Network Map* provides a planar schematic diagram of the objects comprising a water distribution network. The location of objects and the distances between them do not necessarily have to conform to their actual physical scale. New objects can be directly added to the map and existing objects can be clicked on for editing, deleting, and repositioning. A backdrop drawing (such as a street or topographic map) can be placed behind the network map for reference. The map can be zoomed to any scale and panned from one position to another. Nodes and links can be drawn at different sizes, flow direction arrows added, and object symbols, ID labels and numerical property values displayed. The map can be printed, copied onto the Windows clipboard, or exported as a text file (extension MAP), drawing exchange file (DXF) or enhanced metafile (EMF).

Data Browser

The *Data Browser* is accessed from the Data tab on the Browser window. It gives access to the various objects that exist in the network being analysed. The buttons at the bottom are used to add, delete, and edit these objects.

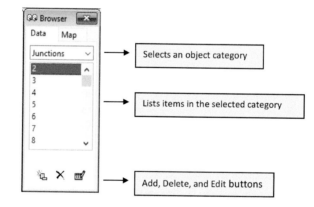

Figure 8.8 Data browser

Map Browser

The *Map Browser* is accessed from the Map tab of the Browser Window. It selects the parameters and time period that are viewed in colour-coded fashion on the Network Map. It also contains controls for animating the map through time.

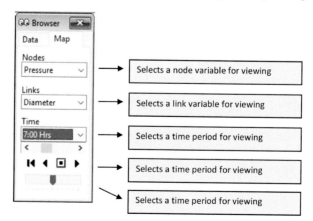

Figure 8.9 Map browser

The work begins by opening a network file, *.NET, by the menu option: **File>>Open** from the Menu Bar or clicking the Standard Toolbar 📂 (If the toolbar is not visible **View>>Toolbars>>Standard** should be selected from the menu bar).

The selected file will be loaded into the computer memory. It is possible to choose to open a file type saved previously as an EPANET project (typically with a .NET extension) or exported as a text file (typically with a .INP extension). EPANET recognizes file types by their content, not their names.

Calculation starts by running the menu option: **Project>>Run Analysis** or clicking the Run button 🏃 on the Standard Toolbar.

If the run was unsuccessful then a *Status Report* window will appear indicating what the problem was. In some situations, the calculation may be completed regardless of the hydraulic boundary conditions. A warning message is displayed in that case. The input data have to be carefully analysed. More information about the calculation progress can be requested in repeated trials. The details about this, together with the description of other programme features, are presented in the full version of the programme manual.

Status Report

If it ran successfully, the user can view the computed results in a variety of ways:

- Select Node Pressure from the Browser's Map page and observe how pressure values at the nodes become colour-coded. To view the legend for the colour coding, the user can select **View>>Legends>>Node** (or right click on an empty portion of the map and select **Node Legend** from the pop-up menu). To change the legend intervals and colours, right click on the legend makes the *Legend Editor* appear.

Legend Editor

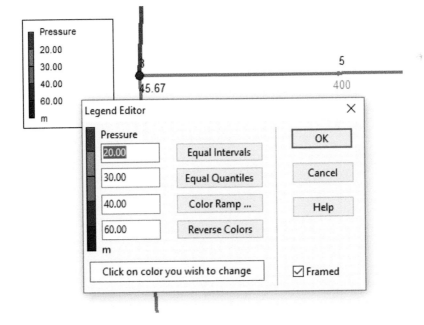

Figure 8.10 Legend editor

• The user can bring up the *Property Editor* (double-click on any node or link) and note how the computed results are displayed at the end of the property list.

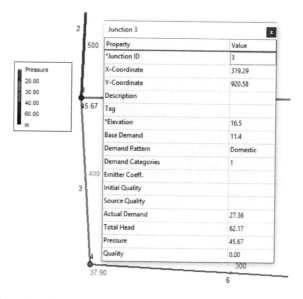

Figure 8.11 Property editor

• A tabular listing of results is created by selecting **Report >>Table** (or by clicking the Table button on the Standard ⊞ Toolbar). It is possible to create the report for all nodes or links for a specific time, or a report for a specific node or link for a time series. In the columns, it is possible to have a report for elevation, base demand, initial quality, demand, head, pressure and quality in the case of a node; and length, diameter, roughness, bulk coefficient, wall coefficient, flow, velocity, unit head loss, friction factor, reaction rate, quality and status in the case of a link.

Node ID	Demand LPS	Head m	Pressure m
Junc 2	10.55	96.38	86.18
Junc 3	6.50	96.30	79.80
Junc 4	9.46	96.23	73.13
Junc 5	19.04	96.33	78.03
Junc 6	13.39	96.17	73.47
Junc 7	15.28	96.01	66.51
Junc 8	6.73	96.28	71.88
Junc 9	10.55	96.09	64.39
Junc 10	6.04	95.96	61.26
Junc 11	0.00	96.28	72.28
Resvr 1	-97.53	10.00	0.00

Figure 8.12 Snapshot report

8.3   Input data

8.3.1   Data preparation

Prior to running EPANET, the following initial steps should be taken for the network being studied:

1. All network components and their connections should be identified. Network components consist of pipes, pumps, valves, storage tanks and reservoirs. The term 'node' denotes a junction where network components connect to one another. Tanks and reservoirs are also considered as nodes. The component (pipe, pump or valve) connecting any two nodes is termed a 'link'.
2. Unique ID numbers should be assigned to all nodes. ID numbers must be between 1 and 2,147,483,647, but need not be in any specific order nor be consecutive.
3. An ID number should be assigned to each link (pipe, pump, or valve). It is permissible to use the same ID number for both a node and a link.
4. The following information should be collected on the system parameters:

    a.   diameter, length, roughness and minor loss coefficient for each pipe,
    b.   the characteristic operating curve for each pump,
    c.   diameter, minor loss coefficient and pressure or flow setting for each control valve,
    d.   diameter and minimum and maximum water depths for each tank
    e.   control rules that determine how pump, valve and pipe settings change with time, tank water levels, or nodal pressures,
    f.   changes in water demands for each node over the time period being simulated
    g.   initial water quality at all nodes and changes in water quality over time at source nodes.

With this information at hand, it is now possible to construct an input file to use with EPANET.

8.3.2   Selecting objects

To select an object on the map:

- The map must be in *Selection mode* (the mouse cursor has the shape of an arrow pointing up to the left). To switch to this mode, the user should either click the Select Object ⬉ button on the Map Toolbar or choose **Edit>>Select Object** from the menu bar.   *Selection mode*
- The mouse has to be clicked over the desired object on the map.

To select an object using the Browser:

- The category of object has to be seleceted from the dropdown list of the Data Browser.
- The desired object has to be selected from the list below the category heading.

8.3.3 Editing visual objects

The Property Editor is used to edit the properties of objects that can appear on the Network Map (Junctions, Reservoirs, Tanks, Pipes, Pumps, Valves, or Labels). To edit one of these objects, the user should select the object on the map or from the Data Browser, then click the Edit button on the Data Browser (or simply double-click the object on the map). The properties associated with each of these types of objects are described in tables 8.1 to 8.7.

Table 8.1 Junction Properties

PROPERTY	DESCRIPTION
Junction ID	A unique label used to identify the junction. It can consist of a combination of up to 15 numerals or characters. It cannot be the same as the ID for any other node. This is a required property.
X-Coordinate	The horizontal location of the junction on the map, measured in the map's distance units. If left blank the junction will not appear on the network map.
Y-Coordinate	The vertical location of the junction on the map, measured in the map's distance units. If left blank the junction will not appear on the network map.
Description	An optional text string that describes other significant information about the junction.
Tag	An optional text string (with no spaces) used to assign the junction to a category, such as a pressure zone.
Elevation	The elevation above some common reference of the junction. This is a required property. Elevation is used only to compute pressure at the junction. It does not affect any other computed quantity.
Base Demand	The average or nominal demand for water by the main category of consumer at the junction, as measured in the current flow units. A negative value is used to indicate an external source of flow into the junction. If left blank then demand is assumed to be zero.
Demand Pattern	The ID label of the time pattern used to characterize time variation in demand for the main category of consumer at the junction. The pattern provides multipliers that are applied to the Base Demand to determine actual demand in a given time period. If left blank then the Default Time Pattern assigned in the Hydraulic Options **(Report>>Default)** will be used.
Demand Categories	Number of different categories of water users defined for the junction. Click the ellipsis button (or press **<Enter>**) to bring up a special Demands Editor which makes it possible to assign base demands and time patterns to multiple categories of users at the junction. Ignore if only a single demand category will suffice.
Emitter Coefficient	Discharge coefficient for emitter (sprinkler or nozzle) placed at junction. The coefficient represents the flow (in current flow units) that occurs at a unit pressure drop. It should be left blank if no emitter is present.
Initial Quality	Water quality level at the junction at the start of the simulation period. Can be left blank if no water quality analysis is being made or if the level is zero.
Source Quality	Quality of any water entering the network at this location. The user should click the ellipsis button (or press **<Enter>**) to bring up the Source Quality Editor.

Table 8.2 Reservoir Properties

PROPERTY	DESCRIPTION
Reservoir ID	A unique label used to identify the reservoir. It can consist of a combination of up to 15 numerals or characters. It cannot be the same as the ID for any other node. This is a required property.
X-Coordinate	The horizontal location of the reservoir on the map, measured in the map's distance units. If left blank the reservoir will not appear on the network map.
Y-Coordinate	The vertical location of the reservoir on the map, measured in the map's distance units. If left blank the reservoir will not appear on the network map.
Description	An optional text string that describes other significant information about the reservoir.
Tag	An optional text string (with no spaces) used to assign the reservoir to a category, such as a pressure zone
Total Head	The hydraulic head (elevation + pressure head) of water in the reservoir. This is a required property.
Head Pattern	The ID label of a time pattern used to model time variation in the reservoir's head. It should be left blank if none applies. This property is useful if the reservoir represents a tie-in to another system whose pressure varies with time.
Initial Quality	Water quality level at the reservoir. Can be left blank if no water quality analysis is being made or if the level is zero.
Source Quality	Quality of any water entering the network at this location. Click the ellipsis button (or press **<Enter>**) to bring up the Source Quality Editor.

Table 8.3 Tank Properties

PROPERTY	DESCRIPTION
Tank ID	A unique label used to identify the tank. It can consist of a combination of up to 15 numerals or characters. It cannot be the same as the ID for any other node. This is a required property.
X-Coordinate	The horizontal location of the tank on the map, measured in the map's unit of scale. If left blank the tank will not appear on the map.
Y-Coordinate	The vertical location of the tank on the map, measured in the map's unit of scale. If left blank the tank will not appear on the map.
Description	Optional text string that describes other significant information about the tank.
Tag	Optional text string (with no spaces) used to assign the tank to a category, such as a pressure zone
Elevation	Elevation above a common reference level of the bottom shell of the tank. This is a required property.
Initial Level	Height of the water surface above the bottom elevation of the tank at the start of the simulation. This is a required property.
Minimum Level	Minimum height of the water surface above the bottom elevation that will be maintained. The tank will not be allowed to drop below this level. This is a required property.

(*Continued*)

Table 8.3 (Continued)

PROPERTY	DESCRIPTION
Maximum Level	Maximum height of the water surface above the bottom elevation that will be maintained. The tank will not be allowed to rise above this level. This is a required property.
Diameter	The diameter of the tank. For cylindrical tanks this is the actual diameter. For square or rectangular tanks it can be an equivalent diameter equal to 1.128 times the square root of the cross-sectional area. For tanks whose geometry is described by a curve it can be set to any value. This is a required property.
Minimum Volume	The volume of water in the tank when it is at its minimum level. This is an optional property, useful mainly for describing the bottom geometry of non-cylindrical tanks where a full volume versus depth curve will not be supplied.
Volume Curve	The ID label of a curve used to describe the relation between tank volume and water level. If no value is supplied then the tank is assumed to be cylindrical.
Mixing Model	The type of water quality mixing that occurs within the tank. The choices include: • MIXED (fully mixed), • 2COMP (two compartment mixing), • FIFO (first-in-first-out plug flow), • LIFO (last-in-first-out plug flow).
Mixing Fraction	The fraction of the tank's total volume that comprises the inlet-outlet compartment of the two-compartment (2COMP) mixing model. Can be left blank if another type of mixing model is employed.
Reaction Coefficient	The bulk reaction coefficient for chemical reactions in the tank. Time units are 1/days. Use a positive value for growth reactions and a negative value for decay. Leave blank if the Global Bulk reaction coefficient specified in the project's Reactions Options applies.
Initial Quality	Water quality level in the tank at the start of the simulation. Can be left blank if no water quality analysis is being made or if the level is zero.
Source Quality	Quality of any water entering the network at this location. Click the ellipsis button (or press **<Enter>**) to bring up the Source Quality Editor.

Table 8.4 Pipe Properties

PROPERTY	DESCRIPTION
Pipe ID	A unique label used to identify the pipe. It can consist of a combination of up to 15 numerals or characters. It cannot be the same as the ID for any other link. This is a required property.
Start Node	The ID of the node where the pipe begins. This is a required property.
End Node	The ID of the node where the pipe ends. This is a required property.
Description	An optional text string that describes other significant information about the pipe.
Tag	An optional text string (with no spaces) used to assign the pipe to a category, e.g. based on age or material

Length	The actual length of the pipe. This is a required property.
Diameter	The pipe diameter. This is a required property.
Roughness	The roughness coefficient of the pipe. It has no units for Hazen-Williams or Chezy-Manning roughness and has units of mm (or millifeet) for Darcy-Weisbach roughness. This is a required property.
Loss Coefficient	Unitless minor loss coefficient associated with bends, fittings, etc. Assumed 0 if left blank.
Initial Status	Determines whether the pipe is initially open, closed, or contains a check valve. If a check valve is specified then the flow direction in the pipe will always be from the Start node to the End node.
Bulk Coefficient	The bulk reaction coefficient for the pipe. Time units are 1/days. Use a positive value for growth and a negative value for decay. It should be left blank if the Global Bulk reaction coefficient from the project's Reaction Options applies.
Wall Coefficient	The wall reaction coefficient for the pipe. Time units are 1/days. A positive value should be used for growth and a negative value for decay. It should be left blank if the Global Wall reaction coefficient from the project's Reactions Options applies.

Table 8.5 Pump Properties

PROPERTY	DESCRIPTION
Pump ID	A unique label used to identify the pump. It can consist of a combination of up to 15 numerals or characters. It cannot be the same as the ID for any other link. This is a required property.
Start Node	The ID of the node on the suction side of the pump. This is a required property.
End Node	The ID of the node on the discharge side of the pump. This is a required property.
Description	An optional text string that describes other significant information about the pump.
Tag	An optional text string (with no spaces) used to assign the pump to a category, e.g. based on age, size or location.
Pump Curve	The ID label of the pump curve used to describe the relationship between the head delivered by the pump and the flow through the pump. It should be left blank if the pump is a constant energy pump.
Power	The power supplied by the pump in horsepower (kW). Assumes that the pump supplies the same amount of energy no matter what the flow is. This information is used when pump curve information is not available. It should be left blank if a pump curve is used instead.
Speed	The relative speed setting of the pump (no units). For example, a speed setting of 1.2 implies that the rotational speed of the pump is 20% higher than the normal setting.
Pattern	The ID label of a time pattern used to control the pump's operation. The multipliers of the pattern are equivalent to speed settings. A multiplier of zero implies that the pump will be shut off during the corresponding time period. It should be left blank if not applicable.
Initial Status	State of the pump (OPEN or CLOSED) at the start of the simulation period.

(Continued)

Table 8.5 (Continued)

PROPERTY	DESCRIPTION
Efficiency Curve	The ID label of the curve that represents the pump's wire-to-water efficiency (in percent) as a function of flow rate. This information is used only to compute energy usage.
Energy Price	The average or nominal price of energy in monetary units per kw-hr. Used only for computing the cost of energy usage.
Price Pattern	The ID label of the time pattern used to describe the variation in energy price throughout the day. Each multiplier in the pattern is applied to the pump's energy price.

Table 8.6 Valve Properties

PROPERTY	DESCRIPTION
ID Label	A unique label used to identify the valve. It can consist of a combination of up to 15 numerals or characters. It cannot be the same as the ID for any other link. This is a required property.
Start Node	The ID of the node on the nominal upstream or inflow side of the valve. (PRVs and PSVs maintain flow in only a single direction.) This is a required property.
End Node	The ID of the node on the nominal downstream or discharge side of the valve. This is a required property.
Description	An optional text string that describes other significant information about the valve.
Tag	An optional text string (with no spaces) used to assign the valve to a category, e.g. based on type or location.
Diameter	The valve diameter. This is a required property.
Type	The valve type (PRV, PSV, PBV, FCV, TCV, or GPV). This is a required property.
Setting	A required parameter that describes the valve's operational setting. Valve Type - Setting Parameter: PRV - Pressure (mwc or psi) PSV - Pressure (mwc or psi) PBV - Pressure (mwc or psi) FCV - Flow (flow units) TCV - Loss Coefficient (no units) GPV - ID of head loss curve
Loss Coefficient	Minor loss coefficient that applies when the valve is completely opened (no units). Assumed 0 if left blank.
Fixed Status	Valve status at the start of the simulation. If set to OPEN or CLOSED then the control setting of the valve is ignored and the valve behaves as an open or closed link, respectively. If set to NONE, then the valve will behave as intended. A valve's fixed status and its setting can be made to vary throughout a simulation by the use of control statements. If a valve's status was fixed to OPEN or CLOSED, then it can be made active again using a control that assigns a new numerical setting to it.

Table 8.7 Map Label Properties

PROPERTY	DESCRIPTION
Text	The label's text.
X-Coordinate	The horizontal location of the upper left corner of the label on the map, measured in the map's unit of scale. This is a required property.
Y-Coordinate	The vertical location of the upper left corner of the label on the map, measured in the map's scaling units. This is a required property.
Anchor Node	ID of node that serves as the label's anchor point. It should be left blank if label is not anchored.
Meter Type	Type of object being metered by the label. Choices are None, Node, or Link.
Meter ID	ID of the object (Node or Link) being metered.
Font	Launches a Font dialogue that allows selection of the label's font, size, and style.

8.3.4 Editing non-visual objects

Curves, Time Patterns, and Controls have special editors that are used to define their properties. To edit one of these objects, the user should select the object from the Data Browser and then click the Edit button 🖉. In addition, the Property Editor for Junctions contains an ellipsis button in the field for Demand Categories that brings up a special Demand Editor when clicked. Each of these specialized editors is described in the following paragraphs.

The *Curve Editor* is a dialogue form as shown below:

Curve Editor

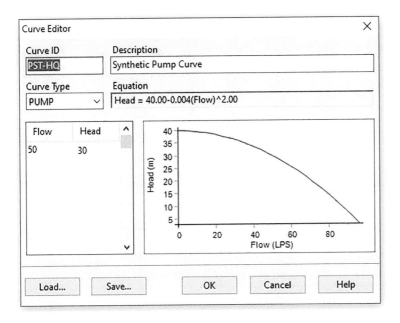

Figure 8.13 Curve editor

The Curve Editor contains the following items:

Item	Description
Curve ID	ID label of the curve (maximum of 15 numerals or characters)
Description	Optional description of what the curve represents
Curve Type	Type of curve
X-Y Data	X-Y data points for the curve

When moving between cells in the X-Y data table (or after pressing the <**Enter**> key) the curve is redrawn in the preview window. For single- and three-point pump curves, the equation generated for the curve will be displayed in the Equation box. The user can click the **OK** button to accept the curve or the **Cancel** button to cancel the entries. It is also possible to click the **Load** button to load in curve data that was previously saved to file or click the **Save** button to save the current curve's data to a file.

Pattern Editor   The *Pattern Editor*, when displayed, edits the properties of a time pattern object.

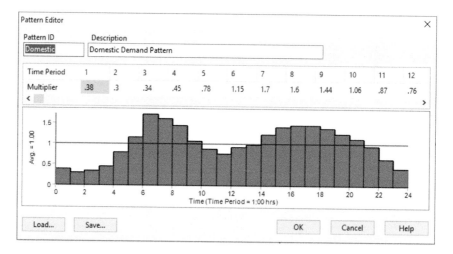

Figure 8.14 Pattern editor

To use the Pattern Editor values for the following items should be entered:

Item	Description
Pattern ID	ID label of the pattern (maximum of 15 numerals or characters).
Description	Optional description of what the pattern represents.
Multipliers	Multiplier value for each time period of the pattern.

Time Periods   As multipliers are entered, the preview chart is redrawn to provide a visual depiction of the pattern. If the end of the available *Time Periods* is reached when entering multipliers, pressing <**Enter**> adds on another period. When finished editing, clicking **OK** accepts the pattern whilst the **Cancel** button cancels the entries. It is also possible to click **Load** to load in pattern data that was previously saved to file or click **Save** to save the current pattern's data to a file.

The *Controls Editor* is a text editor window used to edit both simple and rule-based controls. It has a standard text-editing menu that is activated by right clicking anywhere in the Editor. The menu contains commands for **Undo**, **Cut**, **Copy**, **Paste**, **Delete**, and **Select All**.

Controls Editor

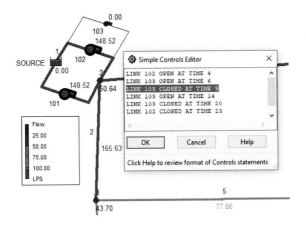

Figure 8.15 Controls editor (simple)

The *Demand Editor* is used to assign base demands and time patterns when there is more than one category of water user at a junction. The editor is invoked from the Property Editor by clicking the ellipsis button (or pressing **<Enter>**) when the Demand Categories field has the focus. The editor is a table containing three columns. Each category of demand is entered as a new row in the table.

Demand Editor

Property	Value
*Junction ID	11
X-Coordinate	1610.70
Y-Coordinate	1205.83
Description	
Tag	Industry
*Elevation	24.0
Base Demand	8.33
Demand Pattern	Factory
Demand Categories	2 ...

Demands for Junction 11

	Base Demand	Time Pattern	Category
1	8.33	Factory	
2	4.21	Domestic	
3			
4			
5			
6			

OK   Cancel   Help

Figure 8.16 Demand editor

The columns contain the following information:

Base Demand

- *Base Demand*: baseline or average demand for the category (required)
- *Time Pattern*: ID label of time pattern used to allow demand to vary with time (optional)
- *Category*: text label used to identify the demand category (optional)

The table is initially sized for 10 rows. Additional rows can be added by pressing **<Enter>** in any cell in the last row. By convention, the demand placed in the first row of the editor will be considered the main category for the junction and will appear in the Base Demand field of the Property Editor.

8.3.5   Editing a group of objects

To edit a property for a group of objects:

- The region of the map that will contain the group of objects to be edited can be selected as follows:

  - Select **Edit>>Select Region**
  - Draw a polygon fence line around the region of interest on the map by clicking the left mouse button at each successive vertex of the polygon.
  - Close the polygon by clicking the right button or by pressing the **<Enter>** key; cancel the selection by pressing the **<Escape>** key.

- Select **Edit>>Group Edit** from the Menu Bar.
- Define what to edit in the Group Edit dialogue form that appears.

The Group Edit dialogue form is used to modify a property for a selected group of objects. The steps to use the dialogue form are as follows:

- Select a category of object (Junctions or Pipes) to edit.
- Check the 'with' box if it is necessary to add a filter that will limit the objects selected for editing. Select a property, relation and value that define the filter. An example might be 'with Diameter below 300 (mm)'.
- Select the type of change to make – 'Replace', 'Multiply', or 'Add To'.
- Select the property to change.
- Enter the value that should replace, multiply, or be added to the existing value.
- Click **OK** to execute the group edit.

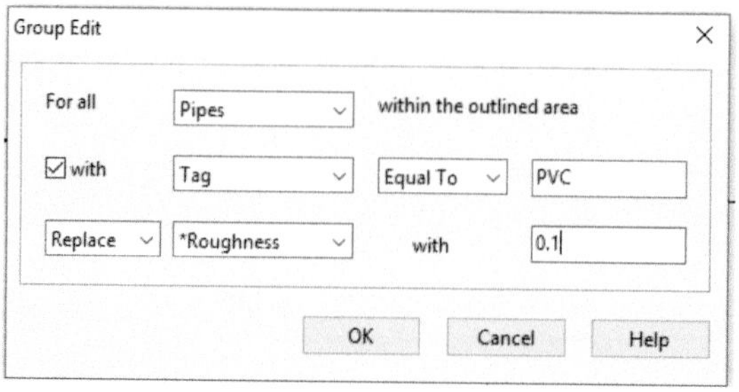

Figure 8.17 Editor for group of objects

8.4   Viewing results

8.4.1   Viewing results on the map

There are several ways in which database values and results of a simulation can be viewed directly on the Network Map:

- For the current settings on the Map Browser the nodes and links of the map will be coloured according to the colour coding used in the Map Legends. The map's colouring will be updated as a new time period is selected in the Browser.
- ID labels and viewing parameter values can be displayed next to all nodes and/or links by selecting the appropriate options on the Notation page of the Map Options dialogue form.
- The display of results on the network map can be animated either forward or backward in time by using the Animation buttons on the Map Browser. Animation is only available when a node or link viewing parameter is a computed value (e.g., link flow rate can be animated but diameter cannot).
- The map can be printed, copied to the Windows clipboard, or saved as a Bitmap, Metafile or Text file.
- Nodes or links meeting a specific criterion can be identified by submitting a *Map Query*; the user should execute the following steps: *Map Query*

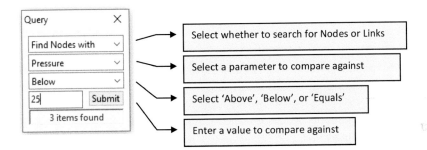

Figure 8.18 Map query editor

1   Select a time period in which to query the map from the Map Browser.
2   Select **View>>Query** or click ?{ on the Map Toolbar.
3   Fill in the following information in the Query dialogue form that appears.
4   Click the **Submit** button. The objects that meet the criterion will be highlighted on the map.
5   As a new time period is selected in the Browser, the query results are automatically updated.
6   It is possible to submit another query using the dialogue box or close it by clicking the button in the upper right corner.

8.4.2   Viewing results with a graph

Analysis results, as well as some design parameters, can be viewed using several different types of graphs. Graphs can be printed, copied to the Windows clipboard, or saved as a data file or Windows metafile. The following types of graphs can be used to view values for a selected parameter:

Type of Plot	Description	Applies To
Time Series Plot	Plots value versus time	Specific nodes or links over all time periods
Profile Plot	Plots value versus distance	A list of nodes at a specific time
Contour Plot	Shows regions of the map where values fall within specific intervals	All nodes at a specific time
Frequency Plot	Plots value versus fraction of objects at or below the value	All nodes or links at a specific time
System Flow	Plots total system production and consumption versus time	Water demand for all nodes over all time periods

The procedure to create a graph is as follows:

- Select **Report>>Graph** or click ![icon] on the Standard Toolbar.
- Fill in the choices on the Graph Selection dialogue box that appears.
- Click **OK** to create the graph.

The Graph Selection dialogue is used to select a type of graph and its contents to display. The choices available in the dialogue consist of the following:

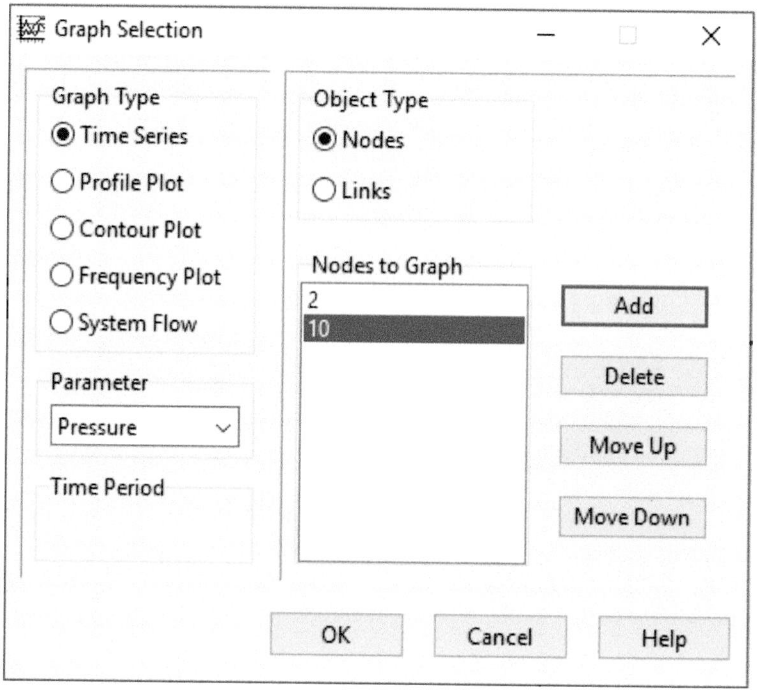

Figure 8.19 Graph selection editor

Item	Description
Graph Type	Selects a graph type
Parameter	Selects a parameter to graph
Time Period	Selects a time period to graph (does not apply to Time Series plots)
Object Type	Selects either Nodes or Links (only Nodes can be graphed on Profile and Contour plots)
Items to Graph	Selects items to graph (applies only to Time Series and Profile plots)

Time Series plots and Profile plots require one or more objects be selected for plotting. The procedure to select items into the Graph Selection dialogue for plotting is as follows:

- Select the object (node or link) either on the Network Map or on the Data Browser. (The Graph Selection dialogue will remain visible during this process).
- Click the **Add** button on the Graph Selection dialogue to add the selected item to the list.

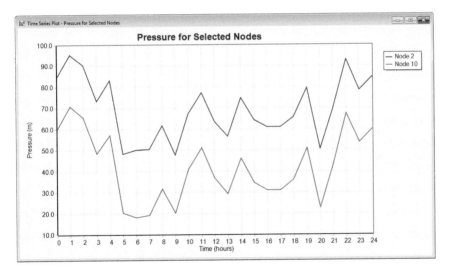

Figure 8.20 Time series diagram

To customize the appearance of a graph, the following steps should be implemented:

- Make the graph the active window (click on its title bar).
- Select **Report>>Options**, or click [icon] on the Standard Toolbar, or right-click on the graph.
- For a Time Series, Profile, Frequency or System Flow plot, use the resulting Graph Options dialogue to customize the graph's appearance.
- For a Contour plot use the resulting Contour Options dialogue to customize the plot.

The Graph Options dialogue form is used to customize the appearance of an X-Y graph. To use the dialogue box, the user can select from among the five tabbed pages that cover the following categories of options:

- General
- Horizontal Axis
- Vertical Axis
- Legend
- Series

The Default box has to be checked if the current settings are also required as defaults for all new graphs. Clicking OK accepts the selections.

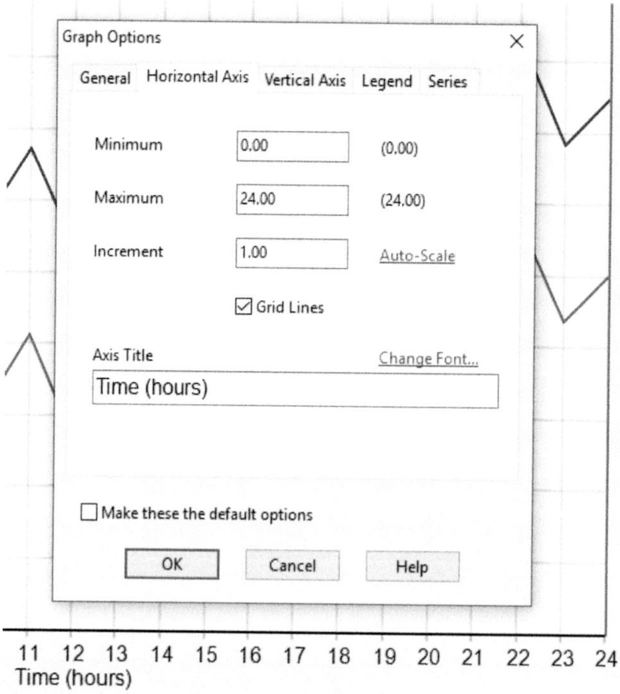

Figure 8.21 Graph options editor

8.4.3 Viewing results with a table

EPANET allows selected project data and analysis results to be viewed in a tabular format:

- A Network Table lists properties and results for all nodes or links at a specific period of time.

- A Time Series Table lists properties and results for a specific node or link in all time periods.

Tables can be printed, copied to the Windows clipboard, or saved to file. To create a table, the user has to:

- Select **View>>Table** or click ▦ on the Standard Toolbar.
- Use the Table Options dialogue box that appears to select:
  - the type of table
  - the quantities to display in each column
  - any filters to apply to the data

The Table Options dialogue form has three tabbed pages. All three pages are available when a table is first created. After the table is created, only the Columns and Filters tabs will appear. The options available on each page are as follows:

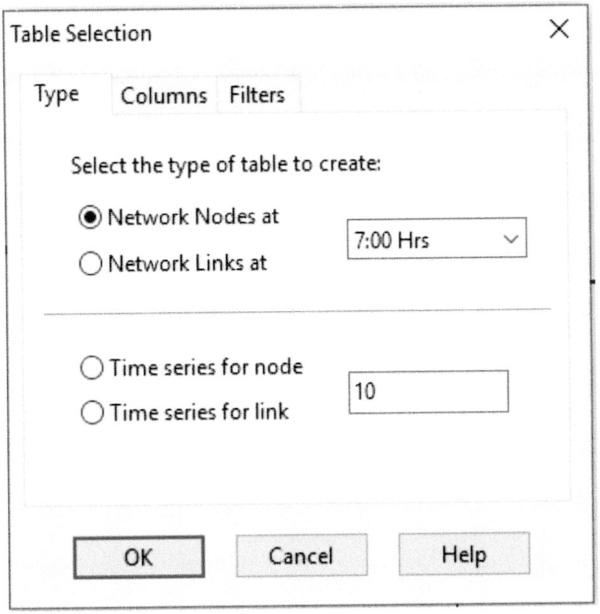

Figure 8.22 Table selection editor

Type Page

The Type page of the Table Options dialogue is used to select the type of table to create. The choices are:

- All network nodes at a specific time period
- All network links at a specific time period

- All time periods for a specific node
- All time periods for a specific link

Data fields are available for selecting the time period or node/link to which the table applies.

Columns Page

The Columns page of the Table Options dialogue form selects the parameters that are displayed in the table's columns. The procedure is as follows:

- Click the checkbox next to the name of each parameter to be included in the table, or if the item is already selected, click in the box to deselect it. (The keyboard's Up and Down Arrow keys can be used to move between the parameter names, and the spacebar can be used to select/deselect choices).
- To sort a Network-type table with respect to the values of a particular parameter, select the parameter from the list and check off the **Sorted By** box at the bottom of the form. (The sorted parameter does not have to be selected as one of the columns in the table.) Time Series tables cannot be sorted.

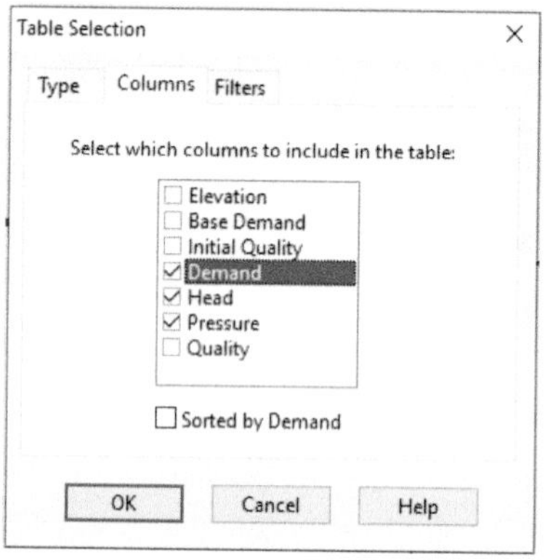

Figure 8.23 Selection of the table columns

Filters Page

The Filters page of the Table Options dialogue form is used to define conditions for selecting items to appear in a table. To filter the contents of a table, the controls at the top of the page should be used to create a condition (e.g. Pressure Below 20). The user can further:

- Click the **Add** button to add the condition to the list.
- Use the **Delete** button to remove a selected condition from the list.

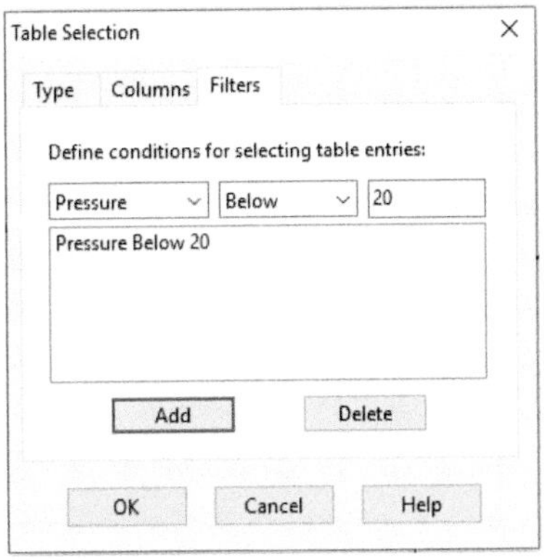

Table Selection ✕

Type   Columns   Filters

Define conditions for selecting table entries:

| Pressure ∨ | Below ∨ | 20 |

Pressure Below 20

Add   Delete

OK   Cancel   Help

Figure 8.24 Filtering of the tabular results

Multiple conditions used to filter the table are connected by ANDs. If a table has been filtered, a re-sizeable panel will appear at the bottom indicating how many items have satisfied the filter conditions.

Once a table has been created it is possible to add/delete columns or sort or filter its data. The user has to:

- Select **Report>>Options** or click 📄 on the Standard Toolbar or right-click on the table.
- Use the Columns and Filters pages of the Table Selection dialogue form to modify the table.

8.5   Copying to the Clipboard or to a File

EPANET can copy the text and graphics of the current window being viewed to both the Windows clipboard and to a file. Views that can be copied in this fashion include the Network Map, graphs, tables, and reports. To copy the current view to the clipboard or to file, the steps are as follows:

- Select **Edit>>Copy To** from the main menu or click.
- Select choices from the Copy dialogue that appears.
- and click **OK**.
- If copy-to-file was selected, enter the name of the file in the Save As dialogue box that appears and click **OK**.

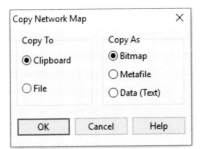

Figure 8.25 Data copying dialogue

Use the Copy dialogue as follows to define how data is to be copied and to where:

- Select a destination for the material being copied (Clipboard or File).
- Select a format to copy in:
  - Bitmap (graphics only)
  - Metafile (graphics only)
  - Data (text, selected cells in a table, or data used to construct a graph)
  - Click **OK** to accept the selections or **Cancel** to cancel the copy request.

8.6 Error and warning messages

Error Number	Description
101	An analysis was terminated due to insufficient memory available.
110	An analysis was terminated because the network hydraulic equations could not be solved. Check for portions of the network not having any physical links back to a tank or reservoir or for unreasonable values for network input data.
200	One or more errors were detected in the input data. The nature of the error will be described by the 200-series error messages listed below.
201	There is a syntax error in a line of the input file created from the network data. This is most likely to have occurred in .INP text created by a user outside of EPANET.
202	An illegal numeric value was assigned to a property.
203	An object refers to undefined node.
204	An object refers to an undefined link.
205	An object refers to an undefined time pattern.
206	An object refers to an undefined curve.
207	An attempt is made to control a check valve. Once a pipe is assigned a Check Valve status with the Property Editor, its status cannot be changed by either simple or rule-based controls.
208	Reference was made to an undefined node. This could occur in a control statement for example.
209	An illegal value was assigned to a node property.
210	Reference was made to an undefined link. This could occur in a control statement for example.
211	An illegal value was assigned to a link property.
212	A source tracing analysis refers to an undefined trace node.
213	An analysis option has an illegal value (an example would be a negative time step value).
214	There are too many characters in a line read from an input file. The lines in the .INP file are limited to 255 characters.
215	Two or more nodes or links share the same ID label.
216	Energy data were supplied for an undefined pump.
217	Invalid energy data were supplied for a pump.
219	A valve is illegally connected to a reservoir or tank. A PRV, PSV or FCV cannot be directly connected to a reservoir or tank. Use a length of pipe to separate the two.

220	A valve is illegally connected to another valve. PRVs cannot share the same downstream node or be linked in series, PSVs cannot share the same upstream node or be linked in series, and a PSV cannot be directly connected to the downstream node of a PRV.
221	A rule-based control contains a misplaced clause.
223	There are not enough nodes in the network to analyse. A valid network must contain at least one tank/reservoir and one junction node.
224	There is no tank or reservoir in the network.
225	Invalid lower/upper levels were specified for a tank (e.g., the lower lever is higher than the upper level).
226	No pump curve or power rating was supplied for a pump. A pump must either be assigned a curve ID in its Pump Curve property or a power rating in its Power property. If both properties are assigned then the Pump Curve is used.
227	A pump has an invalid pump curve. A valid pump curve must have decreasing head with increasing flow.
230	A curve has non-increasing X-values.
233	A node is not connected to any links.
302	The system cannot open the temporary input file. Make sure that the EPANET Temporary Folder selected has write privileges assigned to it.
303	The system cannot open the status report file. See Error 302.
304	The system cannot open the binary output file. See Error 302
308	Could not save results to file. This can occur if the disk becomes full.
309	Could not write results to report file. This can occur if the disk becomes full.

Warning Message	Suggested Action
Pump cannot deliver head.	Use a pump with a larger shut-off head.
Pump cannot deliver flow.	Use a pump with a larger flow capacity.
Flow control valve cannot or provide additionalhead at the valve.	Reduce the flow setting on the valve deliver flow.

8.7   Troubleshooting results

Pumps Cannot Deliver Flow or Head

EPANET will issue a warning message when a pump is asked to operate outside the range of its pump curve. If the pump is required to deliver more head than its shut-off head, EPANET will close down the pump. This might lead to portions of the network becoming disconnected i.e. without any source of supply.

Network is Disconnected

EPANET classifies a network as being disconnected if there is no way to provide water to all nodes that have demands. This can occur if there is no path of open links between a junction with demand and either a reservoir, a tank, or a junction with a negative demand. If the problem is caused by a closed link EPANET will still compute a hydraulic solution (probably with extremely large negative pressures) and attempt to identify the problem link in its Status Report. If no connecting link(s) exist, EPANET will be unable to solve the hydraulic equations for flows and pressures and will return an Error 110 message where an analysis will be made. Under an extended period simulation it is possible for nodes to become disconnected as links change status over time.

Negative Pressures Exist

EPANET will issue a warning message when it encounters negative pressures at junctions that have positive demands. This usually indicates that there is some problem with the way the network has been designed or operated. Negative pressures can occur when portions of the network can only receive water through links that have been closed off. In such cases an additional warning message about the network being disconnected is also issued.

System Unbalanced

A System Unbalanced condition can occur when EPANET cannot converge to a hydraulic solution in some time period within its allowed maximum number of trials. This situation can occur when valves, pumps, or pipelines keep switching their status from one trial to the next as the search for a hydraulic solution proceeds. For example, the pressure limits that control the status of a pump may be set too close together, or the pump's head curve might be too flat causing it to keep shutting on and off.

To eliminate the unbalanced condition it is possible to try to increase the allowed maximum number of trials or loosen the convergence accuracy requirement. Both of these parameters are set with the project's Hydraulic Options. If the unbalanced condition persists, then another hydraulic option, labelled 'If Unbalanced', offers two ways to handle it. One is to terminate the entire analysis once the condition is encountered. The other is to continue seeking a hydraulic solution for another 10 trials with the status of all links frozen to their current values. If convergence is achieved then a warning message is issued about the system possibly being unstable. If convergence is not achieved then a 'System Unbalanced' warning message will be issued. In either case, the analysis will proceed to the next time period.

If an analysis in a given time period ends with the system unbalanced, then the user should recognize that the hydraulic results produced for this time period are inaccurate. Depending on circumstances, such as errors in flows into or out of storage tanks, this might affect the accuracy of results in all future periods as well.

Hydraulic Equations Unsolvable

Error 110 is issued if at some point in an analysis the set of equations that model flow and energy balance in the network cannot be solved. This can occur when some portion of a system demands water but has no links physically connecting it to any source of water. In such a case EPANET will also issue warning messages about nodes being disconnected. The equations might also be unsolvable if unrealistic numbers were used for certain network properties.

Unit conversion table

IMPERIAL (UK) and US ⇒ METRIC*	METRIC ⇒ IMPERIAL (UK) and US
LENGTH: 1 mile = 1.6093 kilometre (km) 1 yard (yd) = 0.9144 metre (m) 1 foot (ft) = 0.3048 m 1 inch (in) = 0.0254 m	LENGTH: 1 km = 0.6214 mile 1 m = 1.0936 yd 1 m = 3.2808 ft 1 centimetre (cm) = 0.3937 in
AREA: 1 square (sq) mile = 2.5898 km² 1 sq yard (yd²) = 0.8361 m² 1 sq foot (ft²) = 0.0929 m² 1 acre = 4047 m² = 0.4047 hectare (ha)	AREA: 1 km² = 0.386 mile² 1 m² = 1.196 yd² 1 m² = 10.764 ft² 1 ha = 2.471 acre
VOLUME: 1 cubic foot (ft³)= 28.32 litre (l) 1 gallon (Imp.) = 4.546 l 1 gallon (US) = 3.785 l	VOLUME: 1 m³ = 35.3 ft³ 1 m³ = 220 Imperial gallon (gal) 1 m³ = 264.2 US gal
WEIGHT/MASS: 1 pound (lb) = 0.4536 kilogram (kg) 1 ounce (oz) = 28.35 gram (g)	WEIGHT/MASS: 1 kg = 2.205 lb 1 kg = 35.27 oz
FLOW: 1 cubic foot per second (ft³/s) = 28.32 litres per second (l/s) 1 Imperial mega-gallon per day (mgd (Imp.)) = 52.62 (l/s) 1 US mega-gallon per day (mgd (US)) = 43.81 l/s 1 Imperial gallon per minute (gpm (Imp.)) = 0.0758 l/s 1 US gallon per minute (gpm (US)) = 0.0631 l/s	FLOW: 1 l/s = 0.0353 ft³/s 1 l/s = 0.019 mgd (Imp.) 1 l/s = 0.0228 mgd (US) 1 l/s = 13.2 gpm (Imp.) 1 l/s = 15.85 gpm (US)
PRESSURE: 1 pound per square inch (psi) = 6895 Pa (0.06895 bar) 1 psi = 0.6765 metres of water column (mwc)	PRESSURE: 1 bar = 14.5 lb/in² (psi) 1 mwc = 1.478 psi
POWER: 1 horse power (hp) = 0.7457 kilo-Watt (kW)	POWER: 1 kW = 1.341 hp

* Some of the units in the table are not metric but are listed because they are in common use.

TEMPERATURE SCALE READINGS:

$$T_F = \frac{9}{5}T_C + 32 \; ; \; T_C = \frac{5}{9}(T_F - 32)$$

The subscripts stand for F (Fahrenheit) and C (Celsius).